International Series in Operations Research & Management Science

Volume 235

More information about this series at http://www.springer.com/series/6161

Christopher W. Zobel • Nezih Altay
Mark P. Haselkorn
Editors

Advances in Managing Humanitarian Operations

Springer

Editors
Christopher W. Zobel
BIT Department
Virginia Tech
Blacksburg
Virginia
USA

Mark P. Haselkorn
HCDE Department
University of Washington
Seattle
Washington
USA

Nezih Altay
Department of Management
DePaul University
Chicago
Illinois
USA

ISSN 0884-8289 ISSN 2214-7934 (electronic)
International Series in Operations Research & Management Science
ISBN 978-3-319-24416-7 ISBN 978-3-319-24418-1 (eBook)
DOI 10.1007/978-3-319-24418-1

Library of Congress Control Number: 2015958654

Springer Cham Heidelberg New York Dordrecht London

Printed on acid-free paper

Springer International Publishing AG Switzerland is part of Springer Science+Business Media (www.springer.com)

Contents

1 Humanitarian Research and Managing Humanitarian Operations 1
Christopher W. Zobel, Nezih Altay and Mark P. Haselkorn

2 Modeling Approaches and Metrics to Evaluate Nonprofit Operations 9
Gemma Berenguer

3 Disaster Response Logistics: Chief Findings of Fieldwork Research 33
José Holguín-Veras, Miguel Jaller, Felipe Aros-Vera, Johanna Amaya, Trilce Encarnación and Tricia Wachtendorf

4 Cash as a Form of Relief 59
Graham Heaslip, Ira Haavisto and Gyöngyi Kovács

5 Measuring Markets That Supply Cash-Based Humanitarian Interventions 79
Laura Rock Kopczak and Gregory Matthews

6 The Partner Proliferation Problem in Disaster Response Networks 111
Sara Hasani, Ramzi El-Haddadeh and Emel Aktas

7 How Humanitarian Culture Informs Change Adoption: A Case Study of Humanitarian Logistics 135
Rebecca Walton, Robin Mays and Mark Haselkorn

8 Information Visibility in Humanitarian Operations: Current State-of-the-Art 159
Natalie Privett

9 Efficient Inventorying and Distribution of Blood Product During Disasters 185
Cigdem Gonul Kochan, Shailesh S. Kulkarni and David R. Nowicki

10 The Vaccine Supply Chain Multathlon: the Reconciliation of Technology, Economy and Access to Medicines 205
Catherine Decouttere, Nico Vandaele, Stef Lemmens and Mauro Bernuzzi

11 Achieving Equity, Effectiveness, and Efficiency in Food Bank Operations: Strategies for Feeding America with Implications for Global Hunger Relief 229
Irem Sengul Orgut, Luther G. Brock III, Lauren Berrings Davis, Julie Simmons Ivy, Steven Jiang, Shona D. Morgan, Reha Uzsoy, Charlie Hale and Earline Middleton

12 Understanding Resilience: A Spatio-temporal Vulnerability Assessment of a Population Affected by a Sudden Lack of Food 257
Thomas Münzberg, Marcus Wiens and Frank Schultmann

Index 281

Contributors

Emel Aktas School of Management, Cranfield University, Bedford, UK

Nezih Altay Department of Management, DePaul University, Chicago, IL, USA

Johanna Amaya Center for Infrastructure, Transportation, and the Environment, and the VREF Center of Excellence for Sustainable Urban Freight Systems, Civil and Environmental Engineering, Rensselaer Polytechnic Institute, Troy, NY, USA

Industrial Engineering, Universidad del Norte, Barranquilla, Colombia

Felipe Aros-Vera Department of Industrial and Systems Engineering Russ College of Engineering and Technology, Ohio University, Athens, OH, USA

Gemma Berenguer Krannert School of Management, Purdue University, West Lafayette, IN, USA

Mauro Bernuzzi GSK Vaccines, Wavre, Belgium

Luther G. Brock III Department of Industrial and Systems Engineering, North Carolina Agricultural and Technical State University, Greensboro, NC, USA

Lauren Berrings Davis Department of Industrial and Systems Engineering, North Carolina Agricultural and Technical State University, Greensboro, NC, USA

Catherine Decouttere Katholieke Universiteit Leuven, Leuven, Belgium

Ramzi El-Haddadeh Brunel University, Uxbridge, UK

Trilce Encarnación Center for Infrastructure, Transportation, and the Environment, and the VREF Center of Excellence for Sustainable Urban Freight Systems, Civil and Environmental Engineering, Rensselaer Polytechnic Institute, Troy, NY, USA

Ira Haavisto HUMLOG Institute, Hanken School of Economics, Helsinki, Finland

Charlie Hale Food Bank of Central & Eastern North Carolina, Raleigh, NC, USA

Sara Hasani Business School, University of Bedfordshire, Luton, UK

Mark P. Haselkorn HCDE Department, University of Washington, Seattle, WA, USA

Graham Heaslip National University of Ireland Maynooth, Ireland & University of New South Wales, Canberra, Australia

José Holguín-Veras Center for Infrastructure, Transportation, and the Environment, and the VREF Center of Excellence for Sustainable Urban Freight Systems, Civil and Environmental Engineering, Rensselaer Polytechnic Institute, Troy, NY, USA

Julie Simmons Ivy Edward P. Fitts Department of Industrial and Systems Engineering, North Carolina State University, Raleigh, NC, USA

Miguel Jaller Civil and Environmental Engineering, University of California, Davis, CA, USA

Steven Jiang Department of Industrial and Systems Engineering, North Carolina Agricultural and Technical State University, Greensboro, NC, USA

Cigdem Gonul Kochan Ohio Northern University, Ada, OH, USA

Laura Rock Kopczak The MIT-Zaragoza Masters in International Logistics Program, Zaragoza, Spain

NYU Wagner School of Public Service, New York, NY, USA

Gyöngyi Kovács HUMLOG Institute, Hanken School of Economics, Helsinki, Finland

Shailesh S. Kulkarni University of North Texas, Denton, TX, USA

Stef Lemmens Katholieke Universiteit Leuven, Leuven, Belgium

Gregory Matthews IRC (International Rescue Committee), New York, NY, USA

Robin Mays Human Centered Design & Engineering, University of Washington, Seattle, WA, USA

Earline Middleton Food Bank of Central & Eastern North Carolina, Raleigh, NC, USA

Shona D. Morgan School of Business and Economics, North Carolina Agricultural and Technical State University, Greensboro, NC, USA

Thomas Münzberg Institute for Nuclear and Energy Technologies, Karlsruhe Institute of Technology, Karlsruhe, Germany

David R. Nowicki University of North Texas, Denton, TX, USA

Natalie Privett Robert F. Wagner Graduate School of Public Service, New York University, New York, NY, USA

Frank Schultmann Institute for Industrial Production, Karlsruhe Institute of Technology, Karlsruhe, Germany

Irem Sengul Orgut Edward P. Fitts Department of Industrial and Systems Engineering, North Carolina State University, Raleigh, NC, USA

Reha Uzsoy Edward P. Fitts Department of Industrial and Systems Engineering, North Carolina State University, Raleigh, NC, USA

Nico Vandaele Katholieke Universiteit Leuven, Leuven, Belgium

Tricia Wachtendorf Department of Sociology and Criminal Justice, Disaster Research Center, University of Delaware, Newark, DE, USA

Rebecca Walton Technical and Professional Communication, Utah State University, Logan, UT, USA

Marcus Wiens Institute for Industrial Production, Karlsruhe Institute of Technology, Karlsruhe, Germany

Christopher W. Zobel BIT Department, Virginia Tech, Blacksburg, VA, USA

About the Authors

Emel Aktas has BSc, MSc and PhD degrees in industrial engineering from Istanbul Technical University, Turkey. She began her career at Istanbul Technical University as a research and teaching assistant. She worked as a visiting researcher at Wayne State University, USA and as a lecturer at Dogus University, Istanbul, Turkey during her PhD studies. She was a researcher in public and private funded projects on location selection, shift scheduling and transportation master plan strategy. Her refereed articles have appeared in a variety of journals including European Journal of Operational Research, Interfaces, and Transportation Research Part A: Policy and Practice. Emel is currently working on minimising waste in the food supply chain, improving urban logistics of food and minimising carbon emissions in maritime logistics. Her research interests are logistics and transportation, disaster relief logistics, supply chain decisions, and mathematical modelling.

Johanna Amaya MSc is a researcher at the Center for Infrastructure, Transportation, and the Environment and at the Volvo Research and Educational Foundation's Center of Excellence for Sustainable Urban Freight Systems at the Rensselaer Polytechnic Institute (RPI) in Troy, NY. She received her BSc in Industrial Engineering from Universidad del Norte, Colombia where she serves as Assistant Professor. She got a MSc in Industrial and Systems Engineering, from University of Florida in Gainesville and currently is pursuing her PhD. in Transportation Engineering at RPI. Ms. Amaya is an Eno Fellow and the recipient of a Fulbright Scholarship sponsored by the U.S. Department of State. Her research interests are in the areas of urban freight transportation systems, humanitarian logistics, and operations management. She has several publications in such areas and has been part of diverse research projects and committees.

Felipe Aros-Vera PhD is an Assistant Professor in the Department of Industrial and Systems Engineering at Ohio University. Felipe holds a PhD in Transportation Engineering from the Rensselaer Polytechnic Institute (RPI), and a MS and BS on Engineering Sciences from the Pontificia Universidad Católica de Chile. His research spans the fields of Disaster Relief Logistics, Transportation and Urban Logistics. Before joining Ohio University, Felipe worked as a Postdoctoral

Research Associate at the Center for Infrastructure, Transportation and the Environment at RPI. His expertise includes transportation operations and planning, operations research, mathematical optimization, and computer programming. His work has been published in prestigious journals such as European Journal of Operational Research (EJOR), Journal of Operations Management (JOM), and Transportation Research Part B: Methodological. Felipe has participated in several research and applied research projects including: Park and Ride Facility Location in New York City, Integrative Freight Demand Management in the New York City Metropolitan Area, and Cyber Enabled Discovery System for Advanced Multidisciplinary Study of Humanitarian Logistics for Disaster Response.

Gemma Berenguer is an Assistant Professor at the Krannert School of Management, Purdue University since June 2012. She received her PhD in Operations Research at the University of California, Berkeley. She also holds an MEng in Logistics and Supply Chain Management from the Zaragoza Logistics Center, an MS in Economics from the Barcelona Graduate School of Economics, and an undergraduate degree in Mathematics from the Universitat Politècnica de Catalunya.

Professor Berenguer's research focuses on integrated supply chain design problems, global health supply chains, and nonprofit operations. She has experience collaborating with public and nonprofit organizations in the global healthcare space and the solar cooking industry. She has published in journals such as Operations Research and Transportation Science. Professor Berenguer teaches Operations Management, Supply Chain Management, and Sustainable and Socially Responsible Operations in the MBA program at Krannert. She is a member of INFORMS, POMS and MSOM.

Mauro Bernuzzi has worked in supply chain operations throughout his professional life. He began his career in the army as a junior lieutenant before joining Glaxo in 1986 where he held various positions both at manufacturing plants and at corporate headquarters. In 2005 he joined GSK Vaccines as Vice President Global Supply Chain Management and he was deeply involved in the supply chain redesign aimed at coping with the challenges triggered by the collaboration with GAVI and UNICEF.

He is currently responsible for the Supply Chain Centre of Excellence for GSK Vaccines and he is a member of the KU Leuven Research Chair on Operations Management scientific committee. In 2007 he was appointed to the adjunct faculty of the Università Cattolica del Sacro Cuore, Milan where he teaches System Dynamics and Simulation. Since 2009 he is also fellow of the KU Leuven Hogenheuvelcollege.

Luther G. Brock III is a member of the Institute for Operations Research and the Management Sciences (INFORMS) and the American Production and Inventory Control Society (APICS). His research interests include the use of optimization models, data mining, and modern heuristics to address issues related to supply chain management. He is particularly interested in addressing logistics challenges in the context of humanitarian relief, public health, and sustainability. Dr. Brock completed his doctoral studies in Industrial & Systems Engineering at North Carolina A&T State University.

Lauren Berrings Davis is an Associate Professor in the Department of Industrial and Systems Engineering at North Carolina A&T State University. She is a member of the Institute for Operations Research and Management Science (INFORMS), the Institute of Industrial Engineers, and served as the president for the INFORMS Minority Issues Forum (2013–2015). Her research interests are in stochastic optimization particularly in the areas of supply chain management and humanitarian logistics. Her research has been funded by NSF and the Department of Homeland Security.

Catherine Decouttere holds a degree Civil Engineering (1995) from the KU Leuven. Sequentially, she was trained in management, R&D management and Leadership at KU Leuven, Vlerick Business School and INSEAD. She conducted further research in coastal engineering at KU Leuven and worked as hydraulic engineer in the consulting business and local government. She worked for 10 years in a large food company, where she expanded her engineering experience with R&D management and Innovation Management. At a Flemish competence center for design and innovation, she managed the research program and was consultant in human-centered design and design management.

Currently she holds a senior research position at the KU Leuven, Center for Operations Management at the GSK Vaccines Research Chair on Operations Management. Her research interests are supply chain design, stakeholder analysis, R&D portfolio management and scenario based planning. She published in international journals and participates regularly in academic and industrial oriented conferences. She is co-founder of Athlycs, a university spin-off company.

Ramzi El-Haddadeh BSc, MSc, PhD is a faculty member at Brunel Business School. Prior to that, Dr. El-Haddadeh was a faculty member at the School of Information Systems, Computing and Mathematics in Brunel University. He is a member of the Information Systems Evaluation and Integration Research Group (ISEing). Dr. El-Haddadeh is on the editorial board of a number of international journals, and has published in a number of leading journals and international conferences in the areas of Telecommunications, Computer Networks and Information Systems.

Trilce Encarnación MSc is a PhD. student in the Department of Civil and Environmental Engineering at Rensselaer Polytechnic Institute. Before joining RPI, Trilce worked as a business analytics consultant in Santo Domingo, Dominican Republic. She holds a BS *cum laude* in Systems Engineering from the Pontificia Universidad Católica Madre y Maestra, in Dominican Republic, and an MSc in Scientific Computing from the University of Puerto Rico at Mayagüez. Her research interests are: sustainable transportation systems, demand modeling, freight transportation and humanitarian logistics.

Ira Haavisto is an Assistant Professor in Supply Chain Management and Social Responsibility at Hanken School of Economics in Helsinki, Finland. She is a co-editor of the Journal of Humanitarian Logistics and Supply Chain Management and the director of the Humanitarian Logistics and Supply Chain Research Institute (HUMLOG Institute) since 2014. Her research interests are in humanitarian logistics and supply chain management.

Charlie Hale is the Vice-President of Information Technology and Operations at the Food Bank of Central and Eastern North Carolina. He has spent more than 7 years in this role and 10+ years at FBCENC. He served on Feeding America's Underserved Counties Task Force in 2010. He received a Bachelor of Science in Meteorology with a minor field of emphasis in Computer Science and Mathematics from NC State University in 1989.

Sara Hasani started her career in various industries including automotive manufacturing and logistics as a supply chain manager, which triggered her interest towards operations management. After completing her BSc in Economics and Msc in business research methods, she obtained her PhD in disaster management from Brunel University. She is currently a lecturer and researcher in Bedfordshire University focusing on various aspects of operations management, systems thinking, disaster management, decision making and serious games. Her most recent piece of research is the development of a predictive framework for disaster response networks as a solution to partner proliferation problems after a disaster strike. She also provides consultancy to various not-for-profit organisations on pro-bono basis.

Graham Heaslip is Associate Professor of Logistics at UNSW, Australia. Prior to joining UNSW Graham was the Deputy Head of the Business School and programme director for the MSc in Humanitarian Logistics and Emergency Management at Maynooth University. Graham completed his PhD studies in the area of Civil Military Cooperation/Coordination at the Logistics Institute, University of Hull, for which he was awarded the James Cooper Memorial Cup for best PhD in Logistics and Supply Chain Management by the Chartered Institute of Logistics and Transport. Prior to entering academia Graham spent 14 years working in the Irish Defence Forces both at home and abroad in a variety of logistical appointments, as well as spending time seconded to Humanitarian agencies in a logistical capacity. Graham's research interests are broadly in the intersections between global logistics/supply chain management, humanitarian logistics and organisational management development.

José Holguín-Veras PhD, PE is the William H. Hart Professor and Director of the Volvo Research and Educational Foundations Center of Excellence for Sustainable Urban Freight Systems, and the Center for Infrastructure, Transportation, and the Environment at Rensselaer Polytechnic Institute. He is the recipient of numerous awards, including the 2013 White House's Transportation Champion of Change Award, the 1996 Milton Pikarsky Memorial Award, and the 2001 National Science

Foundation's CAREER Award. His research interests are in the areas of freight transportation modeling and economics, and humanitarian logistics. His research has led to major changes in transportation policy and substantial improvements in the ability to improve urban freight systems. His work on humanitarian logistics has played an influential role in disaster response procedures, and has led to deeper insight into how best to respond to large disasters and catastrophic events.

Dr. Holguín-Veras is a member of the prestigious National Academy of Sciences' Disaster Research Roundtable, a highly selective group of disaster responders and researchers charged with advising the federal government in disaster policy, and providing a bridge between the research and practitioner communities. Dr. Holguín-Veras' team has pioneered the holistic study of humanitarian logistics by: (1) analyzing response operations as a socio-technical system; (2) conducting research to characterize actual operations and identify lessons learned; and (3) developing suitable mathematical models as decision-support tools. This work has had a transformative effect on disaster policy. His team has conducted extensive fieldwork research and conducted interviews with hundreds of individuals directly involved in the largest disasters of recent times: World Trade Center, Katrina, Joplin, Hurricanes Irene and Sandy, Port-au-Prince and Chile earthquakes, and the Tohoku disasters in Japan, among others. The lessons learned through this fieldwork are routinely shared with disaster agencies. The RPI team develops mathematical models that incorporate the realities and behaviors identified in the fieldwork, such as: inventory allocation models that account for deprivation costs, dynamic control models to manage material convergence, and models of immediate resource requirements, among others.

He is President of the Scientific Committee of the Pan-American Conferences of Traffic and Transportation Engineering, and member of the Scientific Committee of the World Conference of Transport Research. He is a member of numerous technical committees and editorial boards of leading journals. He received his PhD from The University of Texas at Austin in 1996; a MSc from the Universidad Central de Venezuela in 1984; and a BSc from the Universidad Autónoma de Santo Domingo en 1982.

Julie Simmons Ivy is an Associate Professor in the Edward P. Fitts Department of Industrial and Systems Engineering and Fitts Faculty Fellow in Health Systems Engineering. She is a member of Institute of Operations Research and Management Science (INFORMS) and the Institute of Industrial Engineers, Dr. Ivy served as the 2007 Chair (President) of the INFORMS Health Applications Society (HAS) and the 2012–2013 President for the INFORMS Minority Issues Forum. Her research interests are mathematical modeling of stochastic dynamic systems with emphasis on statistics and decision analysis as applied to health care, public health, and humanitarian logistics. Her research has been funded by the NSF and the Centers for Disease Control and Prevention.

Miguel Jaller PhD is an Assistant Professor at the University of California, Davis (UC Davis). He has strong theoretical foundations and practical experience in industrial and transportation engineering. He received his BSc and MSc in Industrial Engineering from Universidad del Norte, Colombia. His ME in Transportation Engineering, MSc in Applied Mathematics, and PhD. in Transportation Engineering from Rensselaer Polytechnic Institute. His research interests are in the areas of humanitarian logistics, disaster response logistics, disaster management, freight transportation, sustainable transportation systems, and operations research. He has scientific and technical publications on these topics, and has presented at different national and international venues. Dr. Jaller has been part of a number of important research projects after Hurricane Katrina, the Port au Prince, Chile and Tohoku Earthquakes, and the storms hitting the Northeast coast of the U.S., among other small and large scale disasters. His research is a combination of field work findings and multidisciplinary perspectives and theories.

Steven Jiang is an Associate Professor in the Department of Industrial and Systems Engineering at North Carolina A&T State University. He is a member of the Institute of Industrial Engineers, Human Factors and Ergonomics Society, and American Society for Engineering Education. His research interests are in visual analytics, human computer interaction as applied to manufacturing and service industries. His research has been funded by NSF, ARO, and ARL.

Cigdem Gonul Kochan is an Assistant Professor of Operations Research and Supply Chain Management at the Ohio Northern University. She has received her MS degree in Information Technology Management from University of Texas at Dallas and her PhD in Logistics and Supply Chain Management from University of North Texas. She is a certified SAP Solution Architect and has 4 years of industry experience in IT and operations management fields. Her research focuses on supply chain resilience, cloud based supply chain management, healthcare supply chains and inventory optimization.

Laura Rock Kopczak is an educator, researcher and consultant in Supply Chain Management, a field in which she has over 25 years of experience. Her current focus is on humanitarian and global health supply chains. Laura teaches courses on humanitarian logistics at NYU Wagner School of Public Service and in the MIT-Zaragoza Masters in International Logistics Program. She also consults and does research with organizations such as MSF, IFRC, Save the Children, UNHCR, UNICEF, UN WASH cluster, GAVI, PSI, JSI, the Rockefeller Foundation, and the Bill and Melinda Gates Foundation. Laura obtained her MBA from Columbia Business School, her MS in Operations Research from Columbia University School of Engineering and her PhD in Industrial Engineering and Engineering Management from Stanford University. In the past, Laura worked as a professor at Stanford University, as one of three co-directors of the Stanford Global Supply Chain Management Forum and in a number of positions at Hewlett-Packard Company.

Gyöngyi Kovács is the Erkko Professor in Humanitarian Logistics at the Hanken School of Economics in Helsinki, Finland. She is a founding editor of the Journal of Humanitarian Logistics and Supply Chain Management, and has also edited a book on Relief Supply Chain Management. She has led the Humanitarian Logistics and Supply Chain Research Institute (HUMLOG Institute) up till 2014 and has been a European regional editor of the International Journal of Physical Distribution and Logistics Management 2008–2013. Her research interests include humanitarian logistics and sustainable supply chains.

Shailesh S. Kulkarni is an Associate Professor of Decision Sciences in the Department of Information Technology and Decision Sciences at the University of North Texas (UNT). Professor Kulkarni's research interests are in the areas of supply chain networks and stochastic modeling and analysis. His research has been published in various academic journals including *Production and Operations Management, IIE Transactions, Decision Sciences Journal, European Journal of Operational Research and the International Journal of Production Research.* He has received teaching and research awards at UNT. He currently serves as the Associate Vice President for Sponsorships for the Production and Operations Management Society (POMS) and on its Board. He also serves on the Editorial Board of the Journal of Industrial Engineering and Engineering Science Letters.

Stef Lemmens is currently pursuing a PhD in Applied Economics at the Katholieke Universiteit Leuven, Faculty of Business and Economics. He holds a master degree Commercial Engineering (2012) and Advanced Business Research (2013). He is a research member of the GSK Vaccines Research Chair on Operations Management.

His primary research focuses on modeling and optimization of real-world supply chains. More specifically, he is interested in the acquisition of (multiple) relevant key performance measures and the integration of internal and external supply chain risks for supply chain design problems. For the development of his research, he particularly uses operations research methods as mathematical programming and factory physics.

Gregory Matthews is the Senior Advisor for Emergency Livelihoods at the International Rescue Committee (IRC). Greg is a food security and nutrition specialist with over 10 years of experience in humanitarian response and coordination. He currently oversees the IRC's cash transfer programming and emergency market analysis efforts, including the promotion and further development of the Emergency Market Mapping and Analysis (EMMA) toolkit. Since joining the IRC in 2012, Greg has primarily focused on starting and supporting emergency programs in response to the Syria crisis and in the Philippines, and is now focused on strengthening capacity to rapidly deliver cash grants at scale in emergencies. Prior to joining the IRC, Greg worked at the Humanitarian Response Department at Oxfam America and the International Disaster Response and Africa units at the American Red Cross. Greg has previously worked in the Philippines, Haiti (several times), Madagascar, Tanzania, Ethiopia, Kenya, Senegal and Gambia, and is currently focused on the Syria crisis response. He holds a MSc in Nutrition from Tufts University, and a BSc in International Politics from Georgetown University.

Robin Mays is an ethnographic researcher who explores the human and contextual factors of disaster and humanitarian response systems that lead to effective response. Her research revolves around understanding contextualized and dynamic meanings of value and effectiveness within humanitarian work; the balance of structure and flexibility in effective rapid response; the role of decision-making and implications for design of technology. She is currently pursuing her PhD with the Department of Human Centered Design & Engineering at the University of Washington. She has worked for over 18 years in rapid response operations and logistics, with an 11-year career as a humanitarian logistician. As a member of the response communities she studies, her research couples an insider perspective with a theoretical framework drawn from human-centered design, understanding hidden work, change adoption, and lowest level empowerment.

Earline Middleton is the Vice-President of Agency Services and Programs at the Food Bank of Central and Eastern North Carolina. She has served more the 15 years in this role and 25 years at FBCENC. She was appointed to serve on the North Carolina State Food Policy Council and continues to serve on North Carolina Local Food Council; she is a founding member of the Southeast Anti-Hunger Consortium. Earline has been recognized by Triangle Business Journal and is a recipient of the Z. Smith Reynolds Foundation Sabbatical. She received a BSEd from Wilberforce University and completed additional studies at the University of Dayton, in Dayton, Ohio and North Carolina State University in Raleigh, NC.

Shona D. Morgan is an associate professor in the School of Business and Economics at North Carolina A&T State University. Her research interests include the analysis and design of heuristics and algorithms for intractable discrete optimization problems, remanufacturing and reverse supply chains, and non-profit supply chains. She has published articles in *IIE Transactions,* the *European Journal of Operational Research,* and *The International Journal of Production Research* to name a few. She is a member of the Decision Sciences Institute and currently serves as the Vice President of Publications for the Southeastern Decision Sciences Institute.

Thomas Münzberg is a research associate at the *Karlsruhe Institute of Technology (KIT).* His research aims at minimizing and preparing for the adverse impacts of critical infrastructure disruptions by providing analytical assistance for decision-makers of local disaster management authorities and critical infrastructure providers. A great interest of his research is the assessment of power outage impacts and community resilience in the context of critical infrastructure disruptions.

Thomas Münzberg is an active member of the *Center for Disaster Management and Risk Reduction Technology (CEDIM),* a joint collaboration between the Helmholtz Centre Potsdam–German Research Centre for Geosciences (GFZ) and the Karlsruhe Institute of Technology (KIT). The *CEDIM* is part of the IRDR International Centres of Excellence on Critical Infrastructures and Strategic Planning (IRDR ICoE-CISP).

Thomas Münzberg has vast practical experiences and corresponding qualifications in safety engineering, business continuity management, disaster management, and emergency medical services. He holds a Master of Science in disaster management and safety engineering from the *Cologne Technical University (CTU), Germany,* where he also worked as a research associate at the *Institute for Rescue Engineering and Civil Protection.*

David R. Nowicki is an Associate Professor at the University of North Texas. He holds a joint appointment in the College of Business' Department of Marketing and Logistics and the College of Engineering's Department of Engineering Technology. Dr. Nowicki received his Masters Degree in Industrial and Systems Engineering from Virginia Tech and both his bachelors and doctorate degrees in Industrial and Systems Engineering from the University of Wisconsin—Madison.

Dr. Nowicki's research efforts focus on applying advanced analytical techniques to solve supply chain management problems from a systems engineering context. Professor Nowicki's research is concentrated on performance based logistics modeling, supply chain management, resiliency and risk, multi-resource optimization, reliability theory, and inventory optimization. Dr. Nowicki brings over 20 years of industry experience holding executive positions at i2 Technologies and the TFD Group with a focus on supply chain management, systems engineering, lifecycle affordability, operations research modeling, reliability, inventory optimization and software engineering.

Irem Sengul Orgut received her PhD in Industrial Engineering in 2015 from the Edward P. Fitts Department of Industrial and Systems Engineering at North Carolina State University. She now works at Lenovo as the Corporate Quality Statistics Project Manager where she uses Big Data and Analytics tools to improve customer engagement. Prior to starting her doctoral studies, she received her BS degrees in Industrial Engineering and Mechanical Engineering from Bogazici University, Istanbul, Turkey in 2010. Her research interests include stochastic modeling of complex supply chains with multiple objectives and conflicting decision makers with application focus on long-term humanitarian issues and public health problems. She received various awards for her teaching and research. She is a member of INFORMS, Alpha Pi Mu and IIE. Her web address is https://iremsengul.wordpress.com/.

Natalie Privett PhD is an Assistant Professor of Management and Policy at the Robert F. Wagner Graduate School of Public Service at New York University. Her research focus is operations and supply chain management in the context of global public health, international public service, and nonprofit and public management. Currently, Privett's multidisciplinary research aims to explore the notion of strategic operational interventions to improve health delivery and markets in developing countries.

Privett received her Masters and PhD from Stanford University's department of Management Science & Engineering, specialized in operations and supply chain management. Privett received her BS in industrial engineering from Texas A&M University and subsequently worked in manufacturing. Prior to joining Wagner, she was a Postdoctoral Research Fellow at the MIT-Zaragoza Logistics Program in Zaragoza, Spain, where she worked on global cold chain pharmaceutical tracking projects supported by the European Space Agency. Privett has most recently been published in the journals Operations Research for Healthcare (ORHC) and Manufacturing & Service Operations Management (MSOM).

Frank Schultmann is Chair Professor at the Karlsruhe Institute of Technology (KIT), Germany, and Director of the KIT's Institute for Industrial Production (IIP) and the French-German Institute for Environmental Research (DFIU). He is also Director of Project Management for the University of Adelaide, Australia.

Professor Schultmann studied Business and Industrial Engineering at the University of Karlsruhe. He completed his doctoral thesis in 1998 at the Faculty of Economics and Business Engineering and his Habilitation in 2003 receiving the venia legendi (teaching authority) in Management Science. Prior to his present positions, he was Professor of Industrial Management at the University of Koblenz-Landau and holder of the Chair of Business Administration, Construction Management and Economics at the University of Siegen.

Professor Schultmann was Editor of an international leading scientific journal and is member of several editorial boards. He served as elected chairman of several scientific boards and is coordinator of international task forces and member of numerous international committees. He conducted various research projects on national and international level and has been awarded with several research grants and prices.

His research interests include sustainable production and logistics, decision support, supply chain management and optimization, project management, technology assessment, construction management, and information and communication technologies. This goes along with various industry collaborations. Among others Professor Schultmann has worked with companies from resource industries, automotive, chemical, construction, security, food, transport and logistics.

Reha Uzsoy is Clifton A. Anderson Distinguished Professor in the Edward P. Fitts Department of Industrial and Systems Engineering at North Carolina State University. He holds BS degrees in Industrial Engineering and Mathematics and an MS in Industrial Engineering from Bogazici University, Istanbul, Turkey. He received his PhD in Industrial and Systems Engineering in 1990 from the University of Florida. His teaching and research interests are in production planning and supply chain management. Before coming to the US he worked as a production engineer with Arcelik AS, a major appliance manufacturer in Istanbul, Turkey. He has also been a visiting researcher at Intel Corporation and IC Delco. He was named a Fellow of the Institute of Industrial Engineers in 2005, Outstanding Young Industrial Engineer in Education in 1997, and has received awards for both undergraduate and graduate teaching.

Nico Vandaele holds a degree Commercial Engineering (1990) and obtained a PhD in Applied Economics, Operations Research and Operations Management from KU Leuven in 1996. He is currently Full Professor at the Research Center for Operations Management at the Katholieke Universiteit Leuven, Faculty of Business and Economics. He is holder of the GSK Vaccines Research Chair on Operations Management. He is also a visiting researcher at CORE and IAG (Université Catholique de Louvain). Nico Vandaele teaches courses in operations research, operations management and supply chain management.

His research interests are situated in modeling of manufacturing and service systems, performance measurement, the design of planning systems, sales and operations management, factory physics, health care management and traffic modeling. Other research deals with decision support systems for product design and development and portfolio management as well as scenario based planning. He published in leading journals and he is active in several executive training programs, both national and international, and has served as consultant/advisor for major global companies as well as for small and medium sized companies. He served as executive director of the Innovation and Incubation Center (2007–2015). He co-founded two KU Leuven spin-off companies, Nyo Alatus and Athlycs.

Tricia Wachtendorf PhD is an Associate Professor of Sociology at the University of Delaware and the Associate Director of the world-renown Disaster Research Center—the oldest center in the world focused on the social science aspects of disaster. Over the past two decades, her research has focused on multi-organizational coordination before, during and after disasters, transnational crises, and social vulnerability to disaster events. Dr. Wachtendorf has engaged in quick response field work after such events as the 2001 World Trade Center attacks, the tsunamis affecting India, Sri Lanka (2004) and Japan (2011), Hurricanes Katrina (2005) and Sandy (2012), as well as the earthquakes in China (2008) and Haiti (2010). With numerous research grants from agencies such as the National Science Foundation, she has published widely on improvisation in disasters as well as disaster convergence. Her most recent funded research projects examine the temporal nature of household and emergency management decision-making during hurricane events, investigate humanitarian logistics during disaster response, and use a visual sociology approach to explore benchmarks of recovery following the 2011 disaster in Japan.

Rebecca Walton is an Assistant Professor of Technical Communication in the Department of English at Utah State University. She studies the role that communication can play in more equitably distributing power. Much of her research investigates how people in resource-constrained environments such as in the Global South and humanitarian organizations create, access, share, evaluate, and use information to meet their personal and professional goals. Her work has appeared in *Technical Communication Quarterly, Journal of Business and Technical Communication, and Information Technologies and International Development,* as well as other journals and edited collections.

Marcus Wiens studied Economics in Bayreuth (GER), Dijon (FRA) and Saarbrücken (GER) and received his PhD in Economics at the Bundeswehr University Munich (GER). He leads the research unit in risk management at the Institute for Industrial Production (IIP) at KIT. His fields of interest are systemic risk management, behavioural risk management, decision theory, game theory and experimental economics. Marcus Wiens is adjunct professor at the International School of Management (ISM), member of the German Operation Research Society, of the German Experimental Economics Society and of the Center for Disaster Management and Risk Reduction Technology (CEDIM), an interdisciplinary research center in the field of disaster management founded by Helmholtz Centre KIT.

His research group participated/participates in projects covering a wide range of risk management topics such as resilience of transportation networks (EU-projects WEATHER and MOWE-IT), resilience of industrial value chains (project KLIMOPASS), safety of the food supply chain (project SEAK), protection of critical infrastructure against adversarial risks (project RIKOV) and other topics such as business continuity management, incentives systems and contracts.

About the Editors

Nezih Altay is an Associate Professor at the Driehaus College of Business of DePaul University. He earned his PhD in Operations Management from Texas A&M University. Dr. Altay is an experienced and highly qualified teacher-scholar. His research specializes in after-sale service operations, disruption management and humanitarian supply chains. He has published his research in leading academic journals and presented in national and international arenas. He co-edited a book titled Service Parts Management: Demand Forecasting and Inventory Control that was published also with Springer. He is the co-Editor-in-chief for the Journal of Humanitarian Logistics & Supply Chain Management, and directs the Master program in Supply Chain Management at DePaul University.

Mark P. Haselkorn is a Professor of Human Centered Design & Engineering at the University of Washington. He is Director of the new university Center on Collaborative Systems for Security, Safety & Regional Resilience (CoSSaR) and currently leads the Maritime Operations Information Sharing (MOISA) project, a research partnership sponsored by three Federal Agencies – DHS Interagency Operations Center (IOC), Program Manager for the Information Sharing Environment (PM-ISE), and National Maritime Intelligence-Integration Office (NMIO) – with the goal of better understanding and enhancing the information sharing requirements for regional maritime safety and security. He also is a lead investigator on an AHRQ R01 to develop work and information centered methods for achieving evidence-based health information technology. Dr. Haselkorn also conducts research for the Red Cross Global Disaster Preparedness Center and has completed an NSF initiative to define the emerging frontier of "Humanitarian Service Science & Engineering." He has worked with the military on a number of projects, including the integration of DOD and VA electronic medical records and the Air Force's strategic management of ICT under the threat of Y2K (a study published by the National Research Council). Dr. Haselkorn has conducted foundational research in the area of intelligent transportation systems, including development of the first Web-based real-time traveler information system (Traffic Reporter, 1990). He is Past President of the IEEE Professional Communication Society, has served on ISO/IEC-JTC1, is a member of the IEEE Medical Technology Policy Committee, and was a founding

Board Member of the International Community on Information Systems for Crisis Response and Management (ISCRAM).

Christopher W. Zobel is the R.B. Pamplin Professor of Business Information Technology in the Pamplin College of Business at Virginia Tech. He earned a Ph.D. in Systems Engineering from the University of Virginia, and an M.S. in Mathematics from the University of North Carolina at Chapel Hill. His primary research interests include disaster operations management and humanitarian supply chains, and he has published his work in journals such as *Decision Sciences, Decision Support Systems*, and the *Journal of Humanitarian Logistics and Supply Chain Management*, among others. Dr. Zobel is one of the Co-Directors of Virginia Tech's Interdisciplinary Graduate Education Program in Disaster Resilience, and he was a 2015 Fulbright Scholar to Germany. He is on the Board of Directors of the ISCRAM Association and an active member of the Decision Sciences Institute (DSI) and the Institute for Operations Research and the Management Sciences (INFORMS).

Chapter 1
Humanitarian Research and Managing Humanitarian Operations

Christopher W. Zobel, Nezih Altay and Mark P. Haselkorn

Abstract Humanitarian operations are a critical and challenging activity. The general class of problems in this area are distinct from those in business operations management for a number of reasons. Among these are the focus on minimizing suffering and empowering affected populations, the requirement for ethical and just treatment, the complex and often politically charged problem setting, the need for transparency and accountability, the uncertain and dynamic environment, the distinctive decision-making processes of non-profit and governmental organizations, and the existence of unusual resource constraints. Each of the chapters in this *Advances in Managing Humanitarian Operations* volume seeks, in different ways, to help us improve our ability to address these problems. This initial chapter emphasizes the importance of the practitioner community to these efforts and it provides a careful overview of the important work contributed to the volume by our many authors.

C. W. Zobel (✉)
BIT Department, Virginia Tech, Blacksburg, VA, USA
e-mail: czobel@vt.edu

N. Altay
Department of Management, DePaul University, Chicago, IL, USA
e-mail: naltay@depaul.edu

M. P. Haselkorn
HCDE Department, University of Washington, Seattle, WA, USA
e-mail: markh@uw.edu

C. W. Zobel et al. (eds.), *Advances in Managing Humanitarian Operations,*
International Series in Operations Research & Management Science 235,
DOI 10.1007/978-3-319-24418-1_1

1.1 Purpose of Volume

Humanitarian operations[1] and humanitarian research have not been close friends. For decades, the humanitarian sector valued one thing—practitioner experience. There was no degree that would be sufficient to get you a job with a major international non-governmental organization (NGO); no conference that every major player had to be at to hear about the latest cross-sector research and developments. Only past practitioner experience qualified you for future work. The disaster environment was viewed as so complex and unique that if you hadn't been there, you couldn't contribute to activities in that environment.

When one of the editors of this volume on current research in humanitarian operations (Haselkorn) attended the joint after-action review of the humanitarian response to the 2004 Indian Ocean tsunami (Brooks and Haselkorn 2005) as the only academic, he was continually greeted with the question, "What disasters have you been at?" Upon hearing that his involvement with disasters did not include field experience, the conversations generally ended abruptly and the questioners moved on to people who they felt really understood the issues. Practitioners and managers with practitioner experience believed that if you hadn't experienced the reality of humanitarian field work then you couldn't really understand, much less help improve, the situation.

Lessons learned documents and after actions reviews were plentiful within humanitarian organizations, but these reports "discovered" the same issues over and over again, and they were not written to actually change the organization or how it operated (Ontko et al. 2007). They were reports that were written to document what had occurred, not research intended to understand the central issues (which often went beyond the single organization) or to answer fundamental questions. Yet, despite this history of devaluing or not even recognizing academic research, the humanitarian sector was changing.

In the late 1990's, senior NGO leaders began organizing to better understand what was happening across the international relief and development sector. One of the leaders was Marc Lindenberg, former Senior Vice President for Programs at CARE USA and then Professor and Dean of the Daniel J. Evans School of Public Affairs at the University of Washington. He helped organize a series of meetings with top NGO leaders and academics to think through the future of international humanitarian efforts, which led to the landmark book, *Going Global: Transforming Relief and Development NGOs* (Lindenberg and Bryant 2001), "the first book-length comprehensive study of the largest of the Northern-based international relief and development NGOs."

The twenty-first century has brought ever increasing efforts by academics to contribute to the humanitarian endeavor and ever increasing acceptance by practi-

[1] By "operations" we mean how humanitarian organizations work. Humanitarian organizations do considerable research on the impact of their programs, but, as Marc Lindenberg put it in 2001, "little is written about how they are doing what they do, and how process affects product." Even more alien is the idea that humanitarian processes ***are*** a product.

tioners of the value of research to their efforts. Humanitarian logistics has been a particularly active area for research, and many of the authors in this volume have contributed in this area (For example Holguín-Veras et al. 2007; Aktas and Ulengin 2005; Heaslip et al. 2012). More recently we have come full circle, with one of the authors in this volume (Mays) being funded by the Red Cross to conduct field-based research on the central elements of successful humanitarian work (Mays et al. 2014). Thus organizations whose culture has been to receive funding and spend it with as little overhead as possible on the direct and immediate needs of those they serve are beginning to understand the value of using resources to conduct research on ways to improve their own work processes.

Research in humanitarian operations generally can be divided into efforts focused on disaster relief (short term), development (long term), and community empowerment (sustainability). In each of these cases, the common thread is the appropriate provision of services in partnership with the affected community. This volume is looking to provide a survey of current research problems and results, to discuss existing applications of humanitarian operations research, and to consider new research efforts that clearly extend existing research and applications. The papers in this volume take both researchers and practitioners further towards a common research basis for successful management of humanitarian operations. The intent is to provide a solid reference for humanitarian operations researchers, practitioners, and policy makers, and thus to encourage further efforts to continue advancement in this area. Accordingly, the chapters contributed to this volume are organized in three sections: overview, theory, and application.

1.2 Overview Section

In the first section of the volume we present four chapters. The first chapter titled "Modeling approaches and metrics to evaluate nonprofit operations" by Gemma Berenguer provides a high-level overview of how performance is measured across the nonprofit spectrum. Since humanitarian organizations (a sub-section of the non-profit sector) have different and more complex objectives than their for-profit counterparts, Dr. Berenguer starts her chapter by proposing several theoretical models of performance measurement that could be used in nonprofit operations. Performance measurement in the humanitarian context is a current topic of discussion among practitioners as well as within the academic community. This initial chapter does an excellent job of bringing the attention of the humanitarian community to different measures that could be used in evaluating performance, and it calls for a greater focus on multi-disciplinary research.

The next chapter titled "Disaster response logistics: chief findings of fieldwork research" by Holguín-Veras et al. focuses on post-disaster relief and recovery operations, one of the main areas within humanitarian operations that has received significant attention. Despite this attention, however, the continuing complexities involved in post-disaster operations, their unique problem characteristics, and the wide range of capabilities and capacities of participating organizations, are all constant reminders of the difficulty in managing humanitarian operations. Based on their extensive field research, Holguín-Veras et al. identify a number of under-studied areas where logistics research could enhance operational effectiveness.

Continuing with the topic of disaster response logistics, an issue that recently has come under scrutiny is the use of cash transfers as a form of relief. Moving cash should be considerably cheaper and faster than moving food, water, or non-food items, but when it comes to implementing such a program numerous questions surface. With this in mind, the next two chapters, "Cash as a form of relief" by Heaslip et al. and "Measuring markets that supply cash-based humanitarian interventions" by Kopczak and Matthews delve into this important topic. These two chapters collectively present an excellent discussion of cash as a form of disaster relief.

In the first of these chapters, Heaslip et al. discuss how funding in humanitarian scenarios is different than funding of for-profit operations. They argue that cash can help to generate local market activity and to restart livelihoods, and emphasize that it is often a more empowering and dignified form of support. They also point out, however, that not only is this type of aid susceptible to theft, corruption, and misuse but also it is prone to targeting errors. Furthermore, they add that cash transfer programs can also cause issues such as inflation. Their discussion provides a very good general overlook at cash transfer programs as a form of relief.

Kopczak and Matthews further extend the discussion on cash transfer programs by specifically examining the measurable impacts on local market systems. Their research focuses on applying commercial and global health supply chain models to help measure the performance of relief markets created by cash transfer programs. They use several different case studies of cash-based market interventions in order to provide insights into such potential measures and they examine three different supply chain models in order to provide insights into what would be appropriate measures at the local market level.

1.3 Theory Section

In the second section of the volume we have three chapters focusing on different aspects of coordination in humanitarian operations. Their common thread is theory building. The chapter titled "The partner proliferation problem in disaster response networks" by Hasani et al. draws on the concept of virtual organizations to provide a structured view of coordination among humanitarian actors after a disaster. The authors propose a theoretical framework that explains how short and long-term collaboration may interact with the disaster management cycle activities. The authors also suggest developing a decision support tool for partner configuration to meet the specific requirements of a disaster response network.

In "How humanitarian culture informs change adoption: a case study of humanitarian logistics," Walton et al. move the focus of the book from disaster response agencies to more general humanitarian organizations. In this qualitative research the authors identify and explain the factors affecting change adoption based on a longitudinal study of emergency logistics teams at a large international humanitarian organization. Issues like technology adoption, coordination and collaboration all represent change to humanitarian organizations which are used to working in silos and generally focus on one task. Walton et al.'s findings have practical implications in facilitating change adoption in humanitarian agencies which consequently could lead to better collaboration.

One of the prerequisites to successful coordination is information exchange (Altay and Labonte 2014). To exchange information, however, that information must first be available and accessible. In "Information visibility in humanitarian operations: current state-of-the-art," Natalie Privett describes information visibility as the clarity and accessibility of information and data throughout the supply chain. While information visibility is still challenging for most commercial supply chains, the problem is even more challenging for the humanitarian sector, since in humanitarian settings supply data and information is often scarce or inadequate. Consequently, Privett argues, increased visibility has the potential to greatly improve humanitarian operations. In order to support this, her contribution lists the tools and techniques available for the humanitarian sector to improve information visibility and it discusses the important characteristics of such technologies.

1.4 Application Section

The final section of the volume contains four chapters of applied research that focus, in particular, on the topics of blood products, vaccines, and food assistance. In "Efficient inventorying and distribution of blood product during disasters," Kochan et al. present a potential application of a quantitative approach for solving the inventory and routing problem for blood products in response to an emergency. Motivated by a *Médecins Sans Frontières* report on the medical care and blood products needs of civilians suffering from the civil war in Syria, the authors develop a two-stage stochastic programming model to represent the storage and distribution of blood products from a single blood collection center to surrounding hospitals.

Decouttere et al. further extend the discussion on medical applications to vaccine supply chains in "The vaccine supply chain multathlon: the reconciliation of technology, economy and access to medicines." In this chapter, the authors take a long-term view, building on their own experience within a vaccine supply chain and presenting a general framework for designing medicinal aid supply chains. Their work clearly illustrates the operative issues in vaccine supply chains and present a multi-level modeling process which is moderated by a group of key stakeholders.

The last two chapters in the section focus on food assistance. The first of the two chapters deals with food assistance for community sustenance through food banking operations. In "Achieving equity, effectiveness, and efficiency in food bank operations: strategies for feeding America with implications for global hunger relief," Orgut et al. take on the problem of hunger, a slow onset and variable impact disaster, by creating a framework for operational decision making at food banks. The result of collaboration between academics and food banking professionals, this chapter presents a collective vision for the future of research in food bank operations. The authors not only list ongoing operations research work in food banking but also identify research needs in food aid distribution.

The second chapter on hunger relief is by Münzberg et al., and is titled "Understanding resilience: a spatio-temporal vulnerability assessment of a population affected by a sudden lack of food." In contrast to the work of Orgut et al., this chapter focuses on understanding the human element in food relief. Based on data from two German cities, the authors investigate food stocking behavior of residents in response to a local emergency which disrupts food supply. This chapter does a great job in presenting both quantitative and qualitative characteristics of the problem, and it provides an overall illustration of the multi-dimensionality of the hunger relief in the developed world.

1.5 Summary

We were very pleased with the quality of the submissions for this edited volume and thank all our authors for their admirable contributions to furthering research in humanitarian operations. Each chapter has been reviewed by at least two referees and has gone through multiple iterations of revisions, and we are grateful to all of our authors for their patience and cooperation throughout the review process. Their perseverance helped to ensure the overall quality of the volume.

The issues to be discussed and the problems to be solved in humanitarian operations are plentiful. In this volume we tried to focus on contemporary issues and unresolved challenges, representing a range of approaches by contribution (overview, theory, application), by topic (disaster/non-disaster) and by focus (money/food/medicine). Our aim with this book was to showcase the latest research in humanitarian operations. We hope that the result is able to address the needs of the several different target communities, and that it will serve both as an information resource for academics and practitioners and as a platform for discussions that lead to even more advances in the field of humanitarian operations.

References

Altay N, Labonte M (2014) Challenges in humanitarian information management and exchange: evidence from Haiti response. Disasters 38(S1):S50–S72

Aktas E, Ulengin F (2005) Outsourcing logistics activities in Turkey. J Enterp Inf Manag 18(3):316–329

Brooks K, Haselkorn M (2005) Joint After-Action Review of our Humanitarian Response to the Tsunami Crisis—Report of Workshop Proceedings, 7–8 April 2005, Bangkok, Thailand, http://www.alnap.org/resource/3297

Heaslip G, Sharif AM, Althonayan A (2012) Employing a systems-based perspective to the identification of inter-relationships within humanitarian logistics. Int J Prod Econ 139(2):377–392

Holguín-Veras J, Pérez N, Ukkusuri S, Wachtendorf T, Brown B (2007) Emergency logistics issues affecting the response to Katrina: a synthesis and preliminary suggestions for improvement. Transp Res Rec: J Transp Res Board 2022:76–82

Lindenberg M, Bryant C (2001) Going global: transforming relief and development NGOs. Kumarian Press, Bloomfield

Mays R, Walton R, Lemos M, Haselkorn M (2014) Valuing what works: success factors in disaster preparedness, International Red Cross, 43 pages. http://www.alnap.org/resource/19219

Ontko M, Williamson S, Kemp R, Haselkorn M (2007) An examination of the effectiveness of lessons-learned reporting within the humanitarian sector. J Inf Technol Soc Chang, Spring 26–46

Chapter 2
Modeling Approaches and Metrics to Evaluate Nonprofit Operations

Gemma Berenguer

Abstract The Operations Research and Management Science (OR/MS) community is paying increasing attention to the analysis of nonprofit practices, which can significantly differ from for-profit ones. While profit is one of the most widely used performance metrics in for-profit settings, it is not appropriate for nonprofit operations. We start the chapter by suggesting a number of theoretical models that could represent nonprofit operations. Next, we present some strategies used by organizations to align the incentives and objectives of different decision makers participating in the same nonprofit operation. The bulk of the chapter is devoted to bringing to the attention of the humanitarian operations community different candidate metrics valid for measuring nonprofit performance. These metrics are not necessarily unique to nonprofit operations but are used by practitioners and scholars in a large variety of nonprofit situations.

Keywords Nonprofit operations · Performance metrics · Modeling approaches · Managerial practices · Input metrics · Output metrics · Efficiency metrics · Costs · Equity · Equality

2.1 Introduction

This chapter is intended to provide information and references to the humanitarian operations community about a fundamental issue in nonprofit operations: how to measure the performance of an operation when the motivation for its implementation is not for profit. Throughout this chapter, the term nonprofit operation refers to any operation whose objective is not to maximize profit and which has a nondistribution constraint that restricts the feasible solution set. The concept of nondistribution constraint is borrowed from the economics literature, where this constraint ensures that "net earnings, if any, must be retained and devoted in their entirety to financing further production of the services that the organization was formed to

---

G. Berenguer (✉)
Krannert School of Management, Purdue University, West Lafayette, IN, USA
e-mail: gemmabf@purdue.edu

C. W. Zobel et al. (eds.), *Advances in Managing Humanitarian Operations*,
International Series in Operations Research & Management Science 235,
DOI 10.1007/978-3-319-24418-1_2

provide" (Hansmann 1980). In other words, the nondistribution constraint prohibits the distribution of residual earnings to individuals who exercise control over the firm. The existence of this constraint for nonprofit operations is a signal of trustworthiness because it dismisses some of the issues that can be brought when there are informational asymmetries under a for-profit setting. In addition to the existence of this nondistribution constraint, the goal of any nonprofit operation should not be profit, but it is not straightforward to convert the goal into measurable performance metrics. Nonprofit operations can be found in different contexts, such as humanitarian disaster relief, health care or community services. In this chapter we argue that context is important but the examples, strategies, and metrics studied in one applied area can be borrowed by other types of nonprofit operations that can learn and adapt the presented tools to their own contexts.

Multiple papers in the OR/MS literature have proposed metrics to evaluate the performance of generic supply chains or operations (Beamon 1999; Chan and Qi 2003; Shepherd and Gunter 2006), but most of them focus on for-profit cases, suggesting metrics and analyzing examples which revolve around the concept of profit or surplus. To find studies related to nonprofit evaluation one can look at the social science literature (e.g., Carman 2009; Alexander 2010), where these analyses naturally lack an operations' perspective.

Going back to the OR/MS literature, there is some work studying specific nonprofit performance metrics such as equity (e.g., Savas 1978) or inequality (e.g., Ogryczak 2000). In addition, there is also some literature related to the measurement of performance in specific nonprofit applied areas such as health care (Jacobs 2001) or humanitarian disaster relief (e.g., Beamon and Balcik 2008; Holguín-Veras et al. 2013). However, to the best of our knowledge, this chapter is the first work to provide an OR/MS overview on ways to measure the performance of any nonprofit operation. Because different nonprofit areas share this same challenge, there is value in collecting examples of tools, strategies, and metrics to measure performance from different applied areas. This provides clear value to the humanitarian operations community that is the targeted audience of this volume because it brings new approaches and metrics that can potentially be used in humanitarian operations.

The chapter has the following outline. Section 2.2 provides approaches employed to address issues related to modeling nonprofit operations and finding a common goal for all stakeholders in the operation. Section 2.2.1 compiles multiple theoretical models from the economics and OR/MS literature. Section 2.2.2 presents the best managerial practices that have been used by organizations to align the incentives and objectives of different decision makers participating in the same nonprofit operation. Section 2.3 presents different candidate metrics, some novel to the OR/MS community, for measuring nonprofit performance. These metrics are classified into three groups: input, output, and efficiency metrics. Finally, Sect. 2.4 concludes the chapter with a final example and a summary.

2.2 Modeling Methods and Strategies

The lack of a single common performance metric to measure nonprofit operations is one of the major and most distinctive challenges of this type of operations[1]. This challenge is two-fold because there is a lack of a common performance metric that can be used for all nonprofit operations and, at the same time, it is very challenging to find a single performance metric to evaluate a specific nonprofit operation. In addition, this challenge is exacerbated for many nonprofit operations that deal with complex supply chain structures. This complexity can be observed in different ways: the number and different type of participants (e.g., donors, suppliers, providers, volunteers), multiple revenue streams (e.g., donations, grants, fees, subsidies), multiple types of beneficiaries (e.g., those socially excluded), and difficult environments (e.g., difficult to reach areas, warzones).

Giving tools that might help find the right unique performance metric for each operation is not the goal of this chapter. In fact, we argue that this is not the best approach due to the nature of nonprofit operations as previously described. This section offers techniques and strategies to address two issues associated with the lack of having a single suitable performance metric. The first issue (2.2.1) consists of finding the most appropriate theoretical model that better represents the nonprofit operation at hand. The second issue (2.2.2) is related to finding best practices that can favor the convergence to the right common set of performance metrics when dealing with multiple myopic goals. These different goals correspond to multiple agents that have their own metrics and incentives to co-participate in the operation. Next we review literature that addresses these problems.

Modeling the Nonprofit Operation

Easley and O'Hara (1983) provide a formal model of a nonprofit firm that includes the nondistribution constraint and compare the performance of a for-profit and nonprofit firm. In their analysis, the nature of a firm is determined by the contract between the firm and society. Society's objective is to maximize expected social welfare given different allocation mechanisms and the firm's problem is modeled according to the manager's utility. If the selected contract fixes the manager's return then the firm is nonprofit. If the manager's return is unspecified, the firm is for-profit.

In economics, the so-called *profit-constrained firms* are those entities that care about other goals besides profit maximization and that restrict the profit to a specific amount. Sometimes this amount is fixed to zero. James and Rose-Ackerman (1986) refer to three different well-known economic models with a profit-constrained firm structure:

[1] Refer to Berenguer et al. (2014) for an overview of the most recurrent challenges in nonprofit operations and managerial strategies that help address each challenge.

- The Baumol's sales-maximizing firm model assumes oligopolistic markets where the firm attempts to maximize sales subject to a minimum profit constraint (Baumol 1967). The model claims that the oligopolist is in equilibrium at the price and output where realized profit is equal to the minimum acceptable profit.
- The Williamson's model with expense preference maximizes managerial utility subject to a minimum profit constraint (Williamson 1964). The utility function is also called expense preference function and is composed by: salary, status, prestige, and power. The model shows the higher preference of managers for staff expenditure as compared to profit.
- The regulated monopoly by Alchian and Kessel (1962) studies profit-constrained firms and compares the regulated versus unregulated monopolistic market. It claims that the profits of a regulated monopoly can be constrained by the regulatory agency's perception of a "fair" rate of return and the profits of an unregulated monopoly can be constrained by anti-trust actions when having excessive profits.

James and Rose-Ackerman (1986) classify models for nonprofit operations into three sets: maximization of the managers' objective (quality, quantity, product type, revenue, or revenue minus fundraising costs), maximization of an ideological goal subject to survival constraints, and maximization of a strong group's interest.

In the OR/MS field, the domain of multi-objective optimization is suitable to study nonprofit operations because of the lack of a single universal performance metric that could be used for most nonprofit operations. By selecting several (two or more) objectives in our analysis we are inherently stating that the analysis will most likely find a set of solutions that are non-dominated (Daskin 2010). Different methods help identify this set of solutions grouped into the trade-off curve. Some of the most well-known techniques are described in the following lines alongside with each technique's tradeoffs in terms of advantages and limitations.

- The weighted sum method converts the multi-objective problem of minimizing the vector of objectives into a scalar problem by defining a weighted sum of all the objectives. Using a standard optimization method the problem can be optimized. This method cannot guarantee to be able to obtain some of the solutions when the boundary of the feasible set of the problem is non-convex.
- The ε-constraint method consists in minimizing one objective and expressing the other objectives in the form of inequality constraints such that the value of each objective has to be smaller or equal than ε. This method can identify some of the solutions of non-convex boundaries that the weighted sum method might miss. Disadvantages of this method are the difficulty of selecting a correct ε and the use of hard constraints.
- Goal programming seeks to obtain a solution that minimizes the total deviation from the objectives set by management. Different types of mathematical programming models can be used to solve goal programming problems, linear, integer, nonlinear, etc. One of the major disadvantages of this technique is that the solutions found might not be Pareto optimal.

Managerial Strategies

According to scholars studying different types of nonprofit operations, there are some effective best practices that help address the challenges related to dealing with multiple myopic goals by different organizations involved in the same nonprofit operation. For example, Tomasini and Van Wassenhove (2009) and Altay and Labonte (2014) analyze this problem in humanitarian logistics and Kraiselburd and Yadav (2012) study it in the area of global healthcare. These are the main strategies:

1. the establishment of a leading coordinator that coordinates the operation from multiple and sometimes conflicting local objectives to a unified common objective.
2. the pooling of resources to avoid tracking each individual organization's resources and, subsequently, to avoid trying to accomplish different myopic goals in favor of a common overall nonprofit goal.
3. Periodic monitoring of the operation by the production and publication of evaluation reports. The performance metrics monitored and published are very important because they ultimately shape the overall objective of the nonprofit operation, which subsistence depends on the values of these performance metrics.
4. the establishment of high quality information management systems that permit communication between different organizations and thus facilitate effective decision-making.

In the next lines we illustrate strategies (1) and (2) with real humanitarian operational examples.

Example 1: The Role of UNJLC in the Aftermath of Natural Disasters Relief activities in the aftermath of large-scale disasters are usually run by multiple organizations, but collaboration between these organizations does not always occur. Samii and Van Wassenhove (2003) review the disasters in which the U.N. Joint Logistics Center (UNJLC) participated. The UNJLC was a collaborative initiative to share logistic capabilities of cooperating agencies[2]. It was initiated with the outbreak of the civil war in Zaire in 1997, when the World Food Program (WFP) and the U.N. Refugee Agency (UNHCR) had spare capacity in their aircraft which was shared with other relief organizations thanks to the coordination services provided by the UNJLC. From then on, the UNJLC was involved in most of the large-scale humanitarian relief efforts, but there were instances in which collaboration was not always smooth. For example, during the aftermath of the Gujarat 7.9 magnitude earthquake in India in 1981, UNJLC deployed late and left the disaster scene only 4 weeks after arriving. In contrast, the International Federation of Red Cross (IFRC) had its own coordination structure with the Indian authorities but the UNJLC was not in sync with them.

Example 2: HIV/AIDS Crisis in Botswana In 2000 a private-public alliance was launched in Botswana, one of the countries with the highest HIV prevalence rate,

[2] The UNJLC does not currently exist.

in response to its HIV/AIDs crisis. The founding partners were a pharmaceutical company (Merck) that provides the drugs and part of the funding, the government of Botswana that provides the health infrastructure and resources, and the Bill and Melinda Gates Foundation that provides the other major part of the funds. Within such an alliance, a third party was required to act as the coordinator represented by a multilateral organization, this partner was named African Comprehensive HIV/AIDS Partnership (ACHAP) and it still exists today. Scholars have studied this partnership and claim that the success of the initiative was linked to the role of ACHAP as catalyst and coordinator to the extent that it relieved the other actors from administrative and implementation burdens and "it tried to ensure the accomplishment of system goals" (Rangan et al. 2006).

2.3 Performance Metrics

We now present a list of performance metrics suitable for measuring nonprofit operations' performance. In measuring this performance, these metrics can be used individually or as part of a multi-objective model. To introduce these metrics we employ a framework based on the simplest representation of an operation: input, output, and efficiency metrics. *Input metrics* refer to the time and the value of the resources needed to run the operation. *Output metrics* are related to the operation's strategic goal and value the quantity, distribution or quality of product or service produced. Finally, *efficiency metrics* refer to the ability of producing maximum outputs with minimum inputs. From this high level classification we group metrics in subcategories that are common in the performance metrics literature but we also add other metrics that can be especially useful to measure nonprofit operations. For example, in the input metrics category we devote some paragraphs to talk about shortage costs as well as donations and in the output metrics category we pay significant attention to equity and equality measures.

Similar approaches as the one employed in this chapter have been used to study performance metrics in supply chain management in general. For example, Neely et al. (1995) identify two types of measures to evaluate supply chain performance: effectiveness and efficiency. Beamon (1999) classifies the measures in three types: resources, outputs, and flexibility. Chan and Qi (2003) employ a similar approach with three types: inputs, outputs, and composite measures. Gunasekaran et al. (2001) prefer to categorize performance depending on its strategic, operational or tactical focus. Shepherd and Gunter (2006) provide an exhaustive review of articles that study supply chain performance systems and metrics.

The metrics presented in this chapter are selected from the OR/MS field and other fields, such as economics and nonprofit management. Thus, this chapter suggests some specific metrics that have not been considered in OR/MS.

Input Metrics

We follow the generic high-level classification from Chan and Qi (2003) that separate input metrics in costs and time. However, we add donations as an additional input metric that is specific for nonprofit operations. In fact, costs and donations are a very good representation of the resources employed to run a nonprofit operation but cannot represent all input aspects of it. Time is the other key input metric that should be considered to evaluate a nonprofit operation, especially for humanitarian relief-oriented operations.

Costs In addition to considering the regular supply chain costs, managers should study the potential reduction of other costs that are more particular to nonprofit operations such as fundraising and volunteering labor costs. Ben-Ner and Van Hoomissen (1991) detail the main costs of entry and operations of a nonprofit organization (NPO):

- identifying and assembling a collection of stakeholders,
- determining whether collective demand is sufficient to cover costs,
- assembling inputs to produce the goods,
- developing and maintaining a control mechanism to cover stakeholders' interests,
- recruiting managers whose values are expected to cause only agency problems,
- identifying and convincing high-demanders (consumers) that its to their benefit to reveal demands and provide financial support,
- designing mechanisms to discourage stakeholders from free-riding on the activities of others.

In humanitarian relief, Beamon and Balcik (2008) state that the three main costs are the costs of supplies, distribution, and inventory holding. Pedraza-Martinez and Van Wassenhove (2013) highlight transportation costs as one of the most important costs and provide a novel humanitarian fleet management policy that differs from the standard commercial vehicle replacement policy. An empirical analysis shows the potential savings of a policy readjustment.

If a stock-out occurred in some of the nonprofit operations introduced so far, the loss and stock-out costs would be dramatically high because the goods and services provided in these settings are meeting "critical needs" (i.e. needs that are essential to the survival of the population, Nagurney et al. 2009). For example, Pierskalla (2005) studies the supply chain management of blood banks. This chain is simply composed of hospitals that act as blood banks (HBBs), in which patients get the transfusions, and community blood centers (CBCs), where the blood is collected from donors. The authors define the blood shortage costs at the CBCs as the set of costs of processing, handling, and transporting blood in an emergency. The shortage costs at the HBBs are based on the cost of keeping a buffer stock at the site or at other sites.

However, some scholars are studying ways to define stock-out cost in the humanitarian relief context, which valuation goes beyond the purely logistics costs. For example, Holguín-Veras et al. (2013) define the concept of deprivation cost as the economic valuation of the human suffering associated with a lack of access to a good or service. The paper provides some estimation foundations and candidate proxies to measure such costs and claims that the preferred objective function for post-disaster humanitarian logistic models should be social costs, which is the summation of logistic and deprivation costs.

Donations Donations are a type of resource used to run nonprofit operations and are one of the most distinctive features of nonprofit operations. Donations can be in cash or in-kind and can also be exogenous or endogenous. From Weisbrod (2000), exogenous donations "come to the organization essentially regardless of its activities" whereas endogenous donations are "influenced by the nonprofit's activities." Managers in nonprofit operations prefer exogenous donations since these funds allow the organization to freely focus on its mission. Volunteering time can be considered an in-kind donation and is typically measured using the estimate wage by which the volunteer labor would be replaced if not donated.

Although donations are freely offered to the nonprofit operation, there are costs related to their management that should be accounted for. For example, volunteering labor might incur costs of recruitment, training, supervision, and retention.

Time Labor is one of the most important resources and it can be measured in time units. This is especially convenient for nonprofit operations because it can indistinctively include paid and volunteer labor. The specific input metric that is most used is hours of labor, which is the number of hours used to produce a good or a service.

Time is a key component in humanitarian operations. These operations can be classified based on whether their main activities are relief or development. Relief activities are short-term activities focused on providing goods and services to minimize immediate risk to human health and survival, whereas development activities are longer-term and focused on community self-sufficiency and sustainability. Thus, time is a critical input factor that should be considered, especially in relief-oriented operations. In particular, speed is the main driver of humanitarian relief operations, where lead time[3] reduction facilitates the possibility of increasing speed (Tomasini and Van Wassenhove 2009).

Output Metrics

Output metrics should refer to the operation's goals and should characterize the quantity, distribution or quality of final product or service produced. Each nonprofit operation is run in a specific context and thus output metrics reflect the nature of

[3] Lead time is defined as the time between when a beneficiary order is placed and when it is received.

this context. For example, in the area of health care in developing countries, Yadav (2010) describes several key public health output measures such as mortality, morbidity, life expectancy, quality-adjusted life year (QALY), and disability-adjusted life year (DALY). Nonetheless, in this section we group the output metrics in sets of general metric categories, which can be used in different forms depending on specific applied areas. To facilitate comprehension, some context-specific examples are provided. The output metrics categories reviewed in this chapter are effectiveness, equity, equality, social welfare, and poverty reduction. We should warn the readers that these different categories of output metrics are not always clearly distinct but we believe are the best way to provide a classification of output metrics. For example, some scholars use the notions of equity and equality interchangeably.

Effectiveness Effectiveness is defined as a measure of the level of satisfaction of a need and the alleviation of its adverse impacts (Savas 1978). This measure can be found in multiple forms depending on the domain of application. Next, the use of effectiveness is illustrated in different applied areas in nonprofit operations:

- In the public service field, Savas (1978) suggest the following measures of effectiveness: expected aggregate utility, level of citizen satisfaction, mortality rates, and environmental factors.
- In healthcare in developing countries, Fleßa (2003) studies the optimal allocation of health care resources in Tanzania and suggests five different metrics to measure effectiveness: number of deaths, years of life lost, incidence of disease, prevalence of disease, and loss of quality of life.
- In the education field, Jauch and Glueck (1975) provide an empirical evaluation of the research performance of university professors. The metric employed is the number of publications in respectable journals.
- In humanitarian relief chains, Beamon and Balcik (2008) claim that effectiveness is correlated with metrics such as response time, number of items supplied, and supply availability. Response time is defined as the time between an order and its corresponding delivery. Short response time can be achieved when there is inventory availability and factors such as procurement and distribution are responsive.

In many situations not all beneficiaries can receive the service they need when they need it. Given this allocation problem, a way of defining assignments is by setting up the goal to maximize demand satisfaction or accessibility.

The OR/MS field has extensive work on location and routing analysis with the main purpose of increasing accessibility. If we restrict it to nonprofit operations, the literature is still abundant. For example, in the field of humanitarian logistics, Balcik and Beamon (2008) present a variant of the maximal covering location model to improve response time and the proportion of demand satisfied. In the context of community services, Francis et al. (2006) study the routing efficiency in interlibrary loan delivery with a model that describes a period vehicle routing problem (PVRP) and incorporates different levels of service. The objective is a combination of travel times, costs, and service benefits, where service benefits increase if higher delivery frequencies are selected. In the context of health care in the developing world, mobile facilities can provide care to rural areas. Doerner et al. (2007) present a multi-objective location-routing problem related to mobile facilities in which one of the objectives is to minimize the percentage of the population unable to reach a tour stop within a predefined maximum distance.

In health care, Griffin et al. (2009) study how OR tools can contribute to access to care. They claim that a good management of location of health care facilities, inventory of health products (e.g., blood platelets), and allocation (e.g., organ allocation) are essential to improving this access. Examples of these works are Griffin et al. (2008) that maximize the total weighted demand coverage of needy population served by studying the optimal location, number, and capacity levels of community health centers. A second example is Daskin and Dean (2004) that review location planning formulations in health care, where by accessibility they refer to the ability of patients to reach the health care facility or the health care providers to reach patients.

Equity According to equity theory, human beings believe that rewards and punishments should be distributed according to recipients' inputs or contributions (see Leventhal 1976). Rawls (1999) studies one of the most popular fairness metrics: the maximum of the minimum individual utility function $\max \min\left(u_{1,\ldots,}u_n\right)$, where u_i is the utility function of individual i. This metric represents the maximum of the poorest individual's utility, but this is one of many other measures of equity proposed in the economics literature.

In the public service literature, Savas (1978) claims that equity is as relevant as efficiency and effectiveness as performance measures. This author states that despite citywide services being efficient and effective, they could be inequitable if all beneficiaries are not treated similarly. Leclerc et al. (2011) review literature that models equity for allocating resources in public service systems with a special focus on the allocation of Emergency Medical Services (EMS). This review contains an interesting discussion on how to construct measures for equity. Balcik et al. (2009) look over the issue of equity in vehicle routing in the nonprofit and public sectors. They provide a summary of equity determinants that is presented Table 2.1.

Table 2.1 Summary of Balcik et al. (2009) sections on applications problems and equity determinants

Application	Equity determinant
Queuing systems	Service waiting time
Resource allocation	Amount of commodities allocated among spatially distributed groups
Workload allocation	Time required to compete tasks
Facility location	Distance between the facility and the group of beneficiaries
Vehicle routing	Arrival time

Equity is considered in facility location problems and Marsh and Schilling (1994) summarize pertinent research. The paper provides a rich theoretical review on equity principles along with twenty different equity measures for facility location.

Campbell (2008) studies vehicle routing in disaster response problems. The model minimizes the maximum arrival time and the average arrival time. These two equity measures are equivalent to minimizing the makespan and the sum of completion times in scheduling systems.

In the community action field, Lien et al. (2014) study food distribution by designing inventory and routing policies for a U.S. food bank. They maximize the minimum expected fill rate (the ratio of amount allocated to demand). A simple example with two nodes is shown here, where $x_i, i = 1,2$ are the amounts allocated and $d_i, i = 1,2$ are the observed demands. The objective is $Max\ E\left[\min\{\beta_1, \beta_2\}\right]$, where $\beta_i = \frac{x_i}{d_i}$ are the node fill rates $i = 1,2$.

In health care, Kaplan (2002) finds a middle-ground solution that addresses both equity rules and cost-efficiency in the study of HIV-prevention strategies. Earmarking a fraction of the total budget for equity alone purposes ensures equity. The remaining funds are allocated according to cost-effectiveness. Su and Zenios (2006) analyze the social benefits of organ transplants by studying the trade-off between the expected aggregate utility (an effectiveness measure) and the minimum utility across all candidates (an equity measure). Zenios et al. (2000) suggest two measures of inequity regarding the allocation of kidneys to patients: the likelihood of transplant for various types of patients and the differences in mean waiting times across patient types.

Related to theoretical OM models, Bertsimas et al. (2011) define the "price of fairness" in resource allocation problems as the relative loss between a fair allocation and the allocation that maximizes the sum of player utilities. They study two fairness criteria, max-min and proportional fairness, and they provide bounds on the "price of fairness" related to each of these measures.

To conclude, we note that most of the studies that employ equity as an objective are accompanied by other measures of efficiency or effectiveness.

Equality Some authors have used the notions of equality and equity interchangeably but these two notions are not synonyms. While equality describes the distribution of goods where one can not obtain more than others, equity is related to fairness and/or justice, i.e. one might obtain more than others but this allocation might still be considered equitable or fair. Atkinson (1970) is an early paper on inequality and income, which notes that conventional inequality measures in empirical studies are:

- variance V^2,
- coefficient of variation $\frac{V}{\mu}$,
- relative mean deviation $\int_0^{\bar{y}} \left| \frac{y}{\mu} - 1 \right| f(y)\,dy$,
- standard deviation of logarithms $\int_0^{\bar{y}} \log\left(\frac{y}{\mu}\right)^2 f(y)\,dy$,

where *f(y)* is the frequency distribution of income *y*, $\overline{y}$ is the upper bound of the income range, μ is the mean of the distribution, and *V* is the standard deviation of the distribution. Theil (1967) suggests a measure derived from Shannon's measure of information entropy, the Theil's (Entropy) index $T = \sum_{i=1,\ldots,n} s_i \left[\log(s_i) - \log\left(\frac{1}{n}\right) \right]$, where s_i is the share of the *ith* group in total income and *n* is the total number of groups.

The Gini coefficient (G) is one of the most popular measures of inequality and it is defined as the ratio of the area between the diagonal line and the *Lorenz curve* $(L(x))$, denoted as area A, and the area under the *Lorenz curve* (this area is $B = \int_0^1 L(x)dx.$). Hence, the Gini coefficient is defined as $G = \frac{A}{A+B} = \frac{A}{0.5} = 1 - 2B = \int_0^1 L(x)dx.$ In some cases, as in Mandell (1991), the Gini coefficient can be applied without referencing the *Lorenz curve*. This index is defined as

$$G = \sum_i \sum_{j>i} \frac{|q_j S_i - q_i S_j|}{\sum_i S_i}$$

where Q_i is the number of individuals contained in area *i*, $q_i = \frac{Q_i}{\sum_j Q_j}$ is the proportion of individuals in area *i* and S_i is the number of service units received in area *i*. This definition of *G* refers to the average "perceived net envy level" associated with the distribution of public services (Fig. 2.1).

In location problems, inequality metrics are often jointly used with other types of metrics. For example, Ogryczak (2000) develops a multiple criteria model where all distances among service recipients are considered as the set of uniform criteria

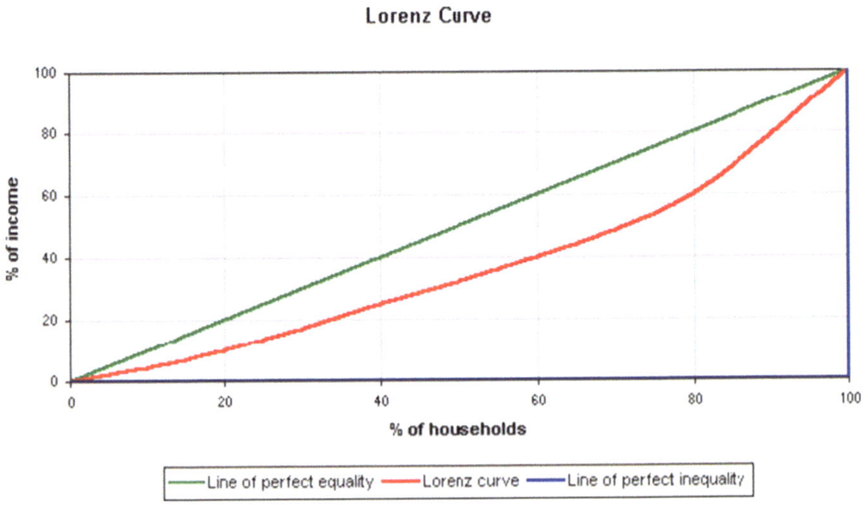

Fig. 2.1 Lorenz curve

to be minimized. A bi-criteria optimization problem is suggested for mean distance and absolute inequality metrics, such as maximum deviation, mean deviation, or mean difference. Some bounded trade-offs are defined to assure that solutions are equitably efficient.

Social Welfare Baumol (1965) is one of the first works to investigate the idea of maximizing social welfare. This economics paper studies the resource allocation problem in terms of an entire community's welfare using two approaches: the maximization of consumers' and producers' surpluses and the Pareto optimality approach. Surplus is understood as the sum of net gains, i.e. the area between the demand curve and the horizontal line representing the price paid for the commodity. The Pareto optimality model maximizes an arbitrary individual's utility while ensuring that no other individual's utility diminishes. Letting x_{ij} be the quantity of commodity i consumed by individual j and $u_j = (x_{1j},\ldots,x_{nj})$ the person j's utility function, Pareto optimality requires:

$$\max u_1 = (x_{11},\ldots,x_{n1})$$

$$\text{s.t.} u_2 = (x_{12},\ldots,x_{n2}) \geq k_2,$$

$$\ldots$$

$$u_m = (x_{1m},\ldots,x_{nm}) \geq k_m.$$

Another example of a welfare measure is the weighted sum of household consumer surpluses (Feldstein 1972). This paper solves a public pricing problem that incorporates customer distributional aspects using the welfare measure $W = N\int_0^{\infty} S(p,y)u'(y)f(y)dy$, where N is the number of households in the population being served, $S(p,y)$ is the consumer surplus of a household with income y and price of good p, $u'(y)$ is the household's marginal social utility of income, and $f(y)$ is the distribution of household's income.

OR/MS researchers have extensively adopted the use of social welfare (or net social benefit) functions as public sector objective metrics. To illustrate, Erlenkotter (1977) solves the uncapacitated facility location problem when prices and location are decided simultaneously. Three solution tools are presented: the private sector solution tool, which maximizes profits, the public sector solution tool, which maximizes net social benefits, and a quasi-public approach, which requires revenues to be above the level required to cover costs. The author defines net social benefit as follows. The net benefit is $(\textit{willingness to pay})-(\textit{price charged for total quantity } D)=p(\eta)-p(D)$. The total consumer surplus is defined as $S(D)=\int_0^D [p(\eta)-p(D)]d(\eta)$. The total social benefit is defined as $S(D)+R(D)=\int_0^D p(\eta)d(\eta)$, where $R(D)=p(D)D$ is total revenue. Among different observations, the work describes how locations with more inelastic demands bear major price increases in the transition from the public to the quasi-public solution.

Poverty Reduction Two simple poverty reduction measures have been proposed in the economics literature:

- the headcount ratio $\left(H = \frac{q}{n}\right)$
- the income gap ratio $\left(I = \sum_{i \in S(Z)} \frac{g_i}{qz}\right)$,

where $z > 0$ is the predetermined poverty line, *S(z)* is the set of people with income no higher than z, $y = (y_1, y_2, \ldots, y_n)$ is the vector of household incomes in increasing order, q is the number of people with income $y_i \le z$, $g_i = z - y_i$, is the income shortfall of the *ith* household, and n is the total population size. Sen (1976) analyzes the shortcomings of these and similar measures and introduces an axiomatic approach to poverty measures. Foster et al. (1984) study a popular class of poverty measures called decomposable poverty measures that accomplish some of Sen's axioms. An example is the normalized weighted sum of the income shortfalls of the poor $P_\alpha(y;z) = \frac{1}{n}\sum_{i-1}^{q}\left(\frac{g_i}{z}\right)^\alpha$, where $\alpha \ge 0$ is a measure of poverty aversion. For detailed analysis of the advantages and shortcomings of several poverty measures refer to Clark et al. (1995), Atkinson (1987), and Kanbur (1987).

There is a body of macroeconomic literature on aid allocation and poverty reduction strategies for governments and international agencies. Collier and Dollar (2002) develop a poverty-efficient aid allocation scheme and compare it with real data. They use three poverty measures: headcount, poverty gap, and squared poverty gap. The corresponding optimization problem maximizes poverty reduction subject to total available aid.

$$\max \sum_i G^i \alpha^i h^i N^i$$

$$s.t. \quad \sum_i A^i y^i N^i = \bar{A}, \quad A^i \ge 0,$$

where G is growth, α is the elasticity of poverty reduction with respect to income, h is one of the three aforementioned measures of poverty, N is population, y is per capita income, $\bar{A}$ is the total amount of aid, and A^i aid in country i.

Mosley et al. (2004) present a similar optimization problem where poverty elasticities vary depending on corruption and inequality. The leverage of public expenditures is explicitly accounted for in the poverty reduction objective function, $\max \sum_i G^i \alpha^i h^i N^i N^{i-\beta} + \Omega^i \gamma^i h^i N^i N^{i-\beta}$. This argument is denoted by Ω and it is called the pro-poor expenditure (PPE) index, $\gamma = \frac{\partial h}{\partial PPE}$, and β is the small-country bias in aid allocation.

There is a specific stream of economics literature on vulnerability-based poverty measures. Ligon and Schechter (2003) define vulnerability as the level of a household's well-being, which depends on income or consumption, and the risks the household faces. They take a utilitarian approach and construct the mathematical definition of vulnerability for household *i* as $V^i(c) = U^i(z) - E\left(U^i\left(c^i\right)\right)$, where c^i is the consumption of household *i*, U^i is a strictly increasing, weakly concave function mapping consumption expenditures into the real line, and z is the poverty line. This measure decomposes into measures of poverty, aggregate risk, and idiosyncratic risk and minimizing this measure is equivalent to maximizing the utilitarian social welfare function $\left(\Sigma_{i=1\ldots n} E\left(U^i\left(c^i\right)\right)\right)$ subject to an aggregate resource constraint.

Another tool to measure vulnerability adapts the standard measures of poverty (P_α) (where α is a measure of risk aversion) to a non-deterministic setting through an expected poverty measure (Foster et al. 1984). For example, Kamanou and Morduch (2004) present the vulnerability measure $E\left(P_{\alpha,t+1}\right) - P_{\alpha,t}$, where the subscripts *t* and *t + 1* denote the current and future time periods.

To the best of our knowledge, poverty reduction measures have not been used in the OR/MS field.

Efficiency Metrics

Efficiency is defined as the ratio of output to total input. Total input is typically measured in dollars, manpower or time while the unit of measurement for output varies and depends on the type of product or service offered. The concept of efficient frontier provides a formal representation of efficiency, where a set of optimal decision making units (DMUs) are efficient compared to other inefficient DMUs. The DMUs can represent different organizations or different departments in the same organization. The efficiency metric measures the ability of a DMU to produce outputs given its inputs (resources). It is challenging to establish the correct level of expected output that a DMU needs to achieve in order to be categorized as efficient. Benchmarking is a good tool to address this challenge, that can be intuitively represented in a graph such as the one in Fig. 2.2, where DMUs A, B, C, and D are on the *efficient frontier* and DMU E is inefficient and thus below the frontier. Note that for E to be on the efficient frontier multiple combinations of inputs and outputs are possible. For example a possibility is to keep the same input value but try to increase Δ its output level. Practitioners might wonder how this improvement can be achieved for DMU E, and this is where the DMUs that are on the efficient frontier located close to E become handy. DMU E can learn from the best practices and strategies implemented by the neighboring efficient DMUs B and C (its reference set) because these DMUs are obtaining higher outputs with similar input levels. In nonprofit operations, a simple example can be found in measuring hospital efficiency, where the input variable could be "resources used in a hospital" and the output variable could be "number of patients treated" (Jacobs 2001).

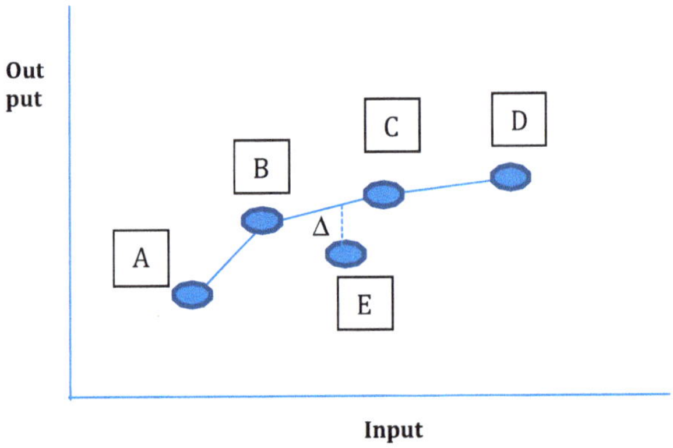

Fig. 2.2 DMUs represented in the input/output space

Data envelopment analysis (DEA) and stochastic frontier analysis (SFA) are two of the most heavily used empirical methods to evaluate efficiency. DEA was introduced by Charnes et al. (1978) and is a nonparametric linear programming method for estimating production frontiers. SFA was developed by Aigner et al. (1977) and is a parametric method for estimating the production frontier via regression techniques. In the nonprofit context, these tools are mainly employed to measure hospital efficiency (e.g. Jacobs 2001), global health programs (Greene 2004; Berenguer et al. 2015), and educational efficiency (e.g., Mancebon and Molinero 2000). DEA and SFA are appropriate candidate tools to measure efficiency in humanitarian settings due to being non-parametric and stochastic methods, respectively. These characteristics provide a more flexibility measure of efficiency because no closed-form or known behavior of the process is required, respectively.

Technical and Allocative Efficiency Economists have studied different types of efficiency measures that focus on different decision aspects. While *technical efficiency* concerns the efficient use of inputs given fixed outputs or the efficient use of outputs given fixed inputs, *allocative efficiency* focuses on producing or distributing the types of goods or services that are more desirable to society. In other words, *technical efficiency* is achieved when organizations combine inputs to produce a given output as inexpensively as possible or, equivalently, producing the maximum output possible from a given set of inputs. The previous paragraphs' discussion on efficiency refers to technical efficiency. *Allocative efficiency* measures the organization's success in choosing an optimal set of inputs to be offered and distributed given a set of input prices. In the context of a nonprofit operation, this will concern the distribution of resources in a population to maximize the nonprofit objective of the operation given resource constraints. The global health area studies these two concepts, where a measure of technical efficiency would be "antiretroviral treatment person-years gained per $1000" and a measure of allocative efficiency would be "malaria cases averted per $1000" (Glassman et al. 2013).

Flexibility Flexibility is the ability to respond to a changing environment in a timely manner. This notion has been traditionally studied in manufacturing systems

Table 2.2 Flexibility types in a supply chain (Beamon 1999)

Flexibility type	Definition (the ability to …)	Measurement
Volume flexibility	Change the output level of products produced	Long-run proportion of demand that can be met by the supply chain
Delivery flexibility	Change planned delivery dates	Proportion of excess slack across all jobs
Mix flexibility	Change the variety of products produced	Changeover time from old to new product mix
New product flexibility	Introduce and produce new products	Time required to add new products

(Gupta and Goyal 1989). Beamon (1999) considers flexibility as one of the three different categories to measure supply chain performance and quantitatively defines four different types of flexibility in the supply chain as summarized in Table 2.2.

In humanitarian relief, Beamon and Balcik (2008) derive its three main flexibility factors:

- ability to respond to different magnitudes of disasters (number of supplies provided during the critical time period for relief),
- time to respond to disasters (minimum response time),
- ability to provide different items (number of different items provided per particular time).

A similar notion to flexibility is resilience, where a company or supply chain is resilient when it has the ability to, and speed at which it can, return to its normal performance level[4] following a disruption (Sheffi 2006). Sheffi explains how resilience can be achieved through redundancy (via underutilization of capacity or safety stock) and through building flexibility. Zobel and Khansa (2014) provide a multi-criteria decision analysis (MCDM) approach to quantitatively measure resilience in the presence of a multi-event disaster situation. The analysis can provide an ordered ranking of disaster scenarios that helps determining more effective mitigation and recovery plans.

Sustainability Sustainability is relevant for development operations and can be defined either as covering all costs or covering all costs except capital. Leonard et al. (2007) describe how a microfinance NPO (Opportunity International) defines sustainability, where they divide the concept in two domains: operational and financial sustainability. *Operational sustainability* is the ability of the NPO to cover its lending expenses with income earned by its lending operations. So, when operational sustainability is 100%, the NPO earns what it spends, without considering inflation or the cost of borrowing at market interest rates. *Financial sustainability* is the capability of an NPO to cover lending expenses and its cost of capital. This definition assumes that the organization takes grants and subsidies as if they were

[4] Performance levels are related to production, service, fill rate, etc.

obtained from commercial sources. If financial sustainability is more than 100 %, the NPO earns a surplus and can expand.

Sustainability in supply chains is reviewed by different publications such as Kleindorfer et al. (2005) and Seuring and Müller (2008). The latter defines sustainable supply chain management as "the management of material, information and capital flows as well as cooperation among companies along the supply chain while taking goals from three dimensions of sustainable development: economic, environmental, and social". These three dimensions are also used as a framework in Clift (2003) to review metrics for supply chain sustainability. For example, the concavity of the segmented curve depicted in Fig. 2.3 is suggested as measure of supply chain unsustainability. This figure is composed by the abscissa that aggregates the Value Added along the supply chain (VA = sales−costs of goods, raw materials and services purchased) and the ordinate axis that refers to any environmental impact indicator appropriated to that supply chain activity (e.g., waste reduction or atmospheric emissions). The origin represents primary resources, point A the finished product, point B the finished product with reused components, the concave segmented curve between O and A shows the supply chain, and the gradient of the chord, *OA*, represents the Overall Business Impact Assessment (OBIA) of a specific environmental impact indicator *i* defined as

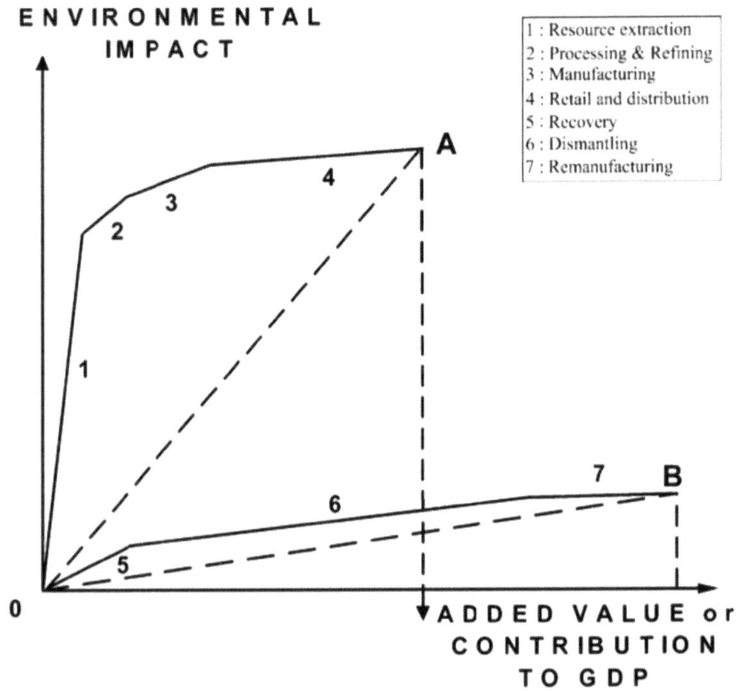

Fig. 2.3 Accumulation of economic value and environmental impact along the supply chain (Clift 2003)

$$\phi_i = \frac{\textit{Impact on category i / Value of business}}{\textit{Total anthropogenic contribution to impact category / Total global economic activity}}$$

Note that when this gradient is reduced the overall eco-efficiency of the product is improved and so when the curve is straight there is more equitable distribution of impacts and benefits along the supply chain.

Sustainable development is another notion defined as "the development that meets the needs of the present without compromising the ability of future generations to meet their own needs" (Brundtland 1987). Pearce and Atkinson (1993) define a sustainable development indicator for a country as $Z_1 = \frac{S}{Y} - \frac{\delta_M}{Y} - \frac{\delta_N}{Y}$, where S is savings, Y is income, δ_M is the value of depreciation on man-made capital, and δ_N is the value of depreciation on natural capital.

2.4 Conclusion

To conclude the chapter we first provide a simple example that illustrates some of the most basic challenges in selecting the best performance metric. Johnson et al. (2005) studied the location of senior center agencies in Allegheny County (PA), a suburban area inaccessible via mass transit, with restrictive governmental budget, and variable demand over time. After having interviewed agency administrators, the authors needed to make a recommendation to policy-makers, so they studied a variety of metrics to represent the objectives of the problem, which are to improve accessibility and respond to changing demands of service. The metrics considered were: total demand served (e.g., meals), demand-weighted distance, facility utilization, demand splits, and proximity of demand to serving facilities. The two metrics selected were demand-weighted distance and total unserved demand. This selection was due to different factors: it provided the most efficient solution based on facility utilization, the most effective solution based on total demand served, and it was the most robust model against parameter changes.

The example illustrates the variety of candidate metrics that a nonprofit operation could have and how the context of the problem and the solution characteristics play a role in the selection process. Nonetheless note that, despite being context specific, the metrics in this example can also be useful to other nonprofit situations such as the location of emergency service facilities. Generally, this chapter expects to be of value to scholars and practitioners in humanitarian operations by providing a number of strategic tools and metrics to measure performance.

The chapter starts discussing the lack of having a single suitable performance metric to evaluate nonprofit operations as "profit" is for for-profit operations. Focusing on finding the most desirable single metric is not the goal of this chapter. We instead devote a section to present theoretical models that can represent a nonprofit operation and best practices to deal with the challenge of having multiple myopic goals from multiple stakeholders of the same nonprofit operation. Most of the models presented in this section consist of multi-goal approaches that are utilized to integrate different relevant performance metrics and constraints coming from the diverse stakeholders and environments. In particular, special focus has been given to economic models, multi-objective optimization models, and managerial best practices.

The bulk of the chapter brings to the attention of the humanitarian operations community different performance metrics suitable for measuring nonprofit operations. These metrics will be used individually or as part of a multi-objective model and are classified as input, output, and efficiency metrics. While some of these metrics are standard performance metrics others, such as the ones related to poverty reduction, are novel to the OR/MS community and especially relevant to humanitarian operations.

The ultimate objective of running a nonprofit operation is to achieve its goal or mission. This will be accomplished by continuous operational improvements obtained by making the right decisions after properly evaluating the operation. Thus, improving the evaluation process is key to the success of any nonprofit operation.

References

Aigner D, Lovell CAK, Schmidt P (1977) Formulation and estimation of stochastic frontier production function models. J Econom 6(1):21–37

Alchian AA, Kessel RA (1962) Competition, monopoly, and the pursuit of pecuniary gain. Asp Lab Econ 157:3–33

Alexander J, Brudney JL, Yang K (2010) Symposium: accountability and performance measurement: the evolving role of nonprofits in the hollow state. Nonprofit Volunt Sect Q 39(4). doi:10.1177/0899764010369662

Altay N, Labonte M (2014) Challenges in humanitarian information management and exchange: evidence from Haiti. Disasters 38(s1):50–72

Atkinson AB (1970) On the measurement of inequality. J Econom Theory 2(3):244–263

Atkinson AB (1987) On the measurement of poverty. Econom: J Econom Soc 55:749–764

Balcik B, Beamon BM (2008) Facility location in humanitarian relief. Int J Logist Res Appl 11(2):101–121

Balcik B, Iravani S, Smilowitz K (2009) A review of equity in nonprofit and public sector: a vehicle routing perspective. Encyclopedia of Operations Research and Management Science

Baumol WJ (1965) Economic theory and operations analysis. Englewood Cliffs, New Jersey

Baumol WJ (1967) Business behavior, value and growth. Harcourt, Brace & World

Beamon BM (1999) Measuring supply chain performance. Int J Oper Prod Manag 19:275–292

Beamon BM, Balcik B (2008) Performance measurement in humanitarian relief chains. Int J Publ Sect Manag 21(1):4–25

Ben-Ner A, Van Hoomissen T (1991) Nonprofit organizations in the mixed economy. Ann Publ Coop Econo 62(4):519–550

Berenguer G, Richter A, Shen Z-J (2014) Challenges and strategies in managing nonprofit supply chains: an OR/MS perspective. Working paper, Purdue University

Berenguer G, Iyer A, Yadav P (2015) Disentangling the efficiency drivers in country-level global health programs: an empirical study. Working paper, Purdue University

Bertsimas D, Farias VF, Trichakis N (2011) The price of fairness. Oper Res 59(1):17–31

Brundtland GH (1987) World commission on environment and development. Our common future. Oxford University Press, Oxford, pp 8–9

Campbell AM, Vandenbussche D, Hermann W (2008) Routing for relief efforts. Transp Sci 42(2):127–145
Carman J (2009) Nonprofits, funders, and evaluation: accountability in action. Am Rev Publ Adm 39(4):374–390
Chan FTS, Qi HJ (2003) An innovative performance measurement method for supply chain management. Supply Chain Manag: Int J 8(3):209–223
Charnes A, Cooper WW, Rhodes E (1978) Measuring the efficiency of decision making units. Eur J Oper Res 2(6):429–444
Clark J (1995) The state, popular participation, and the voluntary sector. World Dev 23(4):593–601
Clift R (2003) Metrics for supply chain sustainability. Clean Technol Environ Policy 5(3):240–247
Collier P, Dollar D (2002) Aid allocation and poverty reduction. Eur Econ Rev 46(8):1475–1500
Daskin M (2010) Service science. Wiley, New York
Daskin MS, Dean LL (2004) Location of health care facilities. Oper Res Health Care: Handb Methods Appl 70:43–76
Doerner K, Focke A, Gutjahr WJ (2007) Multicriteria tour planning for mobile healthcare facilities in a developing country. Eur J Oper Res 179(3):1078–1096
Easley D, O'Hara M (1983) The economic role of the nonprofit firm. Bell J Econ 14(2):531–538
Erlenkotter D (1977) Facility location with price-sensitive demands: private, public, and quasi-public. Manag Sci 24(4):378–386
Feldstein MS (1972) Distributional equity and the optimal structure of public prices. Am Econ Rev 62:32–36
Fleßa S (2003) Priorities and allocation of health care resources in developing countries: a case-study from the Mtwara region, Tanzania. Eur J Oper Res 150(1):67–80
Foster J, Greer J, Thorbecke E (1984) A class of decomposable poverty measures. Econometrica 52(3):761–66
Francis P, Smilowitz K, Tzur M (2006) The period vehicle routing problem with service choice. Transp Sci 40(4):439–454
Glassman A, Fan V, Over M, Silverman R, McQueston K, Duran D (2013) More health for the money: a practical agenda for the global fund and its partners. Technical report, Center for Global Development
Greene W (2004) Distinguishing between heterogeneity and inefficiency: stochastic frontier analysis of the world health organization's panel data on national health care systems. Health Econ 13(10):959–980
Griffin PM, Scherrer CR, Swann JL (2008) Optimization of community health center locations and service offerings with statistical need estimation. IIE Trans 40(9):880–892
Griffin PM, Savelsbergh M, Swann JL (2009) The health care system: operations research and improving access. Technical report, Citeseer
Gunasekaran A, Patel C, Tirtiroglu E (2001) Performance measures and metrics in a supply chain environment. Int J Oper Prod Manag 21(1/2):71–87
Gupta YP, Goyal S (1989) Flexibility of manufacturing systems: concepts and measurements. Eur J Oper Res 43(2):119–135
Hansmann HB (1980) The role of nonprofit enterprise. Yale Law J 89:835–901
Holguín-Veras J, Pérez N, Jaller M, Van Wassenhove LN, Aros-Vera F (2013) On the appropriate objective function for post-disaster humanitarian logistics model. J Oper Manag 31:262–280
Jacobs R (2001) Alternative methods to examine hospital efficiency: data envelopment analysis and stochastic frontier analysis. Health Care Manag Sci 4(2):103–115
James E, Rose-Ackerman SS (1986) The nonprofit enterprise in market economics. Harwood Academic Publishers, London
Jauch LR, Glueck WF (1975) Evaluation of university professors' research performance. Manag Sci 22(1):66–75
Johnson MP, Gorr WL, Roehrig S (2005) Location of service facilities for the elderly. Ann Oper Res 136(1):329–349
Kamanou G, Morduch J (2004) Measuring vulnerability to poverty. Insurance against poverty. Oxford University, Oxford, pp 155–75

Kanbur SMR (1987) Measurement and alleviation of poverty: with an application to the effects of macroeconomic adjustment. Staff Papers-International Monetary Fund, 60–85
Kaplan EH, Merson MH (2002) Allocating HIV-prevention resources: balancing efficiency and equity. Am J Publ Health 92(12):1905–1907
Kleindorfer PR, Singhal K, Van Wassenhove LN (2005) Sustainable operations management. Prod Oper Manag 14(4):482–492
Kraiselburd S, Yadav P (2012) Supply chains and global health: an imperative for bringing operations management scholarship into action. Prod Oper Manag 22(2):377–381
Leclerc PD, McLay LA, Mayorga ME (2011) Modeling equity for allocating public resources. Commun- Based Oper Res 67:97–118
Leonard HB, Epstein MJ, Tritter M (2007) Opportunity international: measurement and mission. Harvard Business School HBS case N9–307–067
Leventhal GS (1976) What should be done with equity theory? New approaches to the study of fairness in social relationships. Department of Psychology, Wayne State University, Detroit, Michigan 48202
Lien R, Iravani SMR, Smilowitz KR (2014) Sequential resource allocation for nonprofit operations. Oper Res 62(2):301–317
Ligon E, Schechter L (2003) Measuring vulnerability. Econ J 113(486):95–102
Mancebon MJ, Molinero CM (2000) Performance in primary schools. J Oper Res Soc 51(7):843–854
Mandell MB (1991) Modelling effectiveness-equity trade-offs in public service delivery systems. Manag Sci 37(4):467–482
Marsh MT, Schilling DA (1994) Equity measurement in facility location analysis: a review and framework. Eur J Oper Res 74(1):1–17
Mosley P, Hudson J, Verschoor A, Down C (2004) Aid, poverty reduction and the 'new conditionality'. Econ J 114(496):F217–F243
Nagurney A, Yu M, Qiang Q (2009) Supply chain network design for critical needs with outsourcing. Pap Reg Sci 90(1):123–142
Neely A, Gregory M, Platts K (1995) Performance measurement system design. Int J Oper Prod Manag 15(4):80–116
Ogryczak W (2000) Inequality measures and equitable approaches to location problems. Eur J Oper Res 122(2):374–391
Pearce DW, Atkinson GD (1993) Capital theory and the measurement of sustainable development: an indicator of "weak" sustainability. Ecol Econ 8(2):103–108
Pedraza-Martinez AJ, Van Wassenhove LN (2013) Vehicle replacement in the international committee of the Red Cross. Prod Oper Manag 22(2):365–376
Pierskalla W (2005) Supply chain management of blood banks. Oper Res Health Care 70(2):103–145
Rangan S, Samii R, Van Wassenhove LN (2006) Constructive partnerships: when alliances between private firms and public actors can enable creative strategies. Acad Manag Rev 31(3):738–751
Rawls J (1999) A theory of justice. Belknap Press, Cambridge
Samii R, Van Wassenhove LN (2003) The United Nations Joint Logistics Centre (UNJLC): the genesis of a humanitarian relief coordination platform. INSEAD case study, Paris Ref 603–010–1
Savas ES (1978) On equity in providing public services. Manag Sci 24(8):800–808
Sen A (1976) Poverty: an ordinal approach to measurement. Econom: J Econom Soc 44(2):219–231
Seuring S, Müller M (2008) From a literature review to a conceptual framework for sustainable supply chain management. J Clean Prod 16(15):1699–1710
Sheffi Y (2006) Resilience reduces risk. Logist Q 12(1):12–14
Shepherd C, Gunter H (2006) Measuring supply chain performance: current research and future directions. Int J Prod Perform Manag 55(3/4):242–258
Su X, Zenios SA (2006) Recipient choice can address the efficiency-equity trade-off in kidney transplantation: a mechanism design model. Manag Sci 52(11):16–47

Theil H (1967) Economics and information theory. North-Holland, Amsterdam
Tomasini R, Van Wassenhove LN (2009) Humanitarian Logistics. Palgrave Macmillan, UK
Weisbrod BA (2000) To profit or not to profit: the commercial transformation of the nonprofit sector. Cambridge University Press, UK
Williamson OE (1964) The economics of discretionary behavior: managerial objectives in a theory of the firm. Prentice-Hall Englewood Cliffs, NJ
Yadav P (2010) Improving public health in developing countries through operations research. Wiley Encyclopedia of Operations Research and Management Science
Zenios SA, Chertow GM, Wein LM (2000) Dynamic allocation of kidneys to candidates on the transplant waiting list. Oper Res 48(4):549–569
Zobel CW, Khansa L (2014) Characterizing multi-event disaster resilience. Comput Oper Res 42:83–94

Chapter 3
Disaster Response Logistics: Chief Findings of Fieldwork Research

José Holguín-Veras, Miguel Jaller, Felipe Aros-Vera, Johanna Amaya, Trilce Encarnación and Tricia Wachtendorf

Abstract This chapter discuses Disaster Response Logistics in connection with the Emergency Functions conducted by relief agencies, and identifies under-studied areas where logistic research could enhance operational effectiveness. The chapter establishes the socio-technical nature of Disaster Response Logistics, and the implications for disaster mitigation, preparation, and response. It relays the chief findings of the fieldwork research conducted by the authors, which support the development of policies to enhance advance preparation and coordination of disaster response activities.

Keywords Disaster management · Disaster plans · Material convergence · Disaster preparedness · Disaster response logistics · Humanitarian logistics · Emergency support functions

J. Holguín-Veras (✉) · J. Amaya · T. Encarnación
Center for Infrastructure, Transportation, and the Environment, and the VREF Center of Excellence for Sustainable Urban Freight Systems, Civil and Environmental Engineering, Rensselaer Polytechnic Institute, 110 8th St, Room JEC 4030, Troy, NY 12180, USA
e-mail: jhv@rpi.edu

F. Aros-Vera
Department of Industrial and Systems Engineering Russ College of Engineering and Technology, Ohio University, Stocker Center 277, 45701 Athens, OH, USA
e-mail: aros@ohio.edu

T. Encarnación
e-mail: encart@rpi.edu

M. Jaller
Civil and Environmental Engineering, University of California, Davis. One Shields Ave, Ghausi Hall, Room 3143, Davis, CA 95616, USA
e-mail: mjaller@ucdavis.edu

J. Amaya
Industrial Engineering, Universidad del Norte, Barranquilla, Colombia
e-mail: amayaj@rpi.edu

T. Wachtendorf
Department of Sociology and Criminal Justice, Disaster Research Center, University of Delaware, 166-F Graham Hall, Newark, DE 19716, USA
e-mail: twachten@udel.edu

C. W. Zobel et al. (eds.), *Advances in Managing Humanitarian Operations,*
International Series in Operations Research & Management Science 235,
DOI 10.1007/978-3-319-24418-1_3

3.1 Introduction

The twenty-first century has provided numerous examples of the devastating effects that disasters can produce on societies. Profound and complex, these effects impact all facets of the socio-technical processes at the core of modern life: social networks, technical activities performed by individuals, and the supporting systems (the operational environment) that rely on other socio-technical systems such as communications and finance. Adding to the challenge, there is great uncertainty about when and where disasters may occur, and the potential severity of their impacts. Disaster mitigation, preparedness, and response plans must be improved to minimize vulnerabilities, and to ensure that what is known can be leveraged to improve the effectiveness of future efforts. Central to these efforts is the field of Disaster Response Logistics (DRL), which is the detailed coordination, management, and planning of activities related to disaster response; the handling and coordination of the associated flows of people, information, equipment, and materials. DRL is a broad field that includes the logistics of a wide range of emergency functions, such as search and rescue, power restoration, volunteer and donation management, and distribution of relief supplies. It is tremendously important to improve DRL operations by means of multi-disciplinary research; such approache has a better chance of designing new paradigms of DRL operations that account for the multi-faceted complexity of societal needs and impacts in disaster situations.

One of the major challenges for developing innovative DRL concepts is the difficulty of studying real-life DRL operations, since they occur with relatively low frequency, and are very complex and dynamic (Holguín-Veras et al. 2014b). Moreover, the transient nature of the response work—done mostly by volunteers who participate for limited amounts of time—prevents the accumulation of experiences central to knowledge creation. As a result, relatively few individuals gain multi-event involvements in disaster response, and those who do typically are not inclined to publish detailed accounts of their experiences. The net effect is that the research community does not have ready access to solid information about the realities of disaster operations, and the corresponding research needs. There are other challenges. If and when lessons are identified in after-action-reports, the authors do not always indicate which agency, or who, must implement the recommendations (Ontko et al. 2007). Moreover, there is always the challenge of overcoming institutional inertia, doubts about the validity of findings, or general apathy. Changing institutional practices is a challenging endeavor. There is a great need for fieldwork research on DRL, and widespread dissemination of detailed accounts of what transpired in DRL operations and the lessons learned. This will help the research community focus on priority areas where their work could make a difference.

The goal of this chapter is to help fill this void by discussing DRL within the context of disaster response literature and practice and summarizing the chief findings from fieldwork research conducted by the authors in the aftermath of dozens of disasters of all sizes. The research reported here is the result of hundreds of formal and informal interviews with disaster responders, and ex-post analyses of relief operations. The chapter has four sections: Sect. 2 defines and discusses DRL in connection with the Emergency Functions conducted by relief agencies; the findings and chief lessons learned are summarized in Sect. 3; and concluding remarks are presented in Sect. 4.

3.2 Disaster Response Logistics (DRL)

To ensure a common base for the discussion and analyses, a basic set of definitions must be adopted. From the dozens of alternative definitions for these terms (Blanchard 2008) the authors include the most widely accepted. These definitions are important because, by categorizing the type of event, they help determine the nature of the response. In the context of this paper:

- An "emergency" is "…an unexpected event which places life and/or property in danger and requires an immediate response through the use of routine community resources and procedures. Examples would be a multi-automobile wreck, especially involving injury or death, and a fire caused by lightning strike which spreads to other buildings. Emergencies can be handled with local resources…" (Drabek 1996).
- A "disaster" is "…a non-routine event that exceeds the capacity of the affected area to respond to it in such a way as to save lives; to preserve property; and to maintain the social, ecological, economic, and political stability of the affected region…" (Pearce 2000). It should be noted that this definition views a disaster as external to the system. Others such as Wisner et al. (2004) hold the view that a disaster is internal to the system itself.
- A "catastrophe" can be defined as "…any disaster that overwhelms the ability of state, local, and volunteer agencies to adequately provide victims with such life-sustaining mass care services as food, shelter, and medical assistance within the first 12–24 h." (General Accounting Office 1993, p. 1). Technically, a catastrophic event is one in which "most or all of the community-built structure is heavily impacted … [and] facilities and operational bases of most emergency organizations are themselves usually hit"; "local officials are unable to undertake their usual work role"; "help from nearby communities cannot be provided"; "most, if not all, of the everyday community functions are sharply and concurrently interrupted"; "the mass media system especially in recent times socially constructs catastrophes even more than they do disasters"; there are "mass out-migrations for protracted periods of time"; and, "because of the previous six impacts, the political arena becomes even more important". (Quarantelli 2006; Holguín-Veras et al. 2012b; Wachtendorf et al. 2013).

It is worth noting that the definition of "disaster" encompasses such events as famines and chronic political crises that threaten vulnerable populations; and that DRL includes all logistical activities, humanitarian or not, dealing with disaster response. The breadth of DRL could be illustrated with the analysis of the Emergency Support Functions (ESFs) defined by the Federal Emergency Management Agency

Table 3.1 Emergency functions (EF)

EF #1: Transportation	*EF #7B: Logistics, management, resource support*
Aviation/airspace management and control	National incident logistics planning/ management…
Transportation safety	Resource support
Restoration/recovery…, [a] debris removal	*EF #8: Public health and medical services*
Movement restrictions	Public health
Damage and impact assessment	Medical
[a] Evacuation planning/operations	Mental health services
EF #2: Communications	Mass fatality management
Coordination with telecom. and IT industries	*EF #9: Search and rescue*
Restoration and repair of telecom. Infrastructure	Life-saving assistance
Protection, restoration, and sustainment of IT	Search and rescue operations
Oversight of communications…	*EF #10: Oil and hazardous materials response*
EF #3: Public works and engineering	Oil and hazardous materials response
Protection and emergency repair	Environmental short—and long-term cleanup
Restoration/recovery, [a] debris removal not in response	*EF #11: Agriculture and natural resources*
Engineering services and construction management	Nutrition assistance
Contracting support for life-saving services	Animal and plant disease and pest response
EF #4: Firefighting	Food safety and security
Coordination of federal firefighting activities	Natural/cultural/historic resources protection/ restoration
Support to wildland, rural, and urban firefighting	Safety and well-being of household pets
EF #5: Emergency management	*EF #12: Energy*
Coordination of incident management response	Energy infrastructure assessment/repair/ restoration
Issuance of mission assignments	Energy industry utilities coordination
Resource and human capital	Energy forecast
Incident action planning	*EF #13: Public safety and security*
Financial management	Facility and resource security
EF #6: Mass care, emergency assistance…	Security planning and technical resource assistance
Mass care, [a] relief distribution planning/ operations	Public safety and security support
Emergency assistance	Support to access, traffic, and crowd control
Disaster housing	*EF #14: Long-term community recovery*
Human services	Social and economic community impact assessment
[a] *EF #7A: Management of convergent help*	Long-term community recovery assistance
Material (solicited, unsolicited) donations	Analysis/review of mitigation program implementation
Monetary (solicited, unsolicited) donations	*EF #15: External affairs*

Table 3.1 (continued)

Service (solicited, unsolicited) donations	Emergency public info and protective action guidance
Information	Media and community relations
Personnel	Congressional and international affairs
	Tribal and insular affairs

[a] Denotes an addition to the original list in Federal Emergency Management Agency (2008). Some of the titles were shortened to make them fit in the table

(FEMA). The ESFs "...provide the structure for coordinating Federal interagency support for a Federal response to an incident" and are "...mechanisms for grouping functions used to provide Federal support to States and Federal-to-Federal support." (Federal Emergency Management Agency 2008). However, since FEMA's list only includes the functions supported by the federal government, it must be complemented with those performed by local responders to obtain a complete list of the tasks conducted in real-life operations. Something to keep in mind is that FEMA's role is to support the local response, hence the emphasis on "Support" in ESFs. However, since the emphasis of the chapter is on actual operations, the word "Support" has been removed to create the term "Emergency Function" or EF. Table 3.1 shows the EFs and tasks.

A number of additions are noteworthy. First, "Evacuation planning and operations" was included as a task under EF #1 (Transportation). Second, "Relief distribution planning and operations" was added to EF #6 (Mass Care, Emergency Assistance, Housing, and Human Services). Third, an EF #7A (Management of Convergent Help) was created to include the tasks related to management of the convergence of the various facets of response. The original EF #7 was renumbered #7B.

Table 3.1 shows the breadth of the activities conducted in disaster response. Some EFs focus on the restoration and management of key economic sectors such as transportation, energy, public works, agriculture, oil and hazardous materials, and communications. Other EFs relate to disaster response tasks such as firefighting, emergency management, mass care, logistics, public health, search and rescue, public safety, and management of convergent help. Of interest to humanitarian logistic researchers is that EF #7B (Logistics, Management, and Resource Support) focuses on the procurement and transportation of supplies to the disaster area, including the prepositioning of critical supplies. Relief distribution, a predominantly local activity, is part of EF#6. Since the activities part of "humanitarian logistics" fall under EF #6 and EF #7B, they are discussed together. The vast majority of the activities in Table 3.1 require fairly sophisticated logistics for maximum effectiveness.

The authors used the Web of Science (Thomson Reuters 2014), with various descriptors of the EFs and the word "logistics" as search keywords. Efforts were made to avoid duplicates and screen out publications incorrectly counted. Three EFs capture in excess of 80 % of the research articles, in roughly equal proportions: EF #1 (Transportation), EF #6/7B (Mass Care, Emergency Assistance, Housing, and Human Services; Logistics, Management, and Resource Support), and EF #8 (Public Health and Medical Services). The articles in EF #1 (Transportation) primarily focus on evacuation logistics. Similarly, the bulk of the articles in EF #6/#7B are related to relief distribution and prepositioning of supplies; while the ones on EF #8 focus on various aspects of health logistics. A distant fourth is EF #7A (Management of Convergent Help) with about 10% of the articles. These results conform to the authors' expectations.

The authors' conjecture is that the popularity of EFs #1, #6/#7B, and #8 reflects the interest of transportation and industrial engineers in disaster research. It seems natural for transportation engineers to do disaster research on evacuation and access restoration, and for industrial engineers to do research on relief distribution and health logistics. These researchers may have simply extended their research to new areas that relate to work in their original disciplines. The influx of researchers from these disciplines has brought to bear their specialized technical skills to the benefit of disaster research. However, in areas such as relief distribution and evacuation logistics, where human and social behaviors can be quite different from those observed in normal conditions, traditional techniques in industrial and transportation engineering are not always applicable. Socio-technical considerations, as discussed in the next section, are indispensable.

In contrast, clearly important EFs have received very little attention from the logistic research community. EFs #2 (Communications), #3 (Public Works and Engineering), and #12 (Energy) could greatly benefit from the development of mathematical models that would create optimal restoration plans for the corresponding infrastructure networks. EFs #4 (Firefighting), #5 (Emergency Management), and #9 (Search and Rescue) could improve operational effectiveness with the development of simulation and optimization tools that enable incident commanders to deploy their limited access in the most effective manner. EF #7A (Management of Convergent Help) needs considerable research, particularly in terms of the management of volunteers and material convergence, given the importance of these activities. EFs #10 and #11 require research developments that account for the environmental aspects of the relief operations. Fostering research in these overlooked subjects is likely to lead to significant contributions to DRL, and to a more balanced use of the talents of the logistic research community.

3.3 Chief Findings from Fieldwork Research

This section discusses the most salient findings from the fieldwork research conducted by the authors. For more than a decade, the authors have been studying real-life DRL operations. As part of the research, the authors routinely visit disaster sites, frequently soon after the event takes place, to interview disaster responders, identify the key lessons offered by the operations and characterize DRL operations. Dozens of disasters of various sizes have been studied, including the most prominent events of recent times (e.g., Indian Ocean tsunami, Hurricane Katrina, Port-au-Prince Earthquake, Tohoku tsunami, Hurricane Sandy). During the fieldwork research, hundreds of formal and informal interviews with disaster responders have been completed. Key findings are routinely published (Holguín-Veras et al. 2007, 2012a, 2013a, 2014b). This fieldwork has provided the authors with a unique perspective into the realities of DRL operations. In selecting which lessons to discuss, emphasis was placed on their importance.

All Facets of Disaster Response Logistics are Socio-Technical Processes

Most human activities are part of socio-technical systems and processes, where a social network of individuals conduct a series of technical activities (Holguín-Veras et al. 2012b). The social side includes aspects such as the interconnections among individuals, cultural and social norms of behavior, the organizational and institutional system, among others. The technical side encompasses the activities conducted by the social networks, the physical systems, and all non-social components. Although there are other useful frameworks, e.g., Bruneau et al. (2003), the authors prefer to use the socio-technical framework as doing so enables the analyses to be positioned in the broader context of the socio-technical systems literature (Kaghan and Bowker 2001; Edwards 2002).

Socio-technical systems rarely work in isolation; they routinely interact with, and use input from, other socio-technical systems such as communications and finance that represent the operational environment surrounding logistical activities. For every socio-technical system, there is an operational environment. The interlinkages among this multitude of interlocking socio-technical systems is such that a disturbance on either the social-network or technical activities in one of the systems is likely to cascade to the other systems that depend on the system impacted, as illustrated in Fig. 3.1. DRL and its different expressions, are no exception. Using Commercial Logistics (CL) as an example, one can see that shippers, carriers, and receivers of supplies are a social network with a commercial orientation. They know each other, understand what is expected of them, have formal ways to communicate with each other, follow structured decision-making rules that define who does what, and perform numerous technical activities such as inventory management, routing, and demand forecasts. Moreover, without a properly functioning environment, the operations cannot perform at their best, and in some cases not at all.

A socio-technical perspective is essential to understand the complexity of DRL. In normal conditions, the social networks of individuals who deliver supplies to population centers are completely functional and intact. If these social networks are severely impacted by a disaster, the socio-technical process could come to a stop, even if the technical systems are intact. Similarly, changes in the behavior of the individuals comprising these social networks will have an effect on the operation of the system. For instance, an inability to contact the receivers of the supplies, together with safety concerns, seems to contribute to the collapse of private sector supply chains after large disasters. The social networks that make commercial supply chains work are easily disrupted by large disasters. Communications are

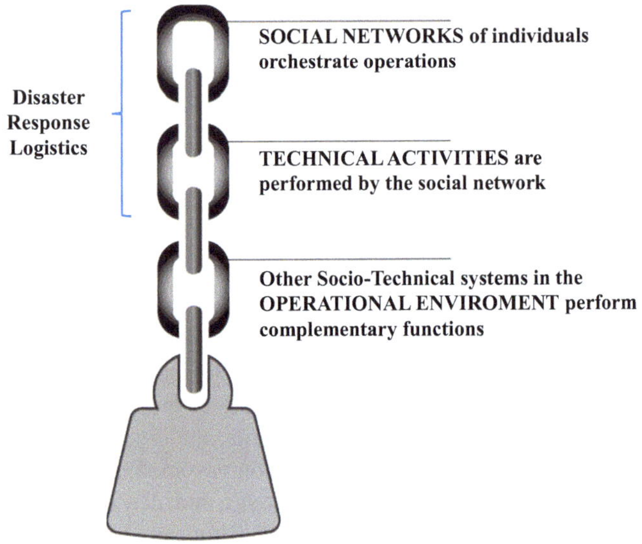

Fig. 3.1 DRL as a socio-technical system

interrupted, access to the area is blocked, transportation networks are severely impacted, members of the social network are injured or dead, or simply decide to abandon their work to help with the response, among many other possibilities. As a result, shippers of supplies located outside the disaster tend to stop operations until the social network is restored. The impacts are tremendous. In both Japan and Haiti, the supply chains delivering goods to areas near the disaster sites stopped functioning because the social connections had been severed (Holguín-Veras et al. 2012a, 2014b). It took about 2 weeks for the normal flow of supplies to resume. From the purely technical point of view, there are no reasons for the supply chains to have stopped. The phenomenon could only be explained if the behavior of the individuals that participate in the social network is taken into account.

The recognition of the importance of the socio-technical nature of DRL has important implications. First, the key to increasing societal resiliency is to enhance the resiliency of social networks, technical activities, and the supporting systems in the operational environment since these crucial components are impacted by disasters. Secondly, the only way to develop a holistic picture of DRL operations is to fuse considerations of the: (1) human and social behaviors that explain decision-making and the performance of the social networks; (2) the technical activities; and (3) the interaction with the other supporting systems in the environment. The importance of considering the social aspects is eloquently described by Rhodes (National Geographic 2011), when he explained the catastrophic impacts of the Hiroshima atomic bomb:

> …of the area's 45 hospitals, only three were usable, but more than the city infrastructure is wiped out. 'When you destroy a city you destroy that intricate web of social and personal connections. One of the horrors for the survivors was that their world has been burned away. They were left with their burns trying to decide where their world was'

A key implication is that developing a thorough understanding of DRL requires multi-disciplinary research involving, among others, psychologists, social scientists, logisticians, and domain experts from the socio-technical systems involved. An inherently socio-technical system cannot be adequately studied by single-perspective approaches that focus either on the social or the technical sides of the phenomenon.

Humanitarian Logistics and Commercial Logistics are Very Different

"Humanitarian logistics" (HL) is the counterpart of Commercial logistics (CL) in the humanitarian domain, i.e., logistics with a humanitarian objective. As stated by Kovacs and Spens (2007), HL *"...encompasses very different operations at different times, and as a response to various catastrophes. All these operations have the common aim to aid people in their survival. Nonetheless, aid to assist the development of a region, famine aid and the running of refugee camps is substantially different from the kind of aid needed after a natural disaster. Thus, two main streams of humanitarian logistics can be distinguished, continuous aid work, and disaster relief..."*

Holguín-Veras et al. (2012b) reached similar conclusions after analyzing the wide range of operational conditions in HL. At one end of the spectrum, one finds humanitarian operations in chronic crisis regions where the work is similar to that in CL, although the objective is humanitarian (termed "Regular Humanitarian Logistics", or R-HL). At the other end, one finds the operations after sudden onset disasters ("Post-Disaster Humanitarian Logistics", PD-HL) where emergent social behaviors create conditions without parallel in the CL case (Holguín-Veras et al. 2012b). Illustrating these differences establishes the need for research that accounts for the unique aspects of DRL.

Holguín-Veras et al. (2012b) identified seven dimensions that define the differences between CL and PD-HL. *First*, CL is concerned with the minimization of either transportation or logistic costs. In contrast, PD-HL attempts to minimize the social costs as defined by various proxy metrics. Recently, welfare economic principles have been brought to bear on HL modeling with the development of the concept of "deprivation cost" (Holguín-Veras et al. 2013b). *Second*, the supplies transported in CL flow from specific vendors to specific customers, and the nature of the cargo is known. In PD-HL, the situation is completely different, particularly in the aftermath of large disasters. In such environments, thousands of donors may send massive amounts of supplies and equipment to the disaster site (quite frequently, the donors do not even specify receiver or consignee), as part of the phenomenon referred to as "Material Convergence", the spontaneous transport of materials and equipment to the disaster site (Fritz and Mathewson 1957). The arrival of such massive flows of cargo produces major complications, as about 60 % of the total is completely useless (Fritz and Mathewson 1957; Holguín-Veras et al. 2014a). Disaster responders refer to this as a "second-tier disaster" (Newsweek 2002). *Third*, in CL the origin and destination of the supplies is known, as are demand patterns over time. The opposite occurs in PD-HL, as there is a high level of uncertainty about supplies needed, quantities, delivery points, and even the origin of the shipments. *Fourth*, CL has pre-established decision-making procedures with a regular set of decision makers. This stands in contrast to PD-HL, where literally thousands of individuals, companies, and relief groups extemporaneously decide to participate in the relief effort, leading to the creation of emergent logistic structures that could, in some cases, outperform much larger established organizations (Holguín-Veras et al. 2012a). *Fifth*, since most operations are regularly scheduled, the participants in CL are able to gradually adjust their operations and optimize their distribution strategies. This is impossible in sudden onset disaster operations, given the highly dynamic and emergent nature of the operations, and because most are once-in-a-lifetime experiences for the responders. *Sixth*, in CL, the social networks behind the operations function at normal capacity, while in PD-HL the social networks are likely to have been impacted, and unable to cohesively identify needs, or coordinate and conduct the required activities. *Finally*, CL activities take place in conditions where the socio-technical systems in the operational environment, such as transportation, are relatively stable and functional. This is in stark contrast with PD-HL operations where these supporting systems are not likely to be fully functional.

The benefits of using multi-disciplinary perspectives on DRL research can be illustrated with recent developments concerning the objective functions of HL analytical models. One of the chief differences between HL and CL is the objective function, which has to do with the presence of markets and the internalization of costs and benefits. In the CL case, there is a functioning market that assigns prices to supplies and services, and where buyers and sellers internalize the costs and benefits of the transaction. In a disaster environment, there are no functioning markets (which is what justifies the humanitarian effort). The recipients of the supplies are in no position to purchase them, and consequently the benefits and costs associated with the transaction become economic externalities (Varian 1992). From the perspective of logistics, the answer to the question of how to take into account the impacts on the beneficiaries is to use the proxy metrics used in similar logistic decisions (e.g., penalties for missed deliveries, delivery constraints, unmet demands). Unfortunately, these approaches are inadequate to properly consider the complex dynamics of the deprivation of critical supplies or services (Holguín-Veras et al. 2013b).

However, if the analysis perspective is broadened to include economics, a different solution emerges: to use welfare economics—the branch of economics that studies the economic welfare impacts of alternative allocation of resources—to decide how best to distribute the available resources. From this perspective, human suffering is an externality that could be valuated using economic valuation techniques, while the logistic costs are the private cost of the effort. Both, private and external costs form the social costs function. The objective is to minimize the social costs function associated with the delivery effort. The use of economic principles sheds new light into a very complex problem. The economic perspective on what initially was a logistic problem enables the use of the arsenal of tools already developed in micro-economics. Using these principles Pérez-Rodríguez and Holguín-Veras (2015) developed an inventory allocation model that provides important insights into the optimal distribution of relief supplies, and the key role of rationing; and Holguín-Veras et al. (2015) produced the first valuations for the economic value of water deprivation. Although far from perfect, the new pathways created by multidisciplinary approaches have great promise.

Size and Complexity Matters: Catastrophes and Disasters Require Radically Different Responses

While it is generally accepted that the nature and complexity of the HL efforts are influenced by the nature and magnitude of the disaster, surprisingly, there are no analytical estimates of the impacts of disaster size on magnitude of the response. To fill this void, the authors decided to use the Point of Distribution (POD) Planner, or POD Planner, developed by Jaller (2011) to estimate the number of staff persons needed to respond to disasters of various sizes (defined by the number of individuals impacted). The POD Planner uses Continuous Approximation models (Beardwood et al. 1959; Daganzo 1984) to compute the optimal number of PODs required to deliver relief supplies to an impacted population uniformly distributed across a convex area. Using the staff requirements from U.S. Army Corps of Engineers (2010), the POD Planner estimates the optimal number of PODs, number of servers at each POD (the number of lines to assist beneficiaries), amount of supplies to be distributed, number of staff members, number of trucks required to supply the PODs, and total amount of supplies needed to minimize the social costs. It is important to mention that these estimates only take into account the local distribution effort, and do not account for the upstream resources needed to procure, gather, and transport the relief supplies to the disaster area. The magnitude of this support force is tremendous. Military logisticians estimate that for every soldier in the field, there are ten additional staff members supporting the soldier. The authors estimate that for every volunteer participating in a relief effort, between five to seven staff members are required.

To estimate the resource requirements, the authors considered two different scenarios for the amount of supplies distributed. The first one, 5 kg/person-day, corresponds to the subsistence level (Sphere Project 2011), which is what may be practically accomplished in the initial days of the response. The second level, 20 kg/person-day, is typical of a well-supplied response once the relief operation is working at full speed and able to fulfill the rest of the beneficiaries' needs (e.g., clothing, cooking utensils, cleaning kits). As a reference, during the Tohoku response, the companies that participated in the relief effort delivered about 20 kg/person-day to the beneficiaries (Holguín-Veras et al. 2014b).

The analyses use the number of individuals impacted as a measure of disaster size. Throughout the numerical experiments, the population density was assumed to be constant and equal to 5544 inhabitants/km^2, which is comparable to the density

Table 3.2 Parameters used in the POD planner

Modeling assumptions		Costs parameters	
Density (inhabitants/km^2)	5544	Fixed cost per POD ($/day)	$1000.00
Supplies distributed (kg/person-day)	5, 20	Daily cost per volunteer	$80.00
Truck capacity (t)	5	Walking cost ($/km)	$1.55
Truck speed (km/h)	24	Waiting cost ($/h)	$1.00
Planning horizon (days)	15	Replenishement cost ($/ration)	$10.00
Service rate per server (beneficiaries/h)	45, 40	Keeping cost ($/ration)	$0.02
Walking speed (km/h)	2.5	Truck cost ($/h)	$60.00

of cities like Madrid, Spain, and London. Since the POD Planner finds the solutions that minimize social costs, i.e., the summation of private and external costs, all impacts are translated into a cost metric. This could be accomplished with the use of economic valuation techniques, as outlined in Bateman et al. (2002). The private costs considered by the POD Planner include the logistics costs of setting the PODs; the resources in terms of the personnel needed to man the PODs; and the costs of transporting the supplies to the PODs from a distribution center. The model explicitly considers the truck travel times and the time required to load and unload the supplies at the PODs. The external costs modeled are the deprivation costs incurred by the beneficiaries, and the impacts of the recipient policy (i.e., who is authorized to receive the relief aid, and whether or not this individual could receive aid for others). Deprivation costs are estimated as the time required to walk to the POD and the waiting time at the POD at the queues. Lastly, the model considers the external cost associated with the risk of theft or violence when transporting the supplies to where they are staying. The parameters used in the computations are shown in Table 3.2. Two service rates were assumed: 45 beneficiaries/h if 5 kg/beneficiary are distributed, and 40 beneficiaries/h for the case of 20 kg/beneficiary (to account for the larger physical effort). The estimated requirements are shown in Table 3.3.

The results show that under the assumption of constant population density, the resources required to perform the relief operations (e.g., manpower, number of delivery trucks to supply the PODs, total amount of supplies) increase in a quasi-linear fashion with the size of the population in need. Interestingly, the POD Planner suggests that the optimal number of servers per POD is practically constant (12 servers, in most cases). The results suggest that the best way to serve the needs is to simply increase the number of PODs with 12 servers.

Table 3.3 provides a solid indication of the magnitude of the challenge associated with responding to large disasters. As shown, fulfilling the subsistence needs of 1 million individuals (5 kg/person) requires 5000 t of supply per day, and about 30,000 staffers manning the PODs. This translates into about 250 semi-trailers transporting the supplies to the disaster area (or 250 C-130 s, or 65 C-17 s if brought by cargo planes); about 189.5-t small trucks to deliver the supplies to the PODs multiple times a day. A disaster impacting a large city—such as Beijing (19.61 million), Tokyo (13.35 million), or Los Angeles (3.79 million) (World Population Review 2014)—will require a response multiple times larger than the one for a mil-

Table 3.3 Resource requirements vs. disaster size

Population (1000 s)	Area (km^2)	Supplies distributed = 5 kg/person					Supplies distributed = 20 kg/person				
		PODs	Servers	Manpower	Trucks	Total supplies (t)	PODs	Servers	Manpower	Trucks	Total supplies (t)
10	1.53	4	11	290	2	50	4	12	317	6	200
20	3.06	8	11	579	3	100	8	12	633	11	400
50	7.61	19	12	1502	8	250	20	12	1581	30	1000
100	15.23	38	12	3003	15	500	41	12	3240	57	2000
200	30.46	76	12	6005	33	1000	81	12	6400	128	4000
500	76.17	189	12	14,932	81	2500	203	12	16,038	373	10,000
1000	152.34	378	12	29,864	189	5000	405	12	31,997	743	20,000
1500	228.54	566	12	44,717	283	7500	607	12	47,956	1113	30,000
2000	304.71	754	12	59,569	453	10,000	809	12	63,914	1780	40,000
2500	380.88	941	12	74,343	565	12,500	1011	12	79,873	2225	50,000
5000	761.77	1874	12	148,053	1406	25,000	2021	12	159,667	5558	100,000
5500	837.94	2060	12	162,748	1545	27,500	2223	12	175,626	6114	110,000
10,000	1523.53	3725	12	294,289	3725	50,000	4041	12	319,254	14,817	200,000
15,000	2285.27	5556	12	438,945	5556	75,000	6061	12	478,842	22,224	300,000
20,000	3047.04	7408	12	585,260	7408	100,000	8081	12	638,429	29,631	400,000
25,000	3808.80	9260	12	731,575	13,890	125,000	10,102	12	798,096	55,561	500,000

Table 3.4 Impacts of disasters and catastrophes. (Adapted from (Holguín-Veras et al. 2012b))

Characteristic	Disaster	Catastrophe
Civic society leadership	Typically survives the disaster and is able to lead the response	In most cases, severely affected and unable to lead an effective response
Local stocks of supplies in businesses and households	Only partly destroyed; surviving supplies may support the response	Mostly destroyed; the role of local supplies in the response is minimal
Demand for supplies	Increases; precautionary/opportunistic buying could be a problem	Increases dramatically because of the impacts; precautionary/opportunistic buying becomes a major problem
Private-sector supply chains	Partly impacted but functional, could help in response	Severely destroyed, cannot help in response
Number of points of entry to the disaster area	Multiple points of entry provide many alternatives to enter the area	Only a few points of entry complicates response efforts
Complexity of the local distribution effort	Challenging but manageable	Exceedingly complex because of size of impacted area and manpower needs
Material convergence of non-priority supplies	A nuisance that can be controlled	A major problem that distracts significant amounts of resources from critical tasks
Net result	Local help is key in initial days; outside help brings additional supplies	Outside help is the primary source of supplies

lion people. These numbers provide an indication of the magnitude of the response required.

In addition to the resource requirements summarized in Table 3.3, it is important to consider the ability of the impacted population to respond to the event. Large disasters and catastrophes not only generate large demands in terms of material and personnel, they also diminish and even completely destroy the local capacity to respond (Government Accountability Office 2006; Quarantelli 2006). The authors' fieldwork research clearly shows that, while in the disaster of typical size, the locals may be able to provide a meaningful response, they are not able to do so after a catastrophe. See the reasons in Table 3.4.

As indicated in Table 3.4, local inventories are impacted very differently in disasters and catastrophes. In disasters, the supplies held by households and businesses largely survive, which enable them to play a key role in the response; in catastrophes they are mostly destroyed. The ability of the locals to muster resources to support the response also depends on their wealth. Vulnerable low-income populations are likely to have less material resources to bring to bear. The demand for critical supplies from the surviving population also changes drastically; while disasters increase the needs of business, people and respondents, after catastrophes the increases in demand are much greater. Precautionary and opportunistic buying is an issue in both cases; though in catastrophes the amount of supplies removed from the markets in the areas surrounding the event could reach huge proportions. This is a major problem because these supplies, given their proximity to the impacted area, are ideally positioned to help the impacted populations. In disasters, private supply chains are disrupted but remain functional and can be integrated in the response effort, while in catastrophes the disruptions to the social networks and the physical damage are so massive that supply chains are interrupted for weeks. Similarly, after disasters there are multiple entry points to the affected area, while in catastrophes, because of the extent of the damage, only a few entry points are typically available. The local delivery effort is also significantly different, as disasters pose complex but manageable challenges. However, catastrophes affect a much larger area, dramatically increasing the complexity of the distribution effort, requiring trips over longer distances on impacted networks, and necessitating large numbers of volunteers. As complexity increases, considerable resources and manpower are required to handle the operations. Material convergence also varies with the nature of the event; as its magnitude increases, so does the flow of non-priority goods to the area, escalating the issue from a nuisance in disasters to a major problem in catastrophes, one that consumes a significant amount of resources.

Essentially, a typical disaster is a localized event in which the local response can play a key role, while a catastrophe is a widespread event that significantly diminishes, or completely destroys, the local capacity to respond. The net result is that in disasters, the locals are able to provide the first wave of response while catastrophic events incapacitate the affected region in such a way that outside help is the only option. The assumption that locals could provide resources needed during the first 48 h of the response is entirely valid in disasters but cannot be expected to hold after a catastrophe. This is a major issue, because most disaster response plans are predicated with the assumption that the local responders can provide a meaningful response during the initial phases (Federal Emergency Management Agency 2015).

It is Crucial to Integrate the Civic Society in the Response Effort

The response to large disasters and catastrophic events requires large amounts of supplies to satisfy the needs of the survivors and the response itself. Another huge challenge is ensuring that both manpower and supplies are deployed to the areas in need in a timely fashion, and that the individuals involved in the response work as a cohesive network following humanitarian standards. Without appropriate local support it is not possible to accomplish these objectives, in the aftermath of large events. This was clearly demonstrated in the initial days of the response to the 2010 Port-au-Prince earthquake, where thousands of tons of critical supplies arrived at the airport and could not be delivered to the population in need. The reason for this response failure, as it was later learned, was that the foreign relief groups were unable to find local interlocutors that could help them connect with the social networks needed to deliver relief supplies to 2 million people (Holguín-Veras et al. 2012a). This was no trivial challenge. The authors estimate that distributing relief supplies in Port-au-Prince after the 2010 earthquake required in excess of 30,000 volunteers. As a reference, the average size of a U.S. Army division is 20,000, which military planners estimate it takes 3–4 weeks to deploy. It is not possible to transplant such a distribution network from the outside, yet assembling such a manpower force in the chaotic conditions prevailing after a large disaster is a huge challenge. The most effective way to do it is to take advantage of existing social networks.

The 2010 Port-au-Prince earthquake provided compelling examples of the role that local social networks could play in the response. Although there is a wide variety of logistic structures, three model types of local deliveries were identified by the authors (Holguín-Veras et al. 2012a): (1) Agency Centric Efforts (ACEs): are relief groups that arrive at the disaster site, have little or no integration with local organizations, and attempt to undertake the local distribution themselves; (2) Partially Integrated Efforts (PIEs): logistics structures in which there is collaboration between a foreign relief group and local organizations, where the local partner undertakes the local distribution; and (3) Collaborative Aid Networks (CANs): large social networks of individuals with relationships established before the disaster, that extend beyond the impacted area, and that, because of their massive size, are able to undertake all aspects of the PD-HL effort including gathering, transporting, and local distribution of supplies. The fact that CANs could easily collaborate with foreign relief groups gives them tremendous flexibility.

The experience in Haiti clearly proved the value of the CANs (Holguín-Veras et al. 2012a). By having access to a massive number of committed individuals, the CANs were able to put in place fairly sophisticated and effective distribution networks that delivered large amounts of relief supplies, under conditions where ACEs and PIEs struggled. Because of the role played by their local collaborators, PIEs did a reasonably good job. The ACEs struggled throughout, and had great difficulties finding equipment and creating the social networks needed to perform the local distribution. Since only the CANs associated with churches and other civic groups can provide the manpower needed after a large disaster, it is essential to engage CANs as proactive partners in disaster preparation and response efforts.

It is also crucial to proactively integrate the private sector, as evidenced by the experience after the Tohoku disasters (Holguín-Veras et al. 2014b). In the aftermath of the tsunami, a handful of companies in the food distribution business played a primary role in distributing relief supplies to the survivors. These companies realized that humanitarian needs were not being met, and on their own and against the stated wishes of the government, deployed hundreds of delivery trucks to deliver food and water. Their timely intervention prevented the Japanese crisis from taking a path similar to the one in Haiti. Their outstanding work was made possible because their operations covered both the impacted areas and the surrounding regions. Essentially, they had the local knowhow and access to regional resources necessary to navigate the streets clogged with debris, and to know the location of disaster shelters where survivors would congregate. This enabled them to bring resources from areas that had not been impacted to reinforce the local operations. Without a doubt, their efforts were central to preventing a humanitarian crisis of serious proportions.

The experiences of Haiti and Japan, briefly discussed here, clearly demonstrate the key role that civic society could play, with effective integration, in response efforts. The logistic industry—particularly the companies involved in the distribution of critical supplies such as food, water and medicines—could play a critical role after a disaster by providing material resources, distribution networks, and local knowhow. The local social networks, e.g., the ones associated with networks of churches, could provide significant amounts of manpower and leadership. The integration between the segments of civic society will create logistics structures with a diverse spectrum of capabilities and strengths.

It is Important to Design Plans for Easy Integration of Outside Help

One of the main lessons from the fieldwork is that the convergence of volunteers, materials, and equipment, brings to the impacted area large amounts of resources that are critical to the response (notwithstanding the problems created by inexperienced volunteers, and non-priority donations). Volunteers typically represent the lion's share of the manpower in post-disaster environments. Similarly, the material convergence of high—and low-priority supplies (both solicited and non-solicited) frequently exceeds the amounts of resources transported by the official response. The same could be said about search and rescue teams that arrive from all over the world, and professional engineers that volunteer to assess the conditions of the impacted infrastructure.

While it is essential to fully exploit these convergent resources, coordinating amorphous groups of people and the material resources they bring is no trivial endeavor. To illustrate the challenge: (1) volunteers, supplies, and equipment tend to arrive at the locations highlighted in media accounts, which leads to over-crowding of resources at some locations, to the detriment of others not mentioned by the media; (2) volunteers may lack familiarity with local customs and practices, or face cultural or language barriers; and (3) equipment or materials may not be appropriate for the needs, or may require inputs not locally available.

Ideally, qualified outside responder groups arrive at the disaster site with the proper materials and equipment, and are assigned tasks to complete that they are qualified to perform and that complement the work of other groups with minimal or no overlap. Achieving this ideal is very difficult, however, as it requires planning and coordination between participants. This coordination needs to be prearranged, to avoid duplication of efforts, interference, and the rest of the problems that arise from lack of coordination.

A number of actions could facilitate the integration of outside help. For instance, disaster response agencies could implement Districting Plans and Volunteer Management Programs. Districting, the subdivision of the overall effort into smaller geographical areas, could help allocate the work among respondent groups. Volunteer management programs could ensure that volunteers are used effectively throughout the effort, and the standardization of tasks could help ensure that individuals from other jurisdictions are able to adjust to the new conditions, with minimal effort. In all of these cases, logistic research can play a very important role. Technical training is crucial. A significant effort should be undertaken to enhance the technical capabilities of key individuals expected to be involved in DRL, such as executives of large companies involved in the distribution of critical supplies (e.g., food, water, medications), leaders of private sector trade groups, and community leaders.

A complicating factor is that disaster response procedures must be effective in diverse conditions. Disaster planners must assess the wide range of needs (from small disasters to large catastrophes) and design scalable response operations, i.e., that could satisfy the needs by adjusting upwards or downwards the input resources to deliver the aid needed at the site. The dual effort of studying the wide range of needs and designing the EFs, in a way that could be seamlessly expanded or contracted to meet the needs in the ground, will ensure that a similar template of action be used in a wide range of scenarios.

Prepositioning Critical Supplies is Required for Effective Response to Catastrophes

To expect that the local response provide the first wave of resources required is unrealistic after a catastrophic event since the bulk of the local inventories will likely be wiped out. Given that resources from outside of the impacted area are the only practical option, it is important to examine the potential sources of relief supplies. Frequently mentioned sources of supplies are the inventory stocks maintained by the private sector at distribution centers and warehouses. However, it is doubtful that the private sector has supplies in the amounts needed to satisfy public needs after a large event. This relates to the lean nature of modern supply chains, which have significantly reduced the large inventories they once had, replacing them with more efficient concepts such as Just-in-Time production systems. Another possibility mentioned is to reroute shipments in transit to private-sector customers, for use in the response. Although this could provide a meaningful amount of relief supplies, its implementation is fraught with challenges. The first may come from the consignees of the supplies, who are waiting for and need the cargo for their economic activities. The second challenge is legal, as the forced requisition of critical supplies is likely to be perceived as infringing on commerce. Although possible, the rerouting of supplies would necessitate the implementation of an appropriate legal framework to ensure that vendors get waivers of their contractual obligations to customers, and that the affected companies are compensated for the losses. Ramping up production to meet the additional demand generated by a disaster could certainly help in the medium—and long-terms. However, most production systems are not designed for rapid escalation of the production level. Moreover, there is always the issue of whether or not the required inputs are available. Without appropriate inputs no escalation of production is possible. Generally speaking, production increases cannot be expected to satisfy the need for critical supplies during the first days of the response, when needs are most acute. Such increases could only be accomplished in a handful of industrial sectors.

The sobering conclusion is that pre-positioning of critical supplies is essential. It is imperative that resources be sited in locations that are accessible to risk-prone areas. These resources, or emergency stocks, could be maintained by the disaster response agencies, relief organizations, or could be integrated with private-sector operations. An interesting model is to develop contracts with private-sector vendors to guarantee delivery of pre-specified supplies within a pre-set delivery window. In exchange, the public sector commits to pay premium prices for the supplies purchased. In any case, the location of these inventories of both supplies and equipment, should be selected to avoid putting them at risk in case of a large disaster. Setting up regional purchase blanket agreements (Holguín-Veras et al. 2007; Holguín-Veras and Jaller 2012) would enable the government and relief organizations to secure the supplies at reasonable prices without having to set up purchasing contracts that activate after the disaster, and could specify delivery timeframes to ensure an expedited relief distribution.

Control of Precautionary/Opportunistic Buying Could Help Provide a First Wave of Resources

Precautionary/opportunistic buying refers to the purchase of local supplies by individuals and/or businesses both before and after the disaster in amounts exceeding normal needs. Although often referred to as "panic buying," this is a misnomer. These purchases are a natural human reaction to concerns about potential shortages that may occur when disasters are expected, or have occurred. This does not mean, however, that the practice is not problematic.

The anecdotal information collected by the authors indicates that, before and after a disaster, the population near the disaster area (including those who have not been directly impacted) tend to purchase critical supplies in excess of actual needs. In the case of the Tohoku disasters in Japan, the manager of one of the largest distribution centers in the Tohoku area indicated that the demand "doubled" after the tsunami (Holguín-Veras et al. 2014b). The logisticians of a large retailer, in the months after Super Storm Sandy, reported to the first author that they had evidence that the closer to the disaster area, the more pronounced precautionary/opportunistic buying becomes. The net effect of these purchases is the removal of the market of critical supplies that, for reasons of proximity, are ideally located to be part of the first wave of resources reaching the impacted area. Supplies that otherwise could have been procured locally must be brought in from the outside, delaying the response. Disaster response agencies must take steps to ensure that precautionary/opportunistic buying does not deplete stocks of critical supplies in the vicinity of the disaster area. Such purchases should be minimized so that the supplies are available for the survivors. Initiatives that could achieve this objective include: agreements with key private-sector vendors to ensure that their inventories of critical supplies become available to the official response; campaigns to educate the public about the negative consequences of precautionary/opportunistic buying; and the implementation of rationing and demand-management schemes.

Control of Material Convergence will Enhance Disaster Response

One of the most unique, overlooked, and poorly understood disaster phenomena is material convergence: the spontaneous flow of supplies, donations, and equipment to the disaster area (Fritz and Mathewson 1957; Jaller 2011; Holguín-Veras et al. 2014a). Although material convergence is vital for the response since it helps to secure needed supplies and equipment, it is also responsible for large amounts of useless supplies and logistical problems. The classification created by the Pan-American Health Organization (PAHO) provides an excellent way to categorize such a heterogeneous flow in: (1) urgent or high-priority supplies are those that are needed for immediate distribution and consumption; (2) non-urgent or low-priority supplies that are those not immediately needed but which could be useful later; and (3) non-priority supplies, which do not have any practical use in the immediate aftermath of a disaster (Pan American Health Organization 2001, p. 70; Holguín-Veras et al. 2012b). The main issue in material convergence is that the vast majority of the supplies that arrive at a disaster site are often non-priority. The fieldwork conducted by the authors suggest that between 50–70 % of these supplies are non-priority, 25–35 % are low-priority, and only 5–15 % are high-priority supplies (Holguín-Veras et al. 2014a). These estimates are in line with the ones produced more than half a century ago by Fritz and Mathewson (1957).

The most problematic of these flows are the non-priority goods, referred to by disaster responders as "a second-tier disaster" (Newsweek 2002). Although typically associated with unsolicited in-kind donations, official donations and other solicited donations could easily become non-priority if they arrive in quantities that exceed needs (Jaller 2011). What really matters is how useful the item is, not whether it was solicited or not. Non-priority goods complicate the logistics response because they: do not reflect the needs of the population; require a vast amount of resources for handling and storage; arrive damaged, expired or in sizes and quantities that render them useless; need to be discarded to make room for other more priority goods or activities; have packaging, directions or labels in languages that are not appropriate; arrive in quantities that exceed actual needs, or at a time when they are not needed; create safety issues because of the requirements for storage or discarding; impose a diplomatic and public relations problem for the recipient organization because they cannot be treated as garbage; and require additional resources for their disposal, or incineration (Fritz and Mathewson 1957; Pan American Health Organization 2001, p. 70). In addition, the vehicles bringing these supplies generate logistics and traffic problems both at the entry points to the disaster area and at the disaster site. These vehicles create congestion and other externalities, consume resources at borders and check points because they need to be inspected, and if they are allowed to continue, transfer these problems to locations where the infrastructure may have been severely impacted or where more critical activities are taking place.

It is important to increase the net benefits of material convergence by expediting the flow of high—priority supplies and reducing or curtailing the non-priority flows. The first step is to explicitly consider material convergence in disaster response plans. This could translate into a number of measures to contend with the flows, such as developing proper control procedures to allow only high-priority supplies to enter the disaster area (Jaller 2011). Another possibility is the implementation of donation and information management systems to let the outside world know what the needs are, and what donors should donate. Essentially, this is what systems such as AidMatrix attempt to do (Federal Emergency Management Agency 2011). Unfortunately, the fieldwork conducted by the authors reveal that the level of penetration of these systems is extremely low as they are only used by large companies with previous experience in disaster response. Individuals and community groups that organize donation drives, and companies with no previous experience in disaster efforts—which are the ones credited with the generation of the bulk of non-priority donations—are not inclined to use such information systems. Moreover, since research suggests that the media's portrayal of disaster impacts and needs, which are often subjective, biased, and based on what is considered newsworthy, is correlated with the generation of non-priority supplies, the media must be engaged in the solution (Wachtendorf et al. 2006; Wachtendorf 2010). Similarly, it is important to understand donor behavior, and to develop educational campaigns to inform the general public and potential donors of the problems created by non-priority donations, and to foster monetary donations (Saber Donar 2011b; Saber Donar 2011a; Center for International Disaster Information 2014). Donation management must be transformed from reactive to a proactive process with active engagement of the media, donors, disaster responders, relief organizations, and governments.

New Supporting Technologies and Systems Could Foster Effectiveness

The experience in large disasters highlights the need to develop and foster the use of supporting technologies to accurately assess the needs on the ground, and help responders use scarce resources in the most effective manner. After large disasters, information systems and supporting technologies can help in facilitating transportation of supplies, assessing infrastructure condition, and ensuring efficient communications among responders. Such systems include platforms that combine satellite imagery, Global Positioning Systems (GPS), remote sensing, and geographic information systems (GIS) to help coordinate the operations and assess conditions on the ground. Technologies such as GPS devices could help guide drivers on the roads of the impacted network even when they are unfamiliar with the area. Performance of these devices should be complemented with information collected using Commercial Remote Sensing (CRS) systems that would provide an assessment of the network conditions and disaster impacts.

However, it is important to recognize that for technologies and systems to be of help to disaster responders, they must take into account the challenging realities of post-disaster conditions. Frequently, power and communication systems have been wiped out, and are unable to function for weeks. Portable, stand-alone technologies and systems able to operate without support from local networks are essential.

The CRS disaster response community has taken huge strides towards more effective implementation of remote sensing for the development of geospatial disaster response products. Researchers and government agencies have initiated efforts to collect airborne imagery and Light Detection and Ranging (LiDAR) data that fed into "Global Earth Observation: Catastrophe Assessment Network" (GEO-CAN), which rapidly identified collapsed buildings using high-resolution aerial imagery in Port-au-Prince (Ghosh et al. 2011). This experience led to the provision of remote sensing and data analysis support in multiple subsequent disasters, e.g., Hurricane Irene and Tropical Storm Lee, and the 2011 Japan disaster (Eguchi et al. 2012). It is important to develop and foster the use of technologies and systems that could enhance disaster response procedures.

3.4 Conclusions

The chapter has presented a comprehensive analysis of the field of Disaster Response Logistics (DRL) and its interconnections with the Emergency Functions (EFs) conducted by disaster response agencies and groups. The analyses indicate that the bulk of the logistics research conducted has focused on three major areas: EF #1 (Transportation), mostly evacuation logistics; EF #6/7B (Mass Care, Emergency Assistance, Housing, and Human Services; Logistics, Management, and Resource Support), primarily relief distribution and prepositioning of critical supplies; and EF #8 (Public Health and Medical Services), for the most part, health logistics.

The rest of the EFs have not received a level of attention commensurate with their importance. These research gaps must be addressed to advance DRL.

The chapter also establishes that DRL is a complex and multi-faceted socio-technical system, where social networks of individuals perform numerous technical tasks related to the EFs. These activities use a host of supporting systems, socio-technical systems in their own right, such as communication and finance that are part of the operational environment. The analyses conducted indicate that, because of the intricate interconnections between the social and the technical aspects, single perspective disciplinary approaches cannot provide a complete picture of the nuanced complexity of DRL operations. Essentially, new paradigms of DRL theory, modeling, and operations that account for the social and technical aspects are needed, which could only be accomplished by multi-disciplinary approaches.

The chapter illustrates the importance and potential benefits of a holistic study of DRL through a discussion of the chief findings of the multi-disciplinary fieldwork research conducted by the authors in dozens of disasters. This research clearly establishes that: the various facets of DRL are socio-technical systems and processes; humanitarian and commercial logistics are very different activities; catastrophes require a response that is fundamentally different from the response required by a typical disaster; it is crucial to integrate the civic society in the response effort; it is important to design plans for easy integration of outside help; prepositioning critical supplies is required for effective response to catastrophes; control of precautionary/opportunistic buying could help provide a first wave of resources; control of material convergence will enhance disaster response; and new technologies/systems could foster effectiveness.

The logistic research community, without doubt, has a crucial role to play in the development of new paradigms of DRL. However, the creation of these paradigms has to be based on a solid understanding of the socio-technical nature of DRL operations, and the identification, formulation, and computation of models that accurately depict them. Multi-disciplinary approaches are central to this quest. Embracing multi-disciplinary perspectives will ensure that the logistics research community can fully leverage the past to improve the future, and contribute its best to advance DRL research and operations.

References

Bateman IJ, Carson RT, Day B, Hanemann WM, Hanley N, Hett T, Jones-Lee M, Loomes G, Mourato S, Özdemiroglu E, Pearce DW (2002) Economic valuation with stated preference techniques: a manual. Edward Elgar, Northampton

Beardwood J, Halton JH, Hammersley JM (1959) The shortest path through many points. Math Proc Camb Philos Soc 55(04):299–327

Blanchard BW (2008, October 22) Guide to emergency management and related terms, definitions, concepts, acronyms, organizations, programs, guidance, executive orders & legislation. http://training.fema.gov/hiedu/docs/terms%20and%20definitions/terms%20and%20definitions.pdf

Bruneau M, Chang SE, Eguchi RT, Lee GC, O'Rourke TD, Reinhorn AM, Shinozuka M, Tierney K, Wallace WA, von Winterfeldt D (2003) A framework to quantitatively assess and enhance the seismic resilience of communities. Earthq Spectr 19(4):733–752

Center for International Disaster Information (2014) Guidelines for giving. http://www.cidi.org/guidelines. Accessed 7 Oct 2014

Daganzo C (1984) The distance traveled to visit N points with a maximum C stops per vehicle: an analytical model and an application. Transp Sci 18(4):331–350

Donar S. (2011a) Afiche #5. http://www.saberdonar.info/materiales/Afiche05.pdf

Donar S. (2011b) Una Iniciativa Interagencial para Promover la Calidad y Eficiencia en las Donaciones Humanitarias en América Latina. http://www.saberdonar.info

Drabek T (1996, September) The social dimensions of disaster (FEMA Emergency Management Higher Education Project College Course Instructor Guide). http://training.fema.gov/EMIWeb/edu/completeCourses.htm

Edwards P (2002) Infrastructure and modernity: force, time and social organization in the history of sociotechnical systems. In: Misa T, Brey P, Feenberg A (eds) Modernity and technology. MIT Press, Cambridge

Eguchi RT, Bevington J, Lin A, Goltz JD, Eguchi MT, Huynh A, Yamazaki F, Koshimura S (2012) The role of urban development patterns in mitigating the effects of tsunami run-up: an early progress report. 10th International Workshop on Remote Sensing and Disaster Management, Sendai, Japan

Federal Emergency Management Agency (2008) Emergency support function annexes: introduction. http://www.fema.gov/media-library-data/20130726-1825-25045-0604/emergency_support_function_annexes_introduction_2008_.pdf

Federal Emergency Management Agency (2011) National Donations Management Network (NDMN). http://sandbox.aidmatrixnetwork.org/FEMA/Index.aspx. Accessed 20 Aug 2011

Federal Emergency Management Agency (2015) National response framework. https://www.fema.gov/national-response-framework. Accessed 7 May 2015

Fritz CE, Mathewson JH (1957) Convergent behavior: a disaster control problem. Special report for the committee on Disaster Studies. Disaster Study 9. pp 1–102

General Accounting Office (1993, May 25) Disaster management: recent disasters demonstrate the need to improve the nation's response strategy. Retrieved GAO/T-RDED-93-46, Testimony before the committee on Armed Services, subcommittee on Nuclear Deterrence, Arms Control and Defense Intelligence, U.S. Senate. http://archive.gao.gov/d43t14/149256.pdf

Ghosh S, Huyck CK, Greene M, Gill SP, Bevington J, Svelka W, DesRoches R, Eguchi RT (2011) Crowd sourcing for rapid damage assessment: The Global Earth Observation Catastrophe Network (GEOCAN). Earthq Spectr 27(Special Issue on the 2010 Haiti Earthquake, Number S1):179–198

Government Accountability Office (2006) Emergency preparedness and response—some issues and challenges associated with major emergency incidents state of California. http://www.gao.gov/new.items/d06467t.pdf

Holguín-Veras J, Jaller M (2012) Immediate resource requirements after hurricane Katrina. Nat Hazard Rev 13(2):117–131

Holguín-Veras J, Pérez N, Ukkusuri S, Wachtendorf T, Brown B (2007) Emergency logistics issues affecting the response to Katrina: a synthesis and preliminary suggestions for improvement. Transp Res Rec 2022:76–82

Holguín-Veras J, Jaller M, Wachtendorf T (2012a) Comparative performance of alternative humanitarian logistic structures after the Port-au-Prince Earthquake: ACEs, PIEs, and CANs. Transp Res Part A Policy Pract 46(10):1623–1640. http://dx.doi.org/10.1016/j.tra.2012.08.002

Holguín-Veras J, Jaller M, Wassenhove LNV, Pérez N, Wachtendorf T (2012b) On the unique features of post-disaster humanitarian logistics. J Oper Manage 30:494–506. http://dx.doi.org/10.1016/j.jom.2012.08.003

Holguín-Veras J, Jaller M, Taniguchi E, Aros-Vera F (2013a). The lessons from catastrophic events for post-disaster humanitarian logistic efforts: the Port Au Prince Earthquake and the Tohoku Disasters. Transportation Research Board 92nd Annual Meeting, Washington DC

Holguín-Veras J, Pérez N, Jaller M, Van Wassenhove LN, Aros-Vera F (2013b) On the appropriate objective function for post-disaster humanitarian logistics models. J Oper Manage 31(5):262–280. http://dx.doi.org/10.1016/j.jom.2013.06.002

Holguín-Veras J, Jaller M, Van Wassenhove LN, Pérez N, Wachtendorf T (2014a) Material convergence: an important and understudied disaster phenomenon. Nat Hazards Rev 15(1):1–12. http://dx.doi.org/10.1061/(ASCE)NH.1527-6996.0000113
Holguín-Veras J, Taniguchi E, Jaller M, Aros-Vera F, Ferreira F, Thompson RG (2014b) The Tohoku disasters: chief lessons concerning the post disaster humanitarian logistics response and policy implications. Transp Res Part A Policy Pract 69(0):86–104. http://dx.doi.org/10.1016/j.tra.2014.08.003
Holguín-Veras J, Amaya-Leal J, Cantillo V, Van Wassenhove LN, Aros-Vera F, Jaller M (2015). Deprivation cost functions for water: results from a contingent valuation experiment. J Oper Manage (JOM) (Under review)
Jaller M (2011) Resource allocation problems during disasters: the cases of points of distribution planning and material convergence handling. Ph.D., Rensselaer Polytechnic Institute
Kaghan WN, Bowker GC (2001) Out of machine age?: complexity, sociotechnical systems and actor network theory. J Eng Tech Manage 18(3–4):253–269. http://dx.doi.org/10.1016/S0923-4748(01)00037-6
Kovacs G, Spens KM (2007) Humanitarian logistics in disaster relief operations. Int J Phys Distrib Logist Manage 37(2):99–114. doi:10.1108/09600030710734820
National Geographic (2011) Hiroshima: the next day. Pamela Caragol Wells—Caragol Wells productions in association with National Geographic Television for National Geographic Channels
Newsweek (2002, February 18th) $75 Million of stuff. pp. 62–63
Ontko M, Williamson S, Haselkorn MP, Kemp RB (2007) An examination of the effectiveness of lessons—learned reporting within the humanitarian sector. J Inf Technol Soc Change Spring Edition, pp 28–48
Pan American Health Organization (2001) Humanitarian supply management and logistics in the health sector. https://books.google.com/books?hl=en&lr=&id=yA1JaZGlv08C&oi=fnd&pg=PR7&dq=Humanitarian+Supply+Management+and+Logistics+in+the+Health+Sector&ots=4uB9rF1uJf&sig=Ubzg_7m05mGCrL1Fj8WsJug5EQ4#v=onepage&q=Humanitarian%20Supply%20Management%20and%20Logistics%20in%20the%20Health%20Sector&f=false
Pearce LDR (2000). An integrated approach for community hazard, impact, risk and vulnerability analysis: HIRV. Doctoral Dissertation, The University of British Columbia
Pérez-Rodríguez N, Holguín-Veras J (2015) Inventory-allocation distribution models for postdisaster humanitarian logistics with explicit consideration of deprivation costs. Transp Sci (Articles in advance) 0(0):1–25. doi:10.1287/trsc.2014.0565
Quarantelli EL (2006, November 15). Catastrophes are different from disasters: implications for crisis planning and managing drawn from Katrina. Understanding Katrina: perspectives from the social sciences. http://understandingkatrina.ssrc.org/Quarantelli/. Accessed 16 Nov 2010
Reuters T. (2014) Web of science. http://www.webofknowledge.com/. Accessed 26 Dec 2014
Sphere Project (2011) 2011 sphere handbook: Humanitarian charter and minimum standards in disaster response. http://www.sphereproject.org/. Accessed 10 July 2011
U.S. Army Corps of Engineers (2010) Florida POD forecast model. http://www.swf.usace.army.mil/pubdata/ppmd/EmerMgt/PDF/FLPODFORECASTMODEL.pdf. Accessed 25 Feb 2010
Varian H (1992) Microeconomic analysis. Norton, New York
Wachtendorf T (2010) When push comes to shove: the framing of need in disaster relief efforts. International research committee on Disasters meeting, XVII World Congress of Sociology, International Sociological Association, Goteborg, Sweden
Wachtendorf T, Kendra JM, Rodriguez H, Trainor J (2006) The social impacts and consequences of the December, 2004 Indian Ocean tsunami: observations from India and Sri Lanka. Earthq Spectr 22(3):693–714
Wachtendorf T, Brown B, Holguin-Veras J (2013) Catastrophe characteristics and their impact on critical supply chains: problematizing material convergence and management following hurricane Katrina. J Homel Secur Emerg Manage 10(2):497–520
Wisner B, Blaikie P, Cannon T, Davis I (2004). At risk: natural hazards, people's vulnerability and disasters. Routledge, London
World Population Review (2014) Population estimates 2014. Accessed 39 Dec 2104. http://worldpopulationreview.com

Chapter 4
Cash as a Form of Relief

Graham Heaslip, Ira Haavisto and Gyöngyi Kovács

Abstract The historical form of humanitarian relief is to provide people in need with goods, however, assistance in the form of physical goods is shifting towards providing cash-based assistance instead of goods (Kovács, Humanitarian logistics: meeting the challenge of preparing for and responding to disasters, pp 275–285, 2014; Heaslip, J Humanit Logist Supply Chain Manag 5(1):2–11, 2015). Cash-based assistance is a new concept, and growing rapidly. While only US$2 million in the form of cash-based assistance was provided as international humanitarian assistance in 2006, the sum increased to US$47.4 million in 2011 (Global Humanitarian Initiative, http://www.globalhumanitarianassistance.org/wp-content/uploads/2012/07/GHA_Report_2012-Websingle.pdf, 2012). Cash-based responses (also known as cash transfer programmes, CTP) are mechanisms to provide resources to a population in two main ways—by providing them directly with cash or by giving them vouchers. Cash and voucher programmes are considered to be a more cost effective solution than the more common in-kind provision of goods and services to beneficiaries and households. A key advantage is that unlike in-kind aid, cash allows households flexibility in deciding their spending needs. Cash can help generate local market activity and restart livelihoods. It is often a more empowering and dignified form of support. However, there is a reluctance to deliver cash transfers. Cash is said to be susceptible to theft, corruption and misuse. It is prone to targeting errors. It can cause inflation and distort local markets. Nevertheless, practitioners are finding solutions to mitigate these risks and have observed that many of the risks also apply to in-kind aid (Creti and Jaspars, Cash-transfer programming in emergencies, 2006). This chapter aims to develop an understanding of cash transfer programmes that will help reduce their risks to humanitarian operations.

Keywords Cash transfers · Cash-based response · Delivery mechanisms · Vouchers · Digital food · Material flow · Financial flow · Purchasing power · Funding · Corruption

G. Heaslip (✉)
National University of Ireland Maynooth, Ireland & University of New South Wales, Canberra, Australia
e-mail: Graham.Heaslip@nuim.ie

G. Kovács · I. Haavisto
HUMLOG Institute, Hanken School of Economics, Helsinki, Finland

C. W. Zobel et al. (eds.), *Advances in Managing Humanitarian Operations*, International Series in Operations Research & Management Science 235, DOI 10.1007/978-3-319-24418-1_4

4.1 Introduction

Providing cash, vouchers, tickets and other similar instruments of aid to the needy and poor has existed for over a century (Devereux 2006). The American Red Cross provided cash assistance during the Franco-Prussian war of 1870–1871 (Harvey and Bailey 2011). During the Second World War ration books were issued to British and Irish citizens so they could acquire basic needs. Although cash and voucher interventions are prominent during conflict and emergencies, the concept of transfer interventions is also widely used in developing countries to provide social protection, also known as a 'social safety net'. In 1948, the British colonial administration in Sudan distributed cash, coffee, and train tickets to famine-affected populations. In Bangladesh there is a long history of cash relief (Harvey 2007). Similarly, many developed countries provide cash transfers as part of their social welfare systems.

In contemporary times, humanitarian cash programmes have been used as far back as the 1972 drought in Maharashtra, India (ACF 2007). Academic support for the concept of cash interventions is derived from Sen's (1976) 'entitlement theory,' from studies of people's coping strategies in response to emergencies, and from experience of the range of livelihood needs that arise following a disaster. Entitlement theory posits that famines are often caused not by lack of food, but by individuals' inability to get access to whatever food exists (Sen 1976, 1986). Entitlement failure could occur through loss of income or loss of employment, or from high food prices or reduced food availability. Famines or food insecurity are therefore as much a result of people's inability to buy food as they are caused by a decline in overall food availability or food production. It is therefore logical to conclude that where famine results from a lack of purchasing power, it can be addressed through income transfers (Sen 1986).

The main difference between the traditional approach of in-kind assistance and cash interventions is: in-kind aid is transported into the area of operations, mainly from abroad; cash interventions allow goods to be bought locally (Barrett et al. 2009). In recent times donors have been more willing to allow agencies to develop alternative programmes for delivering cash programmes in conjunction with, or without in-kind relief (Barrett et al. 2009). The next section discusses the emergence of cash transfer programmes (CTPs) in humanitarian operations.

4.2 Changing Landscape

The World Food Programme's (WFP) shift from food aid to food assistance in 2008 is evidence of the changing humanitarian landscape. WFP's strategy includes, for 2015, a third of all assistance programmes to be implemented using cash, vouchers and other types of 'digital food' (WFP 2013). Many humanitarian agencies have begun to explore cash transfers as an alternative to in-kind aid (Heaslip 2013; Kovács 2014), yet to date they have not been used at scale and in-kind assistance still dominates as a humanitarian response—however, donors are increasingly interested in CTP solutions (Harvey and Bailey 2011). But what exactly is a cash transfer programme? The accepted definition of a cash transfer is: "The provision of cash with or without conditions to a person or household in need of goods or services in order to enable beneficiaries to purchase these goods or services in the market place" (ECHO 2012, p. 12). Essentially instead of delivering aid in terms of material flows, CTPs change the delivery to handing out cash or vouchers.

The Global Humanitarian Initiative (2012) lays out the following arguments for the benefit of moving from providing physical goods towards cash based assistance: Cash-based assistance can (a) address the immediate consequences of humanitarian crisis, (b) be planned in the short term, (c) support households' livelihood better than the provision of physical goods, and (d) support local trade and the provision of basic services better than the provision of physical goods.

The biggest concerns of cash transfers are questions of security. Humanitarian assistance today is heavily regulated, particularly in the form of procurement regulations implemented by donors. These regulations include supplier due diligence and anti-terrorism compliance checks for all suppliers. This form of control is difficult to impose on cash based assistance since the beneficiaries make the choice of whom to purchase goods or services from. The other issue that has been raised is how to ensure that the most vulnerable in the society, such as children, elderly and handicapped, benefit from the cash based assistance (Aker 2013). For example, a pilot case of cash based assistance in Afghanistan uncovered gender issues where women had a hard time accessing banks and could not cash in their entitled vouchers. Furthermore, cash transfer programmes have been questioned for accountability. Cash transfer can be more open for unintended use than direct services or supplies. However humanitarian organisations that do provide for cash transfer programmes are trusting the end-user to purchase what they need the most, thus emphasising downwards accountability (towards the beneficiary) rather than upwards accountability (towards the donors). Although cash transfer might leave more room for misuse, there are other aspects related to the concept of "do no harm", such as, for example, distorting local markets with the influx of imported goods, that can be avoided with cash transfer programmes.

Cash transfers address the 'pull' failure by giving people the means to purchase food. The benefits of cash over food aid are that it allows recipients to make their own consumption decisions and it stimulates local markets (Kovács 2014). This boost to the local economy can facilitate the supply and distribution of food from other regions to the affected area. It may further stimulate the production of food locally and thus address the 'response' failure.

In the humanitarian community, CTPs are gaining greater acceptance in response to emergencies either as an alternative to a particular type of in-kind aid or as a complement or supplement to in-kind aid provision, and the debate around cash-based responses has moved on from whether they are appropriate to how to use them to achieve programme objectives (Bailey et al. 2008). Cash based responses are generally viewed as a tool for providing people with resources in emergencies and while they are not a sector in their own right, they can be considered across all sectors of humanitarian aid (Harvey 2007). Therefore, the use of cash does not alter the aims and objectives of an emergency response, however "cash transfers can encompass a very wide range of objectives related to food security, nutrition, health, education, water and sanitation, livelihood recovery, reintegration, shelter and protection" (Harvey and Bailey 2011, p. 45). The following section discusses the different types of CTPs and provides examples.

4.3 Types of Cash Transfer Programmes (CTPs)

The decision to use cash or vouchers instead of delivering materials is based on the context and the objectives of the intervention. Generally, cash and vouchers can be provided with or without conditions. Whether to attach conditions (including a work component) depends on the objectives of the project. Table 4.1 describes the four types of cash transfer programmes (CTPs), and Table 4.2 summarises various reasons for cash-based interventions, while listing examples of them from the humanitarian context.

The type of cash transfer used, as well as the value of the transfer, will depend on the objectives of the project as well as on a range of factors surrounding the target population and place (Bailey et al. 2008). The types of cash-based interventions can also be mixed in a project aimed at the same target population, where some of the population may receive cash for work or conditional cash transfers while a more vulnerable section of the population may receive unconditional cash transfers (ACF 2007). Cash transfers can be used as an alternative to in-kind aid or they can be used as a complementary or supplementary intervention to in-kind aid (ACF 2007). Cash transfers have been used in a variety of programme contexts. Table 4.2 provides examples of various types of interventions involving CTPs. The table highlights the types of various interventions and provides examples for each of them. The

Table 4.1 Types of CTPs in emergencies. (Source: ODI Good Practice Review 2014)

Type of cash transfer programme	Definition
Unconditional cash transfer	People are given money as a direct grant with no conditions or work requirements. There is no requirement to repay any money, and people are entitled to use the money however they wish
Conditional cash transfer	The agency puts conditions on how the cash is spent, such as reconstructing a home. Alternatively, cash might be given after recipients have met a condition, such as enrolling children in school or having them vaccinated
Vouchers	A voucher is a paper, token or electronic card that can be exchanged for a set quantity or value of goods, the value of the vouchers either as a monetary value or the value of predetermined commodities or services. Vouchers are redeemable with preselected vendors or at 'voucher fairs' set up by the implementing agency
Cash for work	Payment (in cash or vouchers) is provided as a wage for work, usually in public or community programmes

Table 4.2 Examples of CTPs (projects funded by the Directorate-General European Commission's Humanitarian Aid and Civil Protection. (DG ECHO))

Type of intervention	Type of cash transfer programme	Examples	Intents
Providing beneficiaries with cash or vouchers as a direct grant with no conditions or work requirements	Unconditional cash transfer	UNHCR grants to Chechen refugees in Baku, Azerbaijan (primarily to cover a portion of rental expenses)	Avoid refugee camp dependency, sustainability of aid
		Cash for host families (UNHCR Kosovo)	
		Transport allowance for families returning to Liberia from Ivory Coast (UNHCR, IOM and NRC)	
Paying beneficiaries in cash for taking part in a public works programme	Cash for work	Palestine job creation programme (SC, Mercy Corps, UNRWA and others)	Beneficiary empowerment, enhancing feeling of belonging to society during/after complex disaster to avoid continuous fitting, sustainability of aid
		Creation of jobs among traders (PU Palestine)	
		Road clearance (ACF in Afghanistan)	
		Road construction/rehab—DR Congo	
		Water and sanitation projects Somalia	
Paying beneficiaries in vouchers	Cash for work	Vouchers for work, Oxfam Mali and Niger	Beneficiary empowerment, sustainability of aid
Providing a cash grant during training	Conditional cash transfer	Medair, Afghanistan: cash was provided to encourage trainees; the trainers were elderly women with craft-making skills which risked being lost to the community	Support gender equality, beneficiary empowerment, secure cultural heritage
Paying beneficiaries to repair their own houses or rehabilitate farm land	Conditional cash transfer	ACTED Afghanistan	Beneficiary empowerment, sustainability of aid
		ICRC Liberia (clearing of cash crop farm land for returnees)	
Providing beneficiaries with cash on condition that they do something (attend school, plant seeds, demobilise) Destocking	Conditional cash transfer	UNHCR grant for Afghan families returning to Afghanistan from Pakistan and Iran (provision of cash at transit centres in country of return)	Sustainability of aid, avoiding prolonged refugee camp dependency
		Livestock purchase—CORDAID Ethiopia	

Table 4.2 (continued)

Type of intervention	Type of cash transfer programme	Examples	Intents
Providing beneficiaries with vouchers for a particular type of good (e.g. seeds) or a bundle of goods	Vouchers	Seed voucher fairs in DR Congo (AAA, PU)	Sustainability of aid, climate change adaptation, securing means of livelihood, beneficiary empowerment
		Seed/fertilizer/tools voucher schemes by FAO/CRS in Lesotho	
		Seed/tuber replication (e.g. Sweet Potato replication (WV Zimbabwe)	
		Donkey vouchers (Oxfam Sudan)	
		Water vouchers (UNICEF Madagascar—*proposed* for 2007)	
		Vouchers for food (PU DR Congo, ICRC Palestine)	
		Vouchers for fuel—DCA and French Red Cross, Lebanon	
		Vouchers for seed (exchanged through market traders)—ICRC Liberia	

interventions have also been analysed as for the type of CTP they represent, and their (latent) intent.

When humanitarian organizations choose to implement CTPs either as direct cash transfer programmes or via vouchers, the supply side is left to the private sector. The demand side on the other hand is leveraged with a direct provision of cash. While the purchasing power empowers the beneficiary, the downside is distrust that the markets will function for the benefit of the traders. Traders (suppliers) are seen to put their financial interest over the interest of beneficiary in need (Gentilini 2007). Such an example occurred in Ethiopia, where "traders" delivered food either too late or not at all, depending on where a financial gain could be sought (World Food Programme 2013).

In spite of differences amongst donors, aid providers and academics as to when cash transfers are efficient and when not, cash transfers are a growing proportion of humanitarian aid distribution today (Global Humanitarian Initiative 2012). Humanitarian organisations that provide cash transfer have eased up on controlling the supply chain, trusting the end-user to purchase what they need the most. The question here is whether cash transfers are changing the structures and process of humanitarian assistance or whether it is only a new form of program. The next section examines the implications of CTPs on humanitarian supply chains.

4.4 CTPs—Change to Humanitarian Supply Chains

A cash transfer offers greater flexibility to the beneficiaries as it allows them to choose the goods and services that are most appropriate to their personal situation. This has an obvious advantage over in-kind aid which delivers a 'one size fits all' package (Gairdner et al. 2011; Bailey et al. 2008). This is a timely argument as Western 'food-mountains' have been disposed of over the last two decades and surplus food stocks are much smaller today (ACF 2007). Cash transfers intrinsically support local procurement (by beneficiaries themselves), hence ensuring the cultural, technical, and health appropriateness of items bought.

Cash transfers adhere to the oft cited principles of empowerment, dignity and choice for the beneficiaries (Bailey et al. 2008). The beneficiaries need to be consulted and given a participatory role in CTPs (Bailey et al. 2008), similar to the role of customers in a commercial supply chain. Needs assessments in any emergency can be challenging; however, according to Gentilini (2007) "appropriateness cannot be predetermined since programme objectives, the economics of food consumption, market analysis, cost effectiveness and efficiency, capacity requirements and beneficiary preferences all play a role in determining the most appropriate option or combinations of options" (2007, p. 4).

In-kind aid is also affected by rising fuel and transport cost as well as by grain and commodity shortages which cash based interventions do not face (Global Humanitarian Initiative 2012). CTPs also do not incur the transport, warehousing and intensive distribution systems and costs that traditional in-kind aid does.

Another reason that cash transfers are often cited as being more efficient is that beneficiaries receive the full measure of the intended relief aid. This is different than beneficiaries selling in-kind aid that they are not in the need of in order to get cash to be able to purchase supplies to meet their more pressing needs (Bailey et al. 2008). The economic impact of cash transfers on local markets and the wider economy is seen in the multiplier effect (ECHO 2012). This can help to build resilience in the local economy and also diversity.

Cash transfers fundamentally alter the balance of power between the donor and the beneficiary as they increase the freedom of beneficiaries to decide how to use the cash (Aker 2013). Until now, the role of beneficiaries as customers has been disputed, not the least because of the lack of their purchasing power. Cash transfer reinstates the purchasing power of beneficiaries and hence changes their very role in the humanitarian supply chain—as was the case already with voucher systems in Kosovo and FYROM (Matopoulos et al. 2014).

Logistics in disaster relief accounts for about 80% of the total cost (Van Wassenhove 2006) and up to 30% of aid delivered has been identified as wastage in some post-crisis situations (Pettit and Beresford 2009). Managing the humanitarian supply chain in a coordinated way has the potential to reduce unnecessary waste. However, donors and the media are attracted to "agility appeal" where aid agencies can move relief goods with speed so there is little benefit to focusing on cost or leanness (Oloruntoba and Grey 2006). Cash transfers can be viewed as a challenge to the traditional roles established by humanitarian sectors such as nutrition, shelter, etc., as cash can address any of these needs as long as there is market supply

(Bailey et al. 2008). Aid agencies already have the staff and systems in place to deliver in-kind relief items effectively and efficiently whereas cash transfer initiatives can often raise the question of capacity. Since the cash transfer supply chains are not implemented on a large scale, there are still no economies of scale in setting up such a process.

Cash transfers shorten the logistical supply chain, simplify procurement and remove or lessen the need for transport and warehousing consideration, which may shrink the humanitarian sector considerably (WFP 2008). This change would support the objective of most humanitarian actors, who at least state that they want to run themselves out of business. The functionality of cash transfers is however conditional. For cash transfers to work there needs to be an available market. In most development aid contexts this would be the case, whilst in the immediate aftermath of a disaster the first response might still need to rely on physical items.

In short, a shift from material to financial flows diminishes the total cost of aid whilst simultaneously empowering beneficiaries (Kovács 2014). This has meant that beneficiaries have changed from being passive beneficiaries to becoming active members of the humanitarian supply chain (Matopoulos et al. 2007; Heaslip 2015).

Generally speaking, CTPs imply a reconfiguration of the humanitarian supply chain with consequential contributions to the reinstatement of the local economy. The local economy could as well be boosted by the prioritization of procuring local items instead of practising global procurement. However local procurement would not empower the beneficiary, and might in the worst case scenario distort the local market.

In humanitarian supply chains where the main activity is providing physical goods, the actors conducting the activity of distribution are commonly a local partner such as the local authorities or local NGOs. In the distribution of CTPs, there is a shift from this role towards that of an actor who can better handle a financial flow. A pre-condition is still that there are functioning markets on location and that the beneficiary has access to that market.

This form of humanitarian assistance has since become more popular and new telecommunication solutions for cash transfers such as "mobile money" have been launched. For example Safaricom has enabled the use of "mobile money" in various African countries (Kovács 2014). Mobile phones can be used to transfer cash to nomadic or hard-to-reach beneficiaries (Heaslip 2013).

Perhaps the most intriguing change is the impact on financial vs. material flows in the humanitarian supply chain. In the traditional model, financial flows originate from donors to humanitarian organisations, which use these finances to pay for material supplies that they deliver to beneficiaries. In a cash transfer programme, financial flows from donors still come to humanitarian organisations, which then assess the possibility for distributing cash directly to beneficiaries. If this is possible—given that there are items available on a market, for instance—the financial flows go to beneficiaries, who pay themselves for the products and services they need. Humanitarian organisations become the brokers of these flows, and the distributors of cash, but not the providers of the purchased materials (see Fig. 4.1). Their role in delivering materials diminishes to the materials that are not available on the local market.

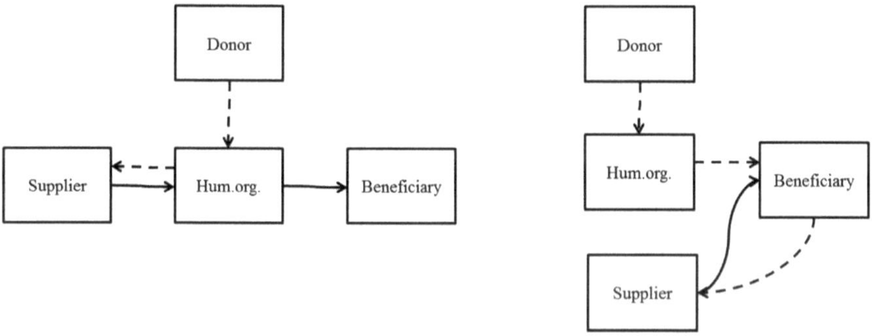

Fig. 4.1 Changing paradigm of humanitarian supply chain—before and after CTPs

Figure 4.1 illustrates the two extremes of before and after supply chains, illustrating material flows with full arrows, and financial flows with dashed arrows. In the original case, humanitarian organisations procure materials and deliver these to beneficiaries. In the ultimate, unconditional cash transfer case, humanitarian organisations would only deliver funding to beneficiaries, who themselves procure from suppliers.

This also implies a significant change in supply chain strategy. The traditional humanitarian supply chain pushes items first and gradually moves towards a pull strategy once more information becomes available. Cash transfer programmes enable a pull strategy to be implemented from the beginning. Through this, arguably, they can meet the actual needs of beneficiaries quicker and more accurately.

4.5 Impact of CTPs on Humanitarian Logisticians (HL)

The ability of logistics and supply chain managers to adapt and operate in unstable and complex conditions, maintaining an agile supply chain, requires professional skills, competence and knowledge (Kovács and Spens 2009; Christopher and Tatham 2011). The skills required for a humanitarian logistician include functional, interpersonal, general management, and problem solving skills and are context specific (Kovács et al. 2012). The humanitarian logistician's skills are systematic throughout the disaster management cycle. The findings of Kovács et al. (2012) demonstrate that significant skills are appropriate and performance enhancing. Problem solving and inter-personal skills (market winning skills) are deemed in the analysed job advertisements as more important than functional logistics (market qualifier skills) and general management, despite market qualifier skills being the backbone of logistics. Interestingly, Kovács et al. (2012) point out that, in relation to specific skills, 'marketing' was considered 'not significantly relevant' to performance.

The focus on HL skills has come about as a result of improving competencies and developing logistics as a profession (Walker et al. 2010). Bölsche et al. (2013), in contrast, contend that improving specific competencies of HL would mitigate the impact of disasters on affected populations and countries. They further contend that gaps still exist and that 74 % of logisticians in 2012 received relevant training, which is effectively unchanged from the Fritz Institute report of 2005, showing 73 % (Thomas and Mizushima 2005).

Allen et al. (2013) carried out a survey with members of the Humanitarian Logistics Association (HLA). They found that a career path in humanitarian logistics requires extra skills and training, particularly with accountancy and finance. They state this shows evidence of a "hierarchy of skills that is important to humanitarian logistics and the needs for these skills depend on the levels of responsibility" (Allen et al. 2013, p. 143). The Allen et al. (2013) study endorses the earlier Harvey (2007) report which posits that logistics is often more straightforward than implementing cash programmes, and that extra skills and competencies are required for cash.

Harvey (2007) further claims that CTPs may diminish certain activities in the humanitarian community as more governments take control through cash. In the CTP arena humanitarian agencies (for example Oxfam, IFRC and UNICEF) have stressed the need for specific competencies when dealing with CTPs (Creti and Jaspars 2006; IFRC 2007). Oxfam, for example, has been developing guidelines and tools for CTPs since 2006 (Creti and Jaspars 2006). The key message from all agencies is that CTPs in response to emergencies are now a common feature of humanitarian action and as such should be morphed into existing procedures, tools, guidelines, and policies. The agencies agree that CTP should be developed in line with training programmes, education, assessment guidelines, employee induction, financial management, contingency, knowledge base, information management, procurement protocols, funding, preparedness planning and policy. In essence CTPs should be embedded into the fabric of agencies and organisations. Figure 4.2 provides a list of competencies required for CTP, and these competencies are discussed below.

Knowledge Management

Knowledge management is an important element for cash transfer programmes as they become increasingly used in humanitarian operations. A knowledge of local markets and the impact cash transfers can potentially have on local economies in positive and negative ways informs the value, type and delivery mechanism of the

No.	Competency
1	*Knowledge management.* An awareness of familiarity gained by situational experience or fact.
2	*System of operations and process management.* A set of systems and processes established to carry out functions in CTPs.
3	*Information management.* The collection and management of information from one or more sources and the distribution and use of that information to one or more audience(s).
4	*Human resources.* 'Human capital,' which is used to describe the knowledge individuals embody and can contribute to an organization.
5	*Funding.* The base and amount of support in a quantifiable monetary value.

Fig. 4.2 Competencies required for CTP. (Source: Adapted from: Meeker et al. 2014; Aker 2013; Kovács et al. 2012; Van Wassenhove 2006)

cash transfer. Knowledge management requires a framework of continuous learning between disasters so that knowledge gained from a past disaster can be used to inform better preparedness for disasters in the future. This would help to increase resilience and ensure that preparedness plans had some insight in disaster relief operations (Kovács and Spens 2007). The problem with achieving this framework is that there is a high turnover of humanitarian workers in field operations so that experience is often lost (Heaslip et al. 2012). There is a lack of a "structured knowledge system that allows information to be shared among people and for them to be transmitted from one occurrence to another" (Cozzolino 2012, p. 28). Furthermore lessons learnt are not, according to Ontko et al. (2007), communicated constructively. This impacts the humanitarian supply chain as there is an absence of information and learning from past disasters and solutions. Back-office management capabilities could potentially be sourced through partnerships with commercial companies. An integrative partnership would allow for cross-sector cooperation to the mutual benefit of both parties (Cozzolino 2012; Heaslip et al. 2012).

Like any relief intervention, cash transfers can be considered in relation to the context in which they will be used. The local market will be affected by an injection of cash so it is important that the market can absorb the extra capacity without causing soaring inflation to occur. (Harvey and Bailey 2011; Bailey et al. 2008; Creti and Jaspars 2006). The market must also be able to supply the goods needed by the beneficiaries and in-kind aid is more appropriate anywhere there is 'supply failure,' which is lack of supply regardless of the existing demand (Harvey and Bailey 2011).

System of Operations and Process Management

The fundamental feature of humanitarian supply chains is that "humanitarians are always faced with the unknown. They do not know when, where, what, how much, where from and how many times. In short, the basics for setting up an efficient supply chain" (Van Wassenhove 2006, p. 480). Technology has improved visibility in commercial supply chains in recent years (Peck 2004); however in humanitarian supply chains there is a lack of uniform systems and they are invariably more "complex interacting networks that link organisations, industries and economies" (Peck 2004, p. 18). Added to the lack of technological industry integration is the nature of data collection and dissemination which involves inventorying and mapping a devastated area and its population. "The humanitarian system is not a homogenous or closely governed entity, and many of the problems associated with the perceived decline in humanitarian space are at least partly attributable to the nature of the system" (Collinson and Elhawary 2012, p. 1). However, as Maiers et al. (2008) points out, an IT-system in itself cannot solve the challenges humanitarians are faced with. For a humanitarian system to function more smoothly both intra and inter-organizational communication needs to be considered.

One of the most integrated global systems in the world is the financial system. There already exists the means and mechanisms to network financial resources to end users through a global system that includes security measures and risk reduction features. Financial networks that have reach or capacity but have been little used in an area can contain the risk of inflation from injecting cash into a local economy. In a market context where goods are not in stock due to a lack of demand but a reliable supply of these goods or stock is available relatively close, then cash transfers can often simulate this supply as the cash transfers create a market demand (Creti and Jaspars 2006; Bailey et al. 2008). This can often help a market recover and stimulate economic activity that was not present before. However markets in developing countries can often be affected by seasonal supply so cash transfers would not work in a context where food supplies are low prior to harvest season, for example, and food aid would be more appropriate in these contexts (Creti and Jaspars 2006).

Another major consideration in assessing if cash transfers are appropriate is the risk of inflation in the price of goods that people need as a direct result of injecting cash into the local economy. "The likelihood of inflation caused by a cash transfer project is connected to the impact of the disaster, the competitiveness, integration and resilience of local markets and the capacity of local traders to respond to the increased business that the cash injection is likely to simulate" (Harvey and Bailey 2011, p. 29). Inflation would increase vulnerability and devalue the buying power of the cash transfer. Inflation is a real concern for any cash based intervention as it not only diminishes the real value of the grant but the cash transfer has the potential to actually cause inflation which not only damages the market but disadvantages recipients and non-recipients alike (Bailey et al. 2008). Agencies usually have fixed budgets and so adjusting the amount of the cash grant can be difficult which can adversely impact the project objectives if inflation erodes the value of the grant. Rising food prices can also be caused by market forces as happened in the 2007–2008 food price crises which created similar problems for fixed cash grants (Wheeler and Devereux 2010). Security considerations for cash transfers go beyond the market, however, and security and safety of the beneficiaries and agency staff delivering the grant also have to be considered.

Information Management

Rapid emergency assessment tools are used by many agencies in order to gather information on the needs of an affected population. While rapid, dependable, initial assessments are always challenging in a major emergency, Bailey et al., argue that they are asking the wrong questions with the entire focus on items needed in various sectors rather than assessing how these needs can best be met in a particular context; "In order to determine the most appropriate form of response, assessments should also include information about local markets, the availability of basic commodities and how these markets would be likely to respond to injections of cash or in-kind goods" (2008, p. 7).

Cash transfer programmes require different types of information to be assessed in needs assessment. The needs of beneficiaries remain in focus, in particular, for vulnerability-capacity assessments and prioritisations of beneficiary (households) for receiving aid. However, other aspects also become important in needs assessment: for example, in order to be able to judge whether a cash transfer programme can be implemented in the first place, it becomes important to assess the availability of vital items and service providers on the market. At the same time, distributing cash typically requires a functioning banking system, and/or other, e.g. mobile, technologies that enable cash distribution. This again requires different types of information to be acquired and assessed by humanitarian organisations than previously. Since many of the cash transfer programs (see Table 4.2) have been implemented as pilot programs the management of information systems can be *ad hoc* and not linked to other programs (Garcia and Moore 2012). What is more, the direction of information flows in the supply chain are also impacted due to changes in the directions of material and financial flows (see Fig. 4.1).

Human Resources

Having a robust human resources policy where you can place properly trained people in a position to effectively plan, act and intervene in an unfolding crisis is essential. Van Wassenhove (2006) argues that preparing, training and recruiting staff from local populations is sustainable and effective. With only 5 % of funding being spent on disaster preparation in 2011 (Global Humanitarian Initiative 2012), there is perhaps some way to go to address preparedness in a more comprehensive way.

The impact of human resource policies on cash transfer supply chains could be an increase in local knowledge, closer ties to local networks, and a more motivated and empowered local population, as well as the cost and logistical benefits of having a team in place when disaster strikes. This type of policy could take the form of sourcing local commercial cash transfer companies, agreements with mobile phone operators or working with local NGOs on planning cash-based programmes. The challenge is to alter industry attitudes and donor attitudes to recognise the importance that preparation of human resources plays in terms of resilience.

Funding and Corruption

Aid for emergency response is different than development aid, which has a more consistent revenue stream. When donors contribute to a disaster appeal, they want their money spent on alleviating the immediate distress in a visible way. Aid agencies compete for donor resources which can be a barrier to collaboration and cooperation. Collaboration in supply chains can be vertical or horizontal. The horizontal partnerships are usually among humanitarian agencies that often compete for the same financial resources (Cozzolino 2012) and this can be a barrier to cash based programmes as they are not as visible as in-kind aid and it is often harder to quantify results to donors.

Preparing a supply chain in the humanitarian sector has recently moved towards diagonal partnerships (cross-sector partnerships), which is a partnership among "actors at different levels on the value chain and from different sectors and industries" (Cozzolino 2012, pp. 26). For cash transfers, this means working with commercial partners like Visa or Western union, mobile phone companies, local NGOs, community groups or local authorities. Working with other partners however raises the question of risks and security as control moves outside of the aid agency and cash based programmes bring their own set of risks and security concerns. However, "a high level of collaborative working across supply chains can help to mitigate risk" (Christopher and Peck 2004, pp. 18) and this is as true for humanitarian supply chains as for commercial ones.

The supply chain for any type of aid always contains the potential risk of corruption and cash transfers are no different. Cash based projects may be more vulnerable to manipulation and corruption of targeting systems as they may be viewed as more attractive than in-kind aid (Bailey et al. 2008). Cash is often considered at greater risk of corruption because it is highly mobile and highly desirable, but corruption risks exist for in-kind aid projects too. Corruption risks for in-kind aid are higher during the phases of procurement, warehousing and transport which are phases that do not apply to cash transfers in the same physical way (Ewins et al. 2006). Targeting an area is difficult regardless of the type of aid being used. This is because targeting the right recipients is a process that often will contain errors of inclusion or exclusion and this can be compounded by different definitions of need and vulnerability between the local population and the aid agency (Taylor et al. 2004). "These difficulties clearly create risks of corruption and opportunities for staff or local elites to influence who is included on the list, potentially in return for a proportion of the aid delivered (a kickback), sexual favours or political support" (Ewins et al. 2006, p. 53).

Corruption risk is present at all stages and in all types of programmes; "However, mitigating many of these corruption risks is fundamentally about good management, and this involves not only having the necessary systems and procedures, but also the necessary management capacity. At a more fundamental level, it is also about greater levels of accountability and transparency to disaster-affected populations" (Ewins et al. 2006, p. 56). Cash transfers usually have close monitoring systems in place because cash transfers systems are designed to make cash visible so there is little evidence of corruption associated with them (Harvey and Bailey 2011). It is worth noting, however, that cash transfers have only been used at a small scale in emergency situations, so it is difficult to compare like with like as in-kind transfers have been used successfully at scale many times, in various geographical, political and social situations.

4.6 CTP—Lean or Agile?

The management of a humanitarian supply chain in the response phase can be divided into three major processes: demand management, supply management, and fulfillment management (Kovács and Spens 2007).

The first process, demand management, is marked by unpredictability of demand resulting in goods being 'pushed' to the location of the disaster (Kovács and Spens 2007). The demand for an immediate response prior to having a clear concept of what goods are needed creates the problem of unsolicited supplies that may be inappropriate and unwanted or arrive in unmanageable forms (Kovács and Spens 2007). These supplies can clog up the supply chain, especially at bottlenecks such as airports and "this creates many redundancies and duplicated efforts and materials" (Kovács and Spens 2007, p. 103). Also cargo movement time can be shorter than information time which can lead to the wrong goods arriving at the wrong time. Cash transfers eliminate many of the problems associated with these response phases; however, cash transfers are rarely thought of in these terms and as a result are under-utilized in the supply chain strategies used to deliver aid.

The second process, supply management, consists of controlling the goods that are being fed into the disaster area in terms of time, amount and cost. Supply management is a process that continues throughout the humanitarian response, whereas the third process, fulfilment management, cannot be executed in the immediacy of the aftermath of a disaster due to a lack of information. This process of supply chain management is probably the closest model to how cash transfers are used in the response phase. It is important at this stage to identify and communicate the needs of the aid recipients as well as to advocate on their behalf and outline agreed and achievable goals for the humanitarian response.

Fulfilment management is concerned with how to deliver aid supplies to the affected area and matching supply with demand. Speed and agility are essential features in the supply chain in the response phase. Agility in the response phase focuses on speed and effectiveness but as the humanitarian response begins to meet the immediate needs of the disaster affected population, the focus shifts from agility to leanness as we move into the recovery phase (Oloruntoba and Grey 2006). Cash transfers are already a lean system with a shorter chain of connections between donors and beneficiaries than in-kind aid; however is a shorter supply chain a sound basis to abandon applying supply chain mechanisms to them?

4.7 Delivery Mechanisms of Cash in Humanitarian Operations

There are a range of mechanisms used to deliver cash transfer aid to recipients. One of the easiest options is to transfer the funds into the beneficiary's bank accounts, but this pre-supposes a well-functioning banking system with branches located in the programme target area. It would also require that the target beneficiaries have bank accounts, which for vulnerable populations is not always the case. However, this is a very attractive option as it simplifies the control and administration, reduces risk of corruption and fraud for donors, and increases safety for the beneficiaries as they withdraw just what they need and leave the rest safely in the bank. The Sri Lankan Government made people open bank accounts to facilitate a cash transfer programme after the 2004 Indian Ocean tsunami as did the Iranian government after the Bam earthquake in 2003 (Doocy et al. 2006). This is not an option for aid agencies so they are limited to supporting target beneficiaries in opening bank accounts which may delay the delivery of relief to an unacceptable level.

Another consideration with banking is access to funds for the beneficiaries especially in conflict areas or remote areas. Mobile banking is an option in these situations which is where mobile ATMs can be driven to a location on a specified day. Beneficiaries use a PIN code or fingerprint recognition to access funds (ACF 2007). Mobile banking needs to be facilitated by banks who assume the risks and responsibility of these operations and since banks are not humanitarian organisations, their motivation is profit and this needs to be considered before implementing this type of delivery mechanism (Farrington and Slater 2006).

Where the banking structure is not a viable option as a delivery mechanism, direct cash transfers may be considered. This can be done by agency staff or it can be outsourced to money transfer agencies like Western Union who already have the expertise and sometimes the presence in the target area. Direct distribution can also be done by the government or NGOs or sometimes local traders can be sub-contracted as they are present and a recognised service provider in the area.

Vouchers are another mechanism for delivering relief; however they are limited as they are usually item specific or valid for one supplier, thus pre-assuming what the beneficiary needs. Pre-paid cards are becoming more popular as they give greater choice to the beneficiary and can be programmed to be valid with more than one supplier which removes the risks associated with monopolies. Mobile phones are another delivery mechanism that is increasing in use due to the high percentage of mobile phone ownership in even the poorest countries. Concern Worldwide pioneered this relief transfer mechanism in a humanitarian context in a remote area of Kenya in early 2008. Funds were transferred to registered numbers which could then be exchanged for actual money or electronically transferred to another phone or used for phone credit through a system they pioneered, the M-PESA system, which does not require a bank account (Datta et al. 2008). This system needs a functioning telecommunication network and an agent who will exchange the digital credit for actual cash, although the mobile phone can serve as a digital wallet with some traders accepting the digital credit as payment (Datta et al. 2008).

Safety is a major factor for consideration in cash transfers, both for the beneficiaries and those physically distributing the cash (Bailey et al. 2008). The argument for gender disadvantage stems from the control that women typically have over food resources and the introduction of cash could disempower them (Bailey et al. 2008); however it is difficult to see how this could not also happen with the sale of in-kind goods. Cash is easier for a recipient to hide; by the same token, it is also easier for a thief to carry away (Harvey and Bailey 2011). Using cash transfers in conflict situations raises concerns not just for secure delivery but also the impact of the cash on the conflict and if it could make the conflict worse. Evidence suggests that methods for safe and secure delivery of cash exist even in conflict situations (Bailey et al. 2008); however it is hard to quantify the impact of humanitarian cash projects on conflict situations. In some cases cash transfers may be considered a more secure alternative to in-kind aid. In Chechnya, cash transfers were considered more appropriate than bulky in-kind aid due to the security situation in this conflict scenario. In this situation there were pre-existing, secure financial systems such as banks and mobile banking available through private/commercial companies (Danish Refugee Council 2008). Some strategies to reduce the risk incurred by agency staff include outsourcing the distribution to a specialised agency or choosing an indirect method of delivery (Harvey and Bailey 2011). A delivery mechanism that is appropriate to the particular emergency context will be required for a cash transfer programme.

All of these delivery mechanisms require logistical planning which needs investment in the planning and preparation phases. Electronic payment schemes can also be expensive in the short term (Smith et al. 2011), which is usually the defining feature of emergency relief. With a more comprehensive approach to cash transfers, it would be possible to reduce these costs through prearranged contracts used like a postponement mechanism or through greater investments in networks that have the potential to be utilised in an emergency.

4.8 Conclusion

In areas where markets are functioning and there is adequate infrastructure and partners to support CTPs, there are good arguments that cash transfers, despite some shortfalls, would work better than the provision of supplies and should, therefore, be prioritized. The benefits of CTPs are however contextually dependent. Use of cash transfers during sudden onset disasters such as floods might not be very effective, unless the response is long-term and markets have normalized.

Despite the fact that CTPs are not new, CTPs as an emergency response are still relatively rare. The challenge for humanitarian agencies in the coming years is to overcome their fears surrounding CTPs, and to implement cash programmes where they are judged to be the most appropriate response. This will require a change in donor policies, but also a fundamental change in the skill set of humanitarian logisticians, who are used to identifying needs and providing commodities and thus to maintaining control over the provision of assistance. A further consequence (and benefit) of CTPs is that agencies must hand over control of aspects of emergency programming to the affected communities themselves.

Do the benefits of cash transfer programmes outweigh their drawbacks? Further understanding is needed of these programmes and of the boundaries within which cash transfer programmes function effectively. Some boundaries were highlighted in this study, such as the availability of food items on the market and the availability of technical and/or banking systems to distribute cash. Apart from the economic and technical limitations these posit, these points also alter the requirements of the abilities of humanitarian organisations to assess food and financial markets in a region that has been affected by a disaster.

The impact of cash transfer programmes on humanitarian operations cannot be understated: They alter supply chain design, the very role of beneficiaries as well as humanitarian organisations, and change the strategy of aid delivery from push to pull. Perhaps the most important factor is the elimination of many logistical activities that needed to be performed by humanitarian organisations. Delivering cash diminishes the needs for lengthy procurement and assessment processes, pre-positioning, transportation and distribution. This bears the potential of significant reductions in costs for delivering humanitarian aid at the same time as it is an important move from aid to trade.

References

ACF—Action Contre la Faim (2007) Implementing cash-based interventions: a guideline for aid workers. http://www.actionagainsthunger.org/publication/2007/10/implementing-cash-based-interventions-guideline-aid-workers. Accessed May 2014

Aker J (2013) Cash or Coupons? Testing the impacts of cash versus vouchers in the Democratic Republic of Congo—Working Paper 320, Center for Global Development. http://www.cgdev.org/publication/cash-or-coupons-testing-impacts-cash-versus-vouchers-democratic-republic-congo-working. Accessed 8 April 2014

Allen AM, Kovács G, Masini A, Vaillancourt A, Van Wassenhove L (2013) Exploring the link between the humanitarian logistician and training needs. J Humanit Logist Supply Chain Manage 3(2):3–3

Bailey S, Savage K, O'Callaghan S (2008) Cash transfers in emergencies: a synthesis of world vision's experience and learning. Humanitarian Policy Group at Overseas Development Institute, London

Barrett CB, Bell R, Lentz EC, Maxwell DG (2009) Market information and food insecurity response analysis. Food Secur 1(2):151–168

Bölsche D, Klumpp M, Abidi H (2013) Specific competencies in humanitarian logistics education. J Humanit Logist Supply Chain Manage 3(2):99–128

Christopher M, Peck H (2004) The five principles of supply chain risk. Logist Eur 12:17–21. (February edition)

Christopher M, Tatham P (2011) Humanitarian logistics: meeting the challenge of preparing for and responding to disaster. Kogan Page Limited, London

Collinson S, Elhawary S (2012) Humanitarian space: trends and issues. Humanitarian Policy Group. [Online] http://www.odi.org.uk/sites/odi.org.uk/files/odi-assets/publications-opinion-files/7644.pdf. Accessed March 2014

Cozzolino A (2012) Humantarian logistics: cross-sector cooperation in disaster relief management. Springer, New York

Creti P, Jaspars S (2006) Cash-transfer programming in emergencies. Oxfam GB, Oxford

Danish Refugee Council (2008) Study of the ECHO cash transfer pilot project. Danish Refugee Council, Copenhagen

Datta D, Ejakait A, Odak M (2008) Mobile phone-based cash transfers: lessons from the Kenya emergency response. Humanit Exch Mag 40(1):37–40

Devereux S (2006) Cash transfers and social protection. In: Cash transfers and social protection. Regional Workshop on Cash Transfer Activities in Southern Africa. pp 9–10

Doocy S, Gabriel M, Collins S, Robinson C, Stevenson P (2006) Implementing cash for work programmes in Post-Tsunami Aceh. Disasters 30(3):277–296

ECHO (2012) The review of cash transfers and vouchers in humanitarian crises. http://ec.europa.eu/echo/policies/sectoral/cash_en.htm. Accessed 9 April 2014

Ewins P, Harvey P, Savage K, Jacobs A (2006) Mapping the risks of corruption in humanitarian action, a report for Transparency International and the U4 Anti-Corruption Resource Centre by the Overseas Development Institute. Overseas Development Institute, London

Farrington J, Slater R (2006) Cash transfers: panacea for poverty reduction or money down the drain? Dev Policy Rev 24(5):499–511

Gairdner D, Mandelik F, Moberg L (2011) We accept cash: mapping study on the use of cash transfers in humanitarian recovery and transitional response. Noras and the Norwegian Ministry of Foreign Affairs, Olso

Garcia M, Moore CMT (2012) The cash dividend, the rise of cash transfer programs in Sub-Saharan Africa. World Bank Publication, Washington, DC

Gentilini U (2007) Cash and food transfers: a primer. Occasional Papers No. 18, World Food Program

Global Humanitarian Initiative (2012) http://www.globalhumanitarianassistance.org/wp-content/uploads/2012/07/GHA_Report_2012-Websingle.pdf. Accessed 5 Feb 2014

Harvey P (2007) Cash-based responses in emergencies. Humanitarian Policy Group at Overseas Development Institute, London
Harvey P, Bailey S (2011) Cash transfer programming in emergencies. Humanitarian Practice Network at Overseas Development Institute, London
Heaslip G (2013) Service operations and humanitarian logistics. J Humanit Logist Supply Chain Manag 3(1):12–26
Heaslip G (2015) Humanitarian operations: an opportunity for service research. J Humanit Logist Supply Chain Manag 5(1):2–11
Heaslip G, Sharif AM, Althonayan A (2012) Employing a systems-based perspective to the identification of inter-relationships within humanitarian logistics. Int J Product Econ 139(2):377–392
International Red Cross and Red Crescent Movement (2007) Guidelines for cash transfer programming. ICRC, Geenva
Kovács G (2014) Where next? The future of humanitarian logistics. In: Christopher M, Tatham P (eds) Humanitarian logistics: meeting the challenge of preparing for and responding to disasters, 2nd edn. Kogan Page, London, pp 275–285
Kovács G, Spens KM (2007) Humanitarian logistics in disaster relief operations. Int J Phys Distrib Logist Manag 37(2):99–114
Kovács G, Spens K (2009) Identifying challenges in humanitarian logistics. Int J Phys Distrib Logist Manage 39(6):506
Kovács G, Tatham P, Larson PD (2012) What skills are needed to be a humanitarian logistician? J Bus Logist 33(3):245–258
Maiers C, Reynolds M, Haselkorn M (2008) Challenges to effective information and communication systems in humanitarian relief organizations. In: Hayhoe GF, Grady HM (eds) Connecting people with technology: issues in professional communication. Baywood, Amityville, pp 119–131
Matopoulos A, Vlachopoulou M, Manthou V (2007) Exploring the Impact of e-business adoption on logistics processes: empirical evidence from the food industry. Int J Logist Res Appl 10(2):109–122
Matopoulos A, Kovacs G, Hayes O (2014) Local resources and procurement practices in humanitarian supply chains: an empirical examination of large scale house reconstruction projects. Decis Sci 45(4):621–646
Meeker J, Perry A, Dolan C, Emary C, Golden K, Abla C, Walsh A, Maclaine A, Seal A (2014) Development of a competency framework for the nutrition in emergencies sector. Public Health Nutr 17(3):689–699
ODI (Overseas Development Institute) (2014) ROMA: a guide to policy engagement and policy influence. ODI, London
Oloruntoba R, Gray R (2006) Humanitarian aid: an agile supply chain. Supply Chain Manage Int J 11(2):115–120
Ontko M, Williamson S, Kemp R, Haselkorn M (2007) An examination of the effectiveness of lessons-learned reporting within the humanitarian sector. J Inf Technol Soc Change Spring 2007: 26–46
Peck H (2004) Resilience-surviving the unthinkable. Logist Manage. [Online] www.som.cranfield.ac.uk/som/scr.co.uk. Accessed Feb 2014
Pettit S, Beresford A (2009) Critical success factors in the context of humanitarian aid supply chains. Int J Phys Distrib Logis Manag 39(6):450–468
Sen A (1976) Famines as failures of exchange entitlements. Econ Polit Wkly XI(31–33):1273–1280
Sen A (1986) Food, economics and entitlements. Lloyds Bank Rev 160:1–20
Smith G, MacAuslan I, Butters S, Tromme M (2011) New technologies in cash transfer programming and humanitarian assistance by cash learning partnership. [Online] file:///C:/Users/maxin_000/Downloads/CaLP_New_Technologies.pdf. Accessed May 2014
Taylor A, Seaman J, Save the Children UK (2004) Targeting food aid in emergencies, emergency nutrition network, ENN special supplement series No 1

Thomas A, Mizushima M (2005) Logistics training: necessity or luxury? Forc Migr Rev 22(22):60–61
van Wassenhove LN (2006) Humanitarian aid logistics: supply chain management in high gear. J Oper Res Soc 57:475–489
Walker P, Russ C (2010) Professionalising the Humanitarian Sector. ELRHA. http://www.elrha.org/uploads/Professionalising_the_humanitarian_sector.pdf
WFP (2008) United Nations World Food Programme (2008), Logistics business plan. http://home.wfp.org/stellent/groups/public/documents/communications/wfp199297.pdf
Wheeler R, Devereux S (2010) Cash transfers and high food prices: explaining outcomes on Ethiopia's Productive Safety Net Programme. Future Agricultures, working paper, 35(4). [Online] http://opendocs.ids.ac.uk/opendocs/bitstream/handle/123456789/2344/FAC_Working_Paper_004.pdf?sequence=1. Accessed May 2014
World Food Programme (2013) Impact evaluation of the targeted food and cash transfer programme (August 2012–March 2013). WFP, Lilongwe

Chapter 5
Measuring Markets That Supply Cash-Based Humanitarian Interventions

Laura Rock Kopczak and Gregory Matthews

Abstract The humanitarian sector is rapidly shifting to a new business model that combines a high reliance on cash-based interventions with traditional goods distributions for humanitarian response. While one reason that agencies are shifting to "cash" is to contribute to recovery, growth and resilience of local markets, measurement of the effect of interventions on markets is not commonly done. This chapter focuses on how to harness lessons from commercial and global health supply chain performance measurement models to better inform how humanitarian actors measure markets and/or supply chains that support new cash-based initiatives in emergencies. We approach this question by first describing the shift to cash and reviewing recent literature on measurement of supply chains and markets. Next we review three existing models of supply chains/markets that may be relevant to this discussion. We then examine and contrast several case studies of cash-based market interventions used in emergency responses in the Philippines, Jordan, Lebanon, Haiti, and the Sahel. Finally, we pull these threads together by revisiting the three supply chain models and providing insights on measures for markets that supply cash-based interventions.

Keywords Humanitarian logistics · Cash programs · Performance measures · KPIs · Case studies · EMMA analysis · Market systems · Cash-based interventions · Value for money

This chapter focuses on markets that supply goods to cash-based humanitarian interventions, and the supply chains behind those markets. We define supply chain as the set of entities, from raw materials suppliers to beneficiaries, which make and sell products. We use the term "markets" to be consistent with the terminology used by humanitarian agencies, to emphasize that beneficiaries are now "going to the

L. R. Kopczak (✉)
The MIT-Zaragoza Masters in International Logistics Program, Zaragoza, Spain
e-mail: kopczak@stanfordalumni.org

NYU Wagner School of Public Service, New York, NY, USA

G. Matthews
IRC (International Rescue Committee), New York, NY, USA

C. W. Zobel et al. (eds.), *Advances in Managing Humanitarian Operations*, International Series in Operations Research & Management Science 235,
DOI 10.1007/978-3-319-24418-1_5

Table 5.1 Four modalities used to meet beneficiary needs

	Goods-based intervention models		Cash-based intervention models	
	Supply "from outside"	*Local purchase*	*Vouchers*	*Cash*
Actor choosing supplier and goods	Donor or agency	Agency	Agency and beneficiary	Beneficiary
Sourcing location	Goods sourced from non-local suppliers or traders or from donors	Goods sourced from suppliers or traders in the country	Vouchers redeemed with local retailers	Cash spent with local retailers
Effect on local or regional markets	New external supply source meets demand at no cost to recipients	Internal supply source meets demand at no cost to recipients Goods may be pushed or pulled	Additional purchasing power pulls goods through supply chain	Additional purchasing power pulls goods through supply chain

market"—shopping to meet their needs, and that economic market forces are in play throughout the supply chain.

The humanitarian sector is rapidly shifting to a new business model that combines a high reliance on cash-based interventions with traditional goods distributions for humanitarian response. Agencies are shifting to "cash" as part of what can be thought of as a balancing of two objectives: (1) To serve beneficiaries well (and cost-effectively) and (2) To pursue "procurement for good" by contributing to recovery, growth and resilience of local economies.

The emerging model combines four "modalities" to meet beneficiary needs (see Table 5.1). The first two are based on distribution of goods such as tents, blankets, food, and medicines to beneficiaries. These goods are either donated or purchased. There has been a strong shift to purchase from in-country or in-region suppliers, based on the realization that bringing in goods "from outside" displaces demand on local suppliers. Buying from in-country or in-region suppliers is consistent with a policy of "Do No Harm" to local markets (Andersen 1999). The last two modalities are based on distribution of cash or vouchers to beneficiaries. Use of cash and vouchers gives beneficiaries greater dignity and choice and enlists local markets and supply chains to serve beneficiaries, supporting economic recovery and/or economic growth in the impact area and adjacent region (Sodhi and Tang 2014).

Before addressing how to measure markets (or supply chains) that support cash-based interventions, consider first *why* to measure these markets. In this new environment, when emergencies occur, humanitarians analyze markets early on in the response in order to choose modalities and design programs for providing goods and meeting beneficiary needs. It seems that a second reason to measure markets is to *monitor* their performance in supporting the cash-based intervention and the response, so that adjustments can be made in programming, if needed. A third reason to measure markets is to understand the effect of the intervention on the markets themselves. A fourth reason to measure markets is to understand "what works" in various settings, and to incorporate learnings into future programming.

Humanitarian agencies *do* conduct "market assessments" early on to inform "cash" program design, and *do* monitor impact of "cash" programs on beneficiaries. However, even though one reason for the use of "cash" is to stimulate and strengthen local markets, most agencies currently do little monitoring or measurement of the performance of supply chains/markets and/or the impact of cash-based interventions on market systems.

This chapter focuses on how to harness lessons from commercial and global health supply chain performance measurement models to better inform how humanitarian actors measure markets and/or supply chains that support new cash-based initiatives in emergencies. We approach this question by first describing the shift to cash and reviewing recent literature on measurement of supply chains and markets. Next we review three existing models of supply chains/markets that may be relevant to this discussion. We then examine and contrast several case studies of cash-based market interventions used in emergency responses in the Philippines, Jordan, Lebanon, Haiti, and the Sahel. Finally, we pull these threads together by revisiting the three supply chain models and providing insights on measures for markets that supply cash-based interventions.

Cash-Based Interventions

Cash-based interventions include approaches such as unconditional cash grants, conditional cash grants, cash for work, and vouchers. Leading agencies argue that when an emergency occurs, and cash-based interventions are being considered, it is imperative to analyze how well markets are functioning in order to minimize (unintended) negative consequences from a cash-based program, such as significant changes in local market prices, a significant drop in demand for local market goods, and market distortions that undermine the future viability of local markets (Austin and Chessex 2013).

'Critical' market systems are defined as those that "played, play or could play a major role in ensuring survival, and/or protecting livelihoods of the target population" (EMMA Toolkit 2010). This chapter focuses on (supply) market systems that provide food, water, shelter materials, essential household items, or services to meet urgent survival needs. When an emergency occurs, agencies consider whether to distribute goods or cash to beneficiaries, following a logic shown in Fig. 5.1 below. Response activities are classified as either direct (provision of assistance to beneficiaries) or indirect (actions to strengthen the market system.)

When beneficiaries spend the cash or vouchers they receive, local sellers pull goods from supply sources in the local or regional area. The situation may be complex. When a disaster occurs, normal sources of supply may not be available in the impact area (e.g. a drought or conflict may affect food supply, an earthquake may destroy stores of goods). In addition, disruption of supply enablers (transport, credit, staffing, etc.) may impede delivery of items to sales points. Thus, it is important to consider what will happen with supply in the areas adjacent to and beyond the impact area and to monitor supply enablers and sources as events unfold (Cuny and Hill 1999).

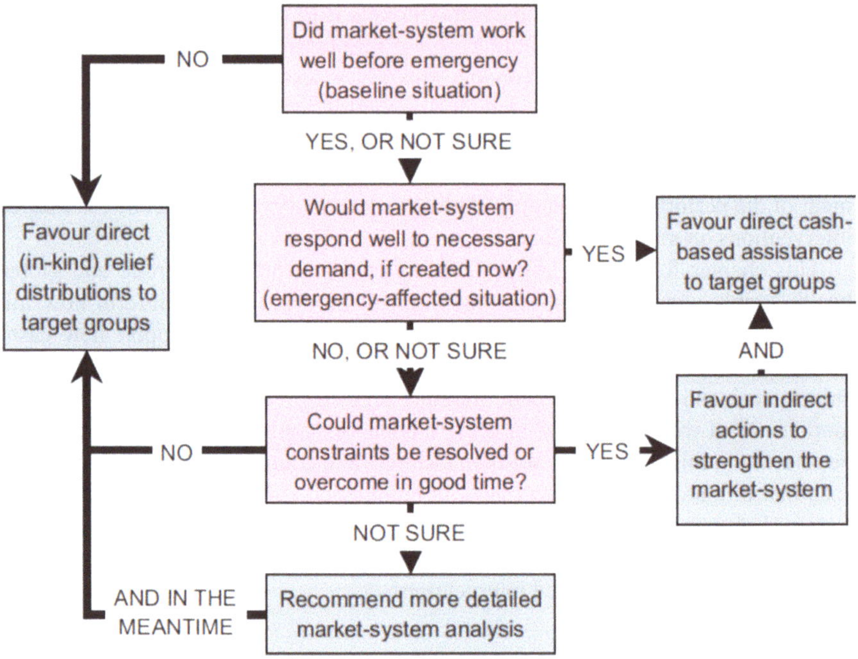

Fig. 5.1 Response analysis logic in a supply system. (Excerpt from EMMA Toolkit, p. 164)

In fact, cash-based vs. goods-based response is not strictly an either/or decision. For example, World Food Program (WFP) addresses hunger and malnutrition through both goods-based interventions (using items such as maize, wheat, rice, oil, pulses) and cash-based interventions. WFP increasingly relies on purchase from in-country or regional suppliers (2013 Annual Statistical Report on UN Procurement).[1] Traditionally WFP has been a logistics operator, delivering very large tonnages of food donated as in kind donations by countries with surplus crops. Recently WFP's "business model" has shifted to one in which a reduced amount of donated food is combined with purchase of large quantities of food from hundreds of thousands of smallholder farmers to meet beneficiary needs. At the same time, WFP has rapidly scaled up its use of vouchers and cash where feasible, appropriate, and advantageous.

WFP's 2013 procurement spend was $1.7 B (2013 Annual Statistical Report on UN Procurement). WFP's logistics and supply department at headquarters looks across the many countries in which it operates and seeks to create the best possible "Value for Money" by globally matching supply sources with beneficiary needs, considering In Kind Donations, harvest seasons and local supply markets, opportunities to use cash, and nutritional aspects of the food it receives and purchases.

[1] The United Nations spent $16.1 billion on goods and services in 2013, 60% of which was sourced from developing countries and those with economies in transition.

However, WFP does not know in advance what new emergencies will occur where and with what impact on markets. Therefore, it works to create flexibility and scalability—to be able to switch back and forth between cash and food, and to draw on the markets that it knows well. Thus, supply for cash programs is implemented against a backdrop of strategic global supply management, to the extent that the global supply strategy is already calling for purchase of these goods from in-country or regional markets.

Use of cash is an extension of the trend to increase procurement from in-country and regional suppliers. Banking technologies such as ATM cards and mobile banking are enabling cash by simplifying cash transfer payments to beneficiaries. With increasing use of cash, donors, agencies and "cash learning groups" are working to share knowledge to capture best practices and to assemble evidence to evaluate what works well and what should be avoided. Reports have recently come out from the Global Logistics Cluster about logistics challenges in cash-based interventions (Global Logistics Cluster 2015) and from the Cash Learning Partnership (CaLP) about the use of cash by various agencies in shelter and in water, sanitation and hygiene (WaSH) interventions (Cash Learning Partnership 2015). The global food security cluster has a working group dedicated to exploring cash and voucher coordination, and the global logistics cluster has begun discussing monitoring and measurement of market performance for cash-based programs, creating in late 2014 a YouTube video to introduce the topic to logistics audiences (see http://www.logcluster.org/cashandmarkets). With all of this activity and evidence being collected, it seems to be a good time for academics to contribute through research on supply chain and market measurement.

5.1 Literature

This section summarizes literature on performance measurement related to commercial supply chains, humanitarian logistics and markets.

Measuring Supply Chain Performance

Much has been written about supply chain measures. A June 2009 literature search of the Emerald database using the keywords "supply" AND "chain" AND "metrics" returned 995 papers (Tatham and Hughes 2011). Recent reviews of supply chain performance measures (Gunasekaran and Kobu 2007; De Toni and Tonchia 2001; Beamon 1999) evaluate and categorize measures being used. Categories mentioned include time, quality, cost, efficiency, flexibility and diagnostic measures. Beamon 1999 stresses that a performance measurement system must include three types of measures: output, resources and flexibility. De Toni and Tonchia (2001) distinguish between traditional cost-based measures and non-cost measures. Fitzgerald et al.

make a distinction between measures that relate to results (e.g. competitiveness and financial performance) and measures that relate to factors that affect results (e.g. quality, flexibility, resource utilization and innovation).

Measuring Humanitarian Supply Chains

Measuring humanitarian supply chain performance is different and more challenging, as a social objective is substituted for the profit objective of commercial supply chains and information may be scarce or difficult to obtain due to the contexts in which these supply chains operate (Beamon and Balcik 2008). Recent literature focuses on what to measure as well as how to construct good measures. In their literature review, Tatham and Hughes (2011) conclude that while measures often focus on operational aspects such as number of boxes delivered and associated cost, there is a need to link these to measures of outcome or effectiveness, to address the question of whether the operations worked effectively towards the program objective. Similarly, Holguin-Veras et al. (2013) argue that the objective function for models seeking to "optimize" logistics decisions should include logistics costs and "deprivation costs"—the costs representing the human suffering resulting from lack of access to a good or service. In contrast, De Leeuw (2010) focuses on how to develop measures and recommends the use of mission maps as a means to derive a complete, useful set of measures that are linked to program and/or organizational objectives. Abidi et al. (2013) examine key success factors and suggest that they should be used as the basis for performance measures.

Agencies are developing supply chain strategies and implementing measures that demonstrate progress linked to those strategies. For example, the International Federation of Red Cross and Red Crescent Societies (IFRC) measures response to sudden onset natural disasters using measures that show that they have become "faster, better, cheaper" (Gatignon et al. 2010; Schulz and Heigh 2009).

The sourcing strategy of the United Nations Children's Fund (UNICEF) for Ready-to-Use Therapeutic Food (RUTF) for treating malnourished children is an example of an evolving, multi-objective supply chain strategy. Since 2008, UNICEF has pursued a sourcing strategy based on a network of "local" and "global" manufacturers (Komrska et al. 2013).[2] As of 2012, UNICEF was purchasing RUTF from "local" suppliers in 11 countries where the product is used, as well as from 10 "global" suppliers in non-using countries. UNICEF measures lead time, timeliness of delivery, product cost and freight cost, and assesses product quality on an ongoing basis. UNICEF also measures the percentage of product supplied by "in-country" and "in-region" suppliers, as an indicator of how well it meets its objective of contributing to local economies. Furthermore, the agency strategizes about how to configure its global network to be agile: UNICEF allocates a steady production

[2] UNICEF is starting to support cash transfers ALONG with RUTF as a way of ensuring the RUTF gets consumed by the child who needs it, rather than sub-divided amongst family members.

volume (order rate) to local suppliers and uses the global suppliers to quickly respond to global emergencies, as the global suppliers can source inputs more quickly and so can adjust output volumes more easily. UNICEF has measured the percent of therapeutic food tonnage that is shipped by air as it has implemented a strategy to be more responsive (Komrska et al. 2013).

Donors are defining their own ways to measure performance of humanitarian supply chains and programs. For example, Department for International Development (DFID) has defined the concept, Value for Money, with four measures related to inputs, processes, outcomes and impact (DFID 2011). The questions they ask agencies to answer as they report on their programs include (verbatim excerpt):

- "Are we or our agents buying inputs of the appropriate quality at the right price?"
- "How well do we or our agents convert inputs into outputs?"
- "How well are the outputs from an intervention achieving the desired outcome on poverty reduction?"
- "How much impact on poverty reduction does an intervention achieve relative to the inputs that we or our agents invest in it?"

The literature described above suggests some interesting questions related to measuring markets that supply cash-based interventions. First, should the same measures be used to measure suppliers to cash-based interventions (retailers, wholesalers, manufacturers) as are used to measure suppliers to goods-based interventions? Second, how much effort should be put into measuring suppliers to cash-based interventions? In a global supply network such as UNICEF's RUTF network, there are 20 suppliers, but in a retail network that sells to beneficiaries, there may be hundreds or thousands of retailers (or there may be just a few). Can the effort and cost to monitor be justified? Last, when managing a global supply network, UNICEF (and other agencies) will consider that in some countries and for some emergencies supply will not go as well as in other locations/emergencies. UNICEF measures its aggregate supply performance, while also acknowledging the realities of individual countries and emergencies. How will this translate to a cash-based environment—will aggregate measures of supply performance be useful in evaluating cash based interventions as a reliable supply channel?

Measuring Markets

Supply chains bring goods to markets. As supply chain practitioners, we generally don't focus on aspects such as market pricing and competition directly. However, as humanitarian organizations increasingly opt for cash-based interventions and also consider the effect on the markets themselves, it becomes important to measure additional aspects of either what is being offered to the consumer (e.g. price, affordability) or how the market is operating (e.g. competition, growth). Nevertheless, supply chain practitioners don't necessarily want to stray too far into the territory of economists.

UNITAID	UNICEF	DALBERG	R4D
Availability	Availability	Appropriate Design	Market Size
Affordability	Affordability	Quality Products	Market Growth Potential
Quality	Competition	Secure Supply	Supplier Power
Acceptability/Adaptability	Quality	Affordable Prices	Buyer Power
Delivery	Acceptability/Adaptability	Availability to the End User	
	Delivery		
	Funding Security		

Fig. 5.2 Measures for global-health markets. (Excerpted from USAID 2014)

A recent primer covering approaches to "market-based interventions" for global health commodities (USAID 2014) highlights measures that four organizations that do work in this area use to measure markets and the impact that market-based interventions have. These "market-based interventions" help improve access to new health technologies (products) by working to remove constraints on market operation and/or by incentivizing supply chain actors to pull or push product through the supply chain. As shown in the figure above (direct excerpt), the organizations vary in the measures used, with the UN organizations (UNITAID and UNICEF) sticking close to supply chain measures and Results for Development (R4D) taking an economist's perspective (Fig. 5.2).

5.2 Three Models to Consider

Measurement is related to the model of the supply chain or market that is chosen. In this section we present three frequently used models that are relevant to this discussion. The first two focus on supply chains, which are also sometimes referred to as "market systems." The third focuses on global health systems and the supply chains that underlie these systems.

Triple-A Supply Chain

The Triple-A Supply Chain approach has been popularized as applied to commercial supply chains. Indeed the Harvard Business Review article (Lee 2004) is one of the best-read supply chain articles. Furthermore, some academic authors have written about applying the concept to humanitarian supply chains (Tomasini and Van Wassenhove 2009; Oloruntoba and Gray 2006).

The article advises that companies should seek to incorporate supply chain strategies that are part of the Triple-A approach, and that doing this will make them more successful. The methodology sets objectives for creating supply chain capability in three areas (excerpted from the article):

- "***Agility***—Respond to short-term changes in demand or supply quickly; handle external disruptions smoothly.
- ***Adaptability***—Adjust supply chain's design to meet structural shifts in markets; modify supply network to strategies, products and technologies.
- ***Alignment***—Create incentives for better performance."

The supply chain strategies that are suggested to achieve agility, adaptability and alignment are very familiar to commercial supply chain experts and range from information sharing and coordination, aligning incentives, management and sharing of risk, managing/optimizing globally and adapting supply chain design and processes as circumstances change.

The EMMA Market Systems Model

The EMMA Emergency Market Mapping and Assessment Toolkit is one of several approaches that have emerged from collaborations among humanitarian agencies. The EMMA Toolkit can be used early on in an emergency response to characterize an impact area market system for a specific commodity and inform the choice of modality and the design and sizing of the market intervention (see: http://emma-toolkit.org/). It is also used to develop a baseline for a market in a place with recurrent emergencies, such as regions with recurrent food emergencies. It has been in use since 2010.

The EMMA Toolkit is becoming institutionalized as more and more people are trained on the methodology. It has been applied in a wide range of sectors, and is an appropriate analytical tool for understanding major constraints in markets which supply goods or services, as well as in income markets from which individuals derive income to meet household needs in an emergency. Often the exercise describes the pre-disaster (baseline or normal) market system and contrasts it with the post-disaster market system to evaluate changes in levels of market functioning following a crisis.

EMMA typically assembles three sets of data that inform program design: the market system maps showing the market chain and limiting factors, the volume of goods or services a household requires from the market, and an analysis of the most appropriate programming options to enable affected households to meet their needs either through the market or through other aid provision channels. An example of an EMMA map is shown in Fig. 5.3 below. This EMMA map was created in the immediate aftermath of the 2010 Haiti Earthquake based on market data gathered by 25 staff from 11 agencies over 10 days.

The map depicts the supply chain in the center of the diagram, focusing on players of interest, including the target beneficiaries and notes changes in flows of goods as well as disruptions to the supply chain. The top band of the map describes institutions, rules, norms and trends influencing the market system. The bottom band describes key infrastructure, inputs and market support services enabling the market to function.

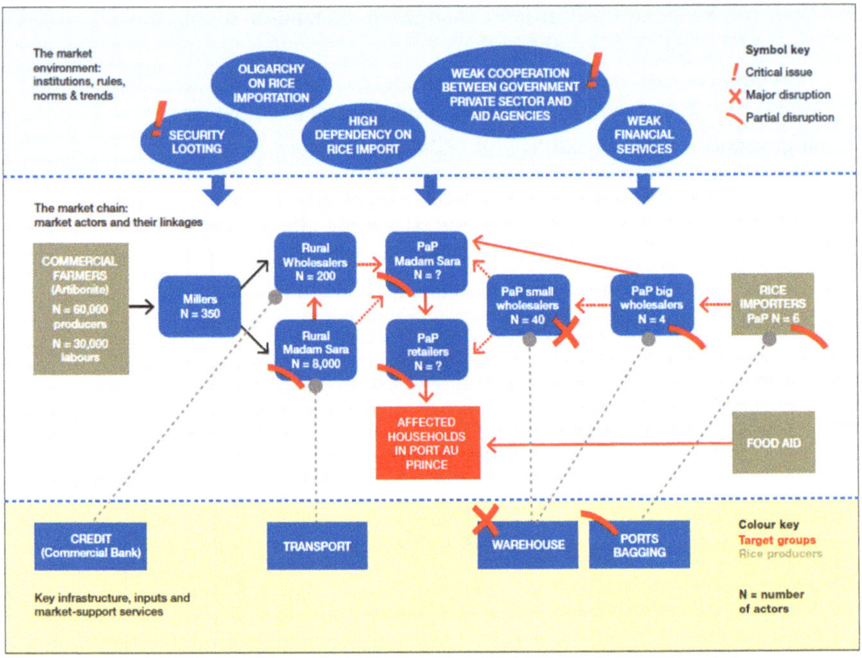

Fig. 5.3 EMMA market map for rice—Haiti. (Excerpt from Brady 2012)

The goal of the exercise is to quickly generate a fairly complete picture of the major factors at play, which can then be used with a narrative to help understand the situation as input into generation of response alternatives and design of the response. In fact, many of the same factors come up in the various EMMAs that are done.

The EMMA map illustrates that markets can be impacted in a large number of ways in a crisis and that markets are highly dynamic and evolve rapidly in post-emergency contexts. Furthermore, market maps depict the various actors and/or factors which can be measured at each level of the map (market chain, environment and infrastructure/inputs), in order to understand how markets are changing in emergencies, and how interventions affect how markets are functioning.

The Access Model

The Access model is perhaps the most common model used to examine supply chain and market issues in global health supply systems, be they private sector, public sector or faith-based. A book written by researchers at Harvard School of Public Health describes this model and applies it to case studies of markets for several commodities including hepatitis B vaccine, rapid diagnostic tests, and the Norplant implant contraceptive (Frost and Reich 2008). The model is appropriate for analyzing an intervention in a single country as well as large-scale interventions that are launched simultaneously in a number of countries.

In the introduction, the authors adapt a common definition of logistics to their area of focus: "Access to a health technology depends on providing the 'right' product at the 'right' place with the 'right' [treatment] protocol at the 'right' time" (Frost and Reich 2008, p. 8). However, their work focuses on helping people who are poor and/or live in remote areas—the underserved—take advantage of new, better health technologies (products) provided through appropriate supply chains. The framework includes four elements:

- ***Architecture***—organizational structure for coordinating and achieving availability, affordability and access;
- ***Availability***—supply logistics activities that ensure provision of a reliable, regular supply of the commodity (technology);
- ***Affordability***—ability of developing country governments and/or purchasers to afford the commodity (technology), given their "pocketbooks";
- ***Adoption***—national, district and community activities that ensure that relevant actors—governments, channel players, health service providers and consumers will adopt and use the commodity (technology), with a focus on demand generation (Fig. 5.4).

5.3 Analysis of Case Studies

In this section we analyze and contrast five recent case studies of cash-based interventions (see Table 5.2 below). The cases have been selected to represent the richness, complexity and variety of settings and interventions, in order to build good theory (Eisenhardt 1989). The Jordan and Lebanon case studies are urban refugee situations in middle income countries, in which local markets must scale up to serve more people. In the case of Jordan, the scale up may not be possible, because of limited water supply. The Philippines case study is a sudden onset natural disaster, with disruption of markets to various extents depending on the location. The scale-up requirement is a temporary surge. The disaster type (typhoon) is recurrent in that country. The Haiti case takes place in a very poor country whose market operations have been devastated by an earthquake. Last, the Sahel case is a recurrent complex food crisis that affects eight low income countries. Market strengthening is critical in order for markets to be more resilient with respect to the next drought/complex emergency.

These cases all include cash or voucher-based interventions. That is, local markets were deemed to be functioning well enough that cash could be used to support affected families to meet their needs through the existing market channels. The last two cases represent situations in which there was some indirect response proposed or implemented to strengthen local markets. It would be useful to do a comple-

Fig. 5.4 The access framework. (Excerpt from Frost and Reich 2008)

mentary study looking at cases where markets were not functioning well enough to support a cash-based intervention. This could be done, for example, using a sample of the 60+ EMMA assessments available on the EMMA Toolkit website, and characterizing what types of disruptions had occurred.

Table 5.2 Summary description of cases

Case title	Syria crisis—Jordan	Typhoon Haiyan	Syria crisis—Lebanon	Haiti—earthquake	Sahel food crisis
Timeframe studied	August–September 2013, with programs following	Nov. 2013–March 2014	Nov. 2013–March 2014	Jan.–Feb. 2010	2012 "lean season"
Number of beneficiaries	6100	xxx	90,000 families	xxx	10 million individuals
Agencies studied	Oxfam	IRC	Interagency	IRC, Oxfam	WFP and its partners
Type of emergency	Urban refugee and host communities	Natural disaster	Urban refugee	Earthquake	Recurrent drought and poverty leading to food emergencies
Temporal nature	Slow onset, ongoing	Rapid onset recurrent with varying location	Slow onset, winter season	Sudden-onset	Slow onset, recurrent with early warning system
Program location(s)	Balqa and Zarqa urban areas	Panay Island, northern Iloilo	Villages in Lebanon	Port au Prince	Eight countries in the Sahel
Program settings	Country of Jordan—cities	Panay Island—cities and towns	Cities above and below 500 ft elevation	Urban, hilly—Port au Prince	Villages where smallholder farmers and pastoralists live
Economic setting for the local markets	Middle income country, low to middle income areas	Low-income	Middle income country	Low income	"Ultra poor", food insecure locations
Commodities studied	Water	Shelter reconstruction materials	Unrestricted cash transfer intended to help people keep warm	Rice, beans, construction labor, corrugated iron	Coarse grains: millet, sorghum, maize, rice, cowpea
Objective of intervention	Ensure that refugees can access existing water systems for drinking and household use in a fair and equitable manner	Enable reconstruction of homes	Enable urban refugees to keep warm through the winter	Reinforce and expand access to goods and services by revitalizing market systems	Stave off hunger and malnutrition, reduce livelihoods impact on farmers and pastoralists

Super Typhoon Haiyan—Cash Grant and Shelter Voucher Program[3]

Typhoon Haiyan, known in the Philippines as Typhoon Yolanda, passed over the Eastern, Central and Western Visayas regions on November 8, 2013, affecting nearly 13 million people and leaving almost 2 million people homeless. The UN classified Haiyan as a Level 3 disaster, similar to the 2004 Indian Ocean tsunami and the Haiti earthquake. Since the typhoon came in from the east, the easternmost islands (and the east side of islands) were most affected. This case focuses on the International Rescue Committee (IRC) shelter rehabilitation program using Cash and Vouchers in the coastal areas of Panay Island, an island that lies on the western end of the storm track and where markets quickly resumed following the storm (Oxfam 2014).

The IRC identified shelter as a priority need of the target population immediately following the typhoon. Following the initial assessment, the IRC conducted a rapid market assessment using the EMMA market mapping tools in late November 2013 to determine the feasibility of using market-based approaches to deliver essential shelter materials to the population. The market assessment focused on markets for essential tools (proxy tool = cross-cut saw); coconut lumber; and Corrugated Iron (CGI) sheeting for rebuilding roofs.

Many other agencies in the same area were procuring materials from overseas or from elsewhere in the country and were facing very long lead-times in receipt of materials. Many agencies were still in the process of receiving and distributing emergency shelter solutions (tarps, tents, etc.) and had not yet shifted into transitional shelter approaches.

The market assessment determined that despite the overwhelming need for construction, demand at retailers had not increased because of a lack of household purchasing power. Following the assessment, the IRC decided to implement a combined cash grant and shelter voucher program to enable affected households to purchase shelter repair materials from local hardware stores and use cash to provide cash grants in order for households to pay for labor, other materials not available from hardware stores, or to meet other essential household needs. The program ran from mid-December 2013 to March 2014.

Based on the market assessment and further structured conversations with hardware store owners and suppliers in Cebu and Manila (major supply hubs), it was concluded that the retail markets and supply chains were functioning well enough (that hardware shops could re-supply key commodities) to support a cash and vouchers program. Nevertheless, several constraints on market functioning were noted:

- Many vendors (small, medium and large hardware stores, wholesalers, etc.) were not sure of future demand, which impeded their ability to plan for purchases and re-stocking. Although people did have a clear need for materials, vendors

[3] The discussion in this section is based on the Oxfam 2013 report included in the references, as well as personal knowledge from one of the co-authors.

were not sure if people would purchase these materials (if given the money) or whether aid agencies would provide those materials in-kind.
- Long lead times on deliveries (particularly for CGI which came mainly by sea from factories in Cebu and Manila), delays in deliveries, and cargo prioritization (one store reported their cargo was offloaded from a vessel because it was not considered a priority).
- Increases in prices since the typhoon for some items (particularly for CGI).
- Stock-outs of key commodities—including CGI, lumber and cross-cut saws for some retailers.
- A possible bottleneck in the number of chainsaw operators was limiting availability of coconut lumber. Lumber retailers were reporting difficulties sourcing coconut timber despite the large number of trees felled by the storm, and many stores had sold out of cross cut saws and other hand tolls.

In addition to providing households with the means to purchase, it was felt that vouchers could address the inability of market actors to predict demand by signaling that the IRC was not buying goods in kind and by indicating the value of cash vouchers to be distributed in the targeted communities. Based on this information, it was expected that retailers would stock up on construction materials, such as CGI. It was expected that they would also place orders with chainsaw operators, so that operators could plan in advance their supply and better organize their capital requirements (such as arranging in advance for trees, etc.).

The IRC contracted four hardware stores to handle the vouchers—three medium-sized hardware retailers and one small retailer. Significant support was provided to each of the shops to assist them in organizing deliveries of materials to all customers, rescheduling deliveries because of delays in receiving goods, and reconciling vouchers against delivery orders. The shops performed at different levels in meeting the needs of voucher customers. One medium retailer could not meet the supply and quality of items demanded, and resorted to delivering sub-quality materials to customers. The IRC terminated this contract. The smallest retailer performed the best and was able to increase the supply of some items.

Syria Refugee Crisis—Water Voucher and Water Equipment Program[4]

Refugees began streaming across the border from Syria into Jordan in March 2011. As of August, 2014, the United Nations had registered 619,000 refugees in Jordan, 80 % of which live in urban areas in the north of Jordan, with the rest living in camps. These Syrian refugees made up about 10 % of the Jordan population (www.unhcr.org).

[4] The discussion in this section is based on the Oxfam 2013 report included in the references.

A key need for these refugees is water. In this context, where water markets are significant, large scale and complex, it was essential to understand how the market functions, what constraints people face in accessing water, and whether the existing water market systems had the capacity to deliver an adequate quantity to refugee populations. From August 22nd to September 3rd 2013, Oxfam conducted an assessment in the urban areas of Balqa and Zarqa Governorates, as well as in the informal tented settlements around factories and agricultural lands (Oxfam 2013). Due to the refugee influx, Balqa and Zarqa experienced an increased demand for water, and as a result the overall per capita consumption dropped. In Balqa, per capita consumption reduced from 88 to 66 l/person/day. Thus, access to water was a problem for both refugee and non-refugee households.

The market assessment analyzed both the water market system and the factors determining access of the vulnerable refugees and host communities. The assessment found that access to water by poor households (in particular refugee populations) is primarily determined by purchasing power and access to adequate water storage at household level. The main constraints on supply of water at household level were:

- Municipal water supply was irregular and did not deliver sufficient quantities to meet all needs
- Limited household income to purchase water, particularly bottled drinking water which was expensive, but also costly to purchase from commercial water truckers
- Lack of household water storage tanks which would have enabled purchase from water truckers (the second cheapest option after municipal supply),
- Lack of social networks of trusted families/friends limited the ability of refugees to form purchasing groups in order to purchase a whole truckload from a commercial water trucker, and divide it among the families.

Water is available through three sources: (1) the municipal water system (piped to houses); (2) water delivered by tank truck, (3) filtered water bought at water shops. The assessment found that refugees had to pay up to $113 a month for filtered water at water shops during the hot summer months. The assessment also found several constraints on the *municipal* supply chain for water:

- The water system was working at 100 % capacity but was not capable of meeting the full need,
- A new pipeline which would increase water supply was being built, but was not yet complete,
- Water delivery was intermittent, particularly in summer, when it was delivered on a rolling basis, once every 3 to 10 days.

The Oxfam program included three interventions: vouchers for water redeemable at local water shops or with commercial water truckers, provision of storage tanks and provision of water filters. The intent was that once water storage tanks and filters were provided, refugees would be able to switch over to use of cheaper municipal water. Since local water shops were already using vouchers, it was easy to adopt these vouchers for the Oxfam program.

Syria Refugee Crisis—Cash-Based Winterization Program in Lebanon[5]

As of the end of 2014, 1.1 million Syrian refugees lived in Lebanon, in almost 2000 communities distributed throughout the country. Syrians made up over one fifth of the Lebanese population.[6] Many of these refugees arrived in 2013.

One challenge facing refugees that live at higher altitudes face is keeping warm and dry during winter cold months. To address this, starting in November 2013, an interagency winterization program began providing assistance to all refugees from Syria, to Lebanese returnees, and to some vulnerable Lebanese families who lived at higher elevations; this was roughly 60 % of the total number of displaced in the country (The IRC 2014). Assistance included cash, tools for improving shelters, non-food items such as blankets, children's clothing and stoves and heating fuel. Because agencies had been using the markets in Lebanon during 2012 and 2013, and had found them to be open, no specific market analysis was done to support the winterization program.

The cash winterization program provided 87,700 registered Syrian refugee households living above 500 m with unconditional cash assistance—US$575 each via ATM cards. Eligibility criteria included geographic (living above 500 m) as well as demographic (vulnerability indicators). The cash was given in monthly installments between November 2013 and March 2014, and was intended for purchase of a stove and fuel. In addition, these families received WFP food assistance through e-vouchers that could be used at specific stores, in the amount of $30/person/month.

UNHCR operated more than half of the cash assistance. According to UNHCR's Winterization Partner Coordination Map (December 2013) implementing and operational partners included: ACTED, AMURT, AVSI, CARE, Caritas, CISP, DRC, Handicap International, Humedica, IOCC, IOM, Makhzoumi, MEDAIR, Mercy Corps, NRC, Oxfam, Save the Children, SHEILD, SIF, Solidar Suisse, and World Vision.

The winterization program was monitored by conducting household interviews of vulnerable households who lived slightly above 500 m and received winterization cash assistance, as well as households who lived slightly below 500 m and did not receive cash assistance. (Both groups received the WFP food vouchers.) The interviews aimed to answer questions both about how cash assistance affected household spending as well as how it affected markets.

Key findings related to household spending were as follows:

- The vast majority of beneficiaries (more than 80 %) preferred cash assistance compared to in-kind assistance (e.g., food parcels).
- Households spent the majority of cash assistance on food and water, and did not spend enough on heating supplies; almost half of beneficiaries reported that they were unable to keep warm. Thus, even with the food vouchers from WFP, they

[5] The discussion in this section is based on the IRC 2014 report that is included in the references.

[6] http://reliefweb.int/report/lebanon/refugees-syria-lebanon-march-2015

did not have enough income to meet all of their basic needs. (Heating supplies *were* available in the market.)
- There was no evidence of beneficiaries spending cash assistance "irresponsibly" or meaningfully reducing labor supply.
- Lack of household labor income resulted in an average value of currently outstanding cash loans of about $500 per household.

Key findings related to markets were as follows:

- Markets were able to provide sufficient quantities of the goods and services demanded, and the program had no meaningful impact on prices. Across approximately 50 consumer goods no meaningful trend toward higher prices in higher altitude communities was observed.
- It was estimated that each dollar that beneficiaries spent generated $2.13 of GDP for the Lebanese economy.
- There was no evidence that cash assistance incentivized additional refugees to settle in higher altitude communities.

Haiti Earthquake—Potential for Indirect Response[7]

The earthquake that hit Haiti in January 2010 killed 220,000 people and displaced 1.5 million people. It was estimated that as a result, poverty levels rose by 30–50% in Port au Prince and by 80% in the commune of Carrefour. With reduced purchasing power, people had less access to food.

An interagency EMMA assessment was carried out three weeks after the earthquake (Brady 2012). In choosing which markets to assess, the team considered markets that supply to meet basic needs, markets that facilitate economic recovery and income markets. The markets selected for evaluation were two food markets (rice and beans), one income market (construction labor) and one shelter material market (corrugated iron sheeting). We focus here on the rice market and a cash intervention for helping restauranteurs rebuild. In both of these markets, agencies considered indirect interventions that would help markets recover and function, and so hasten the time when agencies could shift from in kind food distribution to cash distribution to meet food needs.

While the EMMA informed actions of the 11 agencies that were represented on the EMMA team, and other international actors as well, in this case study we focus on consideration of indirect response interventions by the IRC and Oxfam.

As can be seen in Fig. 5.3 in Sect. 5.2.2 above, there was significant disruption to the rice market as a result of the earthquake. Rice is a commodity that most Haitians eat every day. Imported rice was impacted by the port closure, damage to warehouses and loss of stocks, looting (insecurity) and lack of operating cash/access to finance for the smaller wholesalers. Furthermore, expectations that humanitarian

[7] The discussion in this section is based on the Brady 2012 report included in the references.

agencies would embark on a very large in kind food distribution program that would include rice most likely discouraged importers from working to restart the market. The supply chain for local rice (to be harvested in June-August) was affected in similar ways; the effect was probably more severe, as one tier—the Madame Saras—consists of very small players who are likely to suffer more from looting (insecurity) and lack of access to credit than do the small wholesalers.

Brady (2010) notes that the EMMA assessment did influence decision-making by agencies, and highlighted how it affected their understanding of the needs of small supply chain actors. Specifically, the case mentioned that (direct excerpt):

- "In all market chains studied, the most affected actors were those with fewer assets and more restricted access to credit: middlemen and retailers in the 'quartiers.'
- Small retailers lost their working capital, their stock, and at times, their shops. … Out of necessity, people had to travel much further afield for basic supplies, incurring higher transport costs.
- Storage facilities were damaged and led to increased insecurity.
- The insecurity and unpredictability of sales meant that credit from wholesalers to middlemen and retailers were substantially reduced. … Many micro-finance institutions were no longer issuing credits, limiting the capacity of micro-enterprises to restart or continue their activities."

In deciding how to intervene, agencies must choose the extent to which they will seek funds for direct response (e.g. programs that provide cash or income to target populations) vs. indirect response (i.e. programs that help reestablish market chain actors and get markets going again. Brady (2010) argues for pursuing both direct and indirect response, stating, "Organisations may seek to target the most vulnerable; however, to effectively and sustainably meet the needs of this group, other actors within the larger market system, including traders, wholesalers and importers, might need support to enable them to function properly." The report also makes reference to the concept of "Do No Harm," with regards to in kind distributions displacing demand from markets, thus delaying recovery of local markets.

While most humanitarian assistance focused on meeting the needs of beneficiaries in camps, some indirect assistance was also proposed. The IRC planned for an asset replacement scheme, to refurbish and restock a number of key warehouses, and did further assessments to determine needs and barriers to rebuilding small warehouses. Although funding for replacement and rebuilding of warehouse assets was not secured, there was significant interest in the proposal by donors and major international actors.

Oxfam Great Britain (OGB) sought to understand and assist both local populations and traders to reestablish market systems. Specifically, Oxfam sought to strengthen neighborhoods market systems, so hasten economic recovery. As shown in Fig. 5.5 below, OGB programming aimed to enable small "restauratrices" and grocers to restart their businesses, through unconditional livelihoods grants, provision of containers as shops, business grants and linkages to micro-finance institutions.

Category	Response
Very poor, with no limited work capacity	Hot meals for 2 months Basic needs grants (unconditional) Livelihoods grants (unconditional)
Very poor with work capacity	Cash for work Basic needs grant + livelihoods grant (unconditional)
Poor with work capacity	Cash for work
Medium	Trades people – unconditional livelihoods grant / voucher for tool kit "Restauratrices" – unconditional livelihoods grants
Medium, with higher investment capacity	Grocery stores: Provision of containers as shops Grant for business investment combined to loan (linkage with a MFI)

Fig. 5.5 Oxfam Great Britain Haiti response. (Excerpted from Brady 2012)

Sahel Food Crisis—Food for Assets and Cash for Work[8]

The Sahel region is subject to recurrent food emergencies, driven by drought, poverty and other factors. Food crises have occurred in the Sahel three times in the past decade. During food emergencies, populations who raise crops and/or animals have less to eat from their own production and take in less agricultural and pastoral income because there is less output to sell. Rates of hunger, malnutrition and disease rise (Cuny and Hill 1999). Furthermore, future livelihoods of smallholder farmers and pastoralists are affected, to the extent that seed is eaten by farmers and animals of pastoralists die or are sold at low prices (Cuny and Hill 1999).

Food emergencies are often complex crises, whose effects last multiple years, even with intervention. The 2012 crisis increased food insecurity and malnutrition for 18 million people (United Nations 2013). The 2012 humanitarian response provided assistance to 10 million people in 8 countries: Burkina Faso, Cameroon, Chad, Mali, Mauritania, Niger, Senegal and the Gambia. The 2013 UN Sahel Regional Strategy report mentioned the following aspects of the 2012 crisis: food insecurity and malnutrition; chronic cholera, measles and meningitis; floods; locust infestations; and lack of access to and displacement of populations due to conflict in Mali.

The humanitarian community responded through complex programming that includes cash for work, food for work, unconditional cash assistance, food aid, blanket supplementary feeding, and treatment of acute malnutrition (see Table 5.3 below).

In the Sahel, World Food Programme (WFP) seeks to source food locally or regionally where advantageous, and to use cash and vouchers to encourage the flow of locally or regionally-sourced food through local markets. Cash and voucher programs allow affected people to purchase food while supporting the local economy. Through cash and voucher programs, WFP aims to:

[8] The discussion in this section is based on the United Nations 2013 report.

Table 5.3 Food assistance and nutrition statistics for the Sahel response for 2012. (From Sahel Regional Strategy 2013)

Indicator	Achievement as of Sept. 2102 (% of target)
Food Assistance	
Conditional transfers (Cash/Voucher based)	Monthly average 325,000 (Apr–June) (96 %)
Conditional transfers (Food based)	Monthly average 477,000 (Apr–Sept) (62 %)
Unconditional transfers (Cash/Voucher based)	Monthly average 1,221,000 (May–Sept) (88 %)
Unconditional transfers (Food based)	Monthly average 2,102,000 (Apr–Sept) (70 %)
Nutrition	
Children 6–59 months with severe acute malnutrition admitted to therapeutic care	Monthly average 705,243 (64 %) (as of Sept)
Blanket feeding in high risk areas (preventative)	Monthly average 779,000 (Apr–Sept) (58 %)
Targeted supplementary feeding program—malnourished women	Monthly average 198,000 (May–Sept) (97 %)

- "Reinforce the capacity of communities to prepare for the next agricultural season and strengthen their resilience.
- Strengthen local markets and empower recipients to select the food commodities most appropriate to their needs.
- Allow communities to choose fresh, nutritious food options when food is locally available." (excerpted from Famine prevented: a success story… 2013)

Pricing and availability of basic food commodities at local, regional and global food markets change month-to-month and harvest-to-harvest. WFP and FAO analyze markets for food commodities that are relevant to the Sahel, including maize, millet, rice, and sorghum. Prices paid in various cities in the region are noted and compared to the average over the past 5 years. The bulletin also notes trading prices for livestock and other indicators of the level of food insecurity in various locations. The information in these bulletins is used to detect and characterize emerging food crises, develop program design and raise funds, and to monitor (and adjust) interventions as they are implemented.

An opportunity and challenge in the Sahel region is the network of traders, which is quite amorphous. The most successful traders are flexible and quickly adjust where they buy and to whom they sell, which can result in un-served populations when rural household purchasing power typically falls during a food crisis. On the other hand, these traders can often move food commodities in conflict areas where aid agencies are unable to reach.

Another aspect is the substitution of commodities that are more available for those that are not, and will result in the substitution of a different supply chain, for example, imported rice for locally grown millet. This reliance on purchased imported commodities in place of own-produced goods during food crises forces households to take on additional expenses, and has implications on local livelihoods.

Aside from the emergency response in these 8 countries, there are many development projects being implemented by WFP and other organizations to strengthen agricultural supply chains, by, for example, increasing acreage under cultivation and crop yields, improving warehousing to reduce loss, improving information flow, and improving credit and transport.

Summary of Case Studies

These case studies represent a variety of settings, disaster types, commodities and responses. (see Table 5.4). The types of markets involved vary, although all included local retailers. The scale-up requirement varies—some required replacement of lost agricultural output, some required recovery to pre-disaster levels, and some required a permanent scale-up. While all were situations in which beneficiaries lacked income, some also addressed shortages of equipment and assets related to the situation. The nature of the disruption to the setting also differs—while for Jordan and Lebanon the disruption was related to new demand from arriving refugees, for Haiti

Table 5.4 Comparison of markets represented in the cases

	Syria crisis—Jordan	Typhoon Haiyan—Philippines	Syria crisis—Lebanon	Haiti earthquake	Sahel food crisis
Type of markets	Public water distribution networks; private water market	Two local manufacturers, local distributors and retailers	Village shops selling heaters, ovens; fuel sellers	Importers, small farmers, traders, small grocery stores, small restaurants	Smallholder farmers, local traders
Scale-up requirement	Increase water usage by >10% for the foreseeable future	Increase sales of home recon-struction materials by 5–10x for 3–6 months	One-time sale of heating stoves. Scale-up of fuel market by 25%	Replacement of last season's farm output, recovery of grocery and restaurant businesses	Replacement of typical farm output for duration of the drought
Shortage addressed	Income, household water storage capacity, lack of filtration. Water shortage at country level	Income	Income	Income, business assets that had been destroyed by earthquake	Income; own-farm-production (for consumption and/or income)
Market disruption	Arrival of refugees	Physical destruction	Arrival of refugees	Physical destruction	Both

and the Philippines the disruption was physical destruction or degradation related to the natural disasters. The Sahel case included both.

In three case studies—Jordan, Lebanon and the Philippines, the programmatic *expectation* for the markets was that they would supply the goods that beneficiaries would purchase with vouchers or cash. In contrast, in the Haiti and Sahel case studies, there were more explicit programmatic *objectives* for the markets themselves—that restaurants and grocers would reestablish themselves (Haiti), that agricultural and pastoral producers would recover and recommence activities quickly (Sahel) and that food markets would recover and support cash purchases everywhere that programs were operating, including in remote areas (Sahel). This difference is in part because markets were only partially functioning in Haiti and in the Sahel, and also because the very fragmented markets for food are intertwined with livelihoods in these locations. Recovery of livelihoods often does entail an indirect intervention in the market (e.g. to replace assets or to initiate credit mechanisms).

5.4 Discussion—Measures and Models

In this section we revisit the three models and consider how they may be used to understand cash-based interventions and develop performance measures. Based on the case studies and literature review, we note that for a given context/response, in effect there are three inter-linked systems in operation simultaneously and/or in turns (see Fig. 5.6). Two of these systems are driven by humanitarian agencies acting as the coordinators, decision makers and actors. This includes goods-based aid provision, which relies on aid agencies' own supply chains to source, transport and distribute relief commodities, as well as cash-based aid models which both enable crisis-affected populations to access local markets and incentivize local market actors to scale their supply networks to meet the increased demand. The third system—the local market system—is a constellation of independent actors operating according to market principles and is not coordinated directly by humanitarian agencies. Nevertheless, emergency relief efforts do impact on markets (whether as in-kind relief or cash-based programming), and in the case of cash-based in-

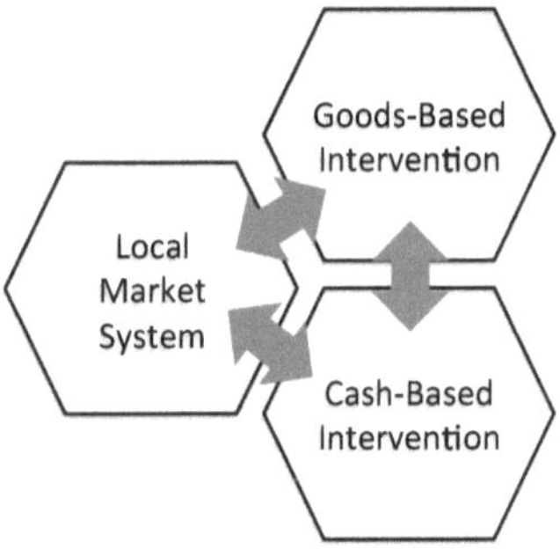

Fig. 5.6 Interaction of humanitarian system and local market system

terventions, may actually serve to strengthen the functioning of local market systems. Measuring that impact on local markets (both positive and negative) is not as straightforward as measuring an organization's or business's own supply chain, and as such this research focuses on identifying opportunities for applying commercial and global health supply chain models to measure the impact of cash-based interventions on local market systems.

Current Measures

As can be seen in Table 5.5 below, current measures focus mainly on beneficiary behavior and benefit to the beneficiary and the extent to which the program has met its objectives. Much of this information is collected through post-intervention household surveys. In addition, the "Who-what-where" analyses put together by clusters for monitoring and coordination are also helpful in describing actual *coverage* of beneficiary needs in terms of number of beneficiaries receiving cash with various objectives, and number of beneficiaries receiving goods.

What is measured and the use of that information varies by humanitarian agency and by intervention. Some agencies do monitor commodity markets in locations that are subject to recurrent crises, such as the Sahel for famine. For example, WFP monitors prices and availability of grains in many locations and characterizes their supply chains (through value chain analysis and market assessment approaches) on an ongoing basis, and uses the information not only to monitor program effects, but also to detect and react to new food crises and to inform indirect interventions in supply chains and markets ("development work"). While WFP does monitor prices and availability, disciplined, consistent monitoring and analysis of information on market system performance and the impact of cash-based interventions on supply networks is rare in the sector. Similarly, while a study of the multiplier effect of cash programming was recently conducted in Lebanon, only very rarely (and after exhaustive economic analysis) is multiplier effect analysis conducted for cash-based interventions.

Table 5.5 Current indicators for cash-based interventions

	Indicator	Comments
Household-oriented	Availability of commodities	Post-distribution household survey
	Quality/acceptability	Observation, post-distribution household survey
	Source of purchase	Post-distribution household survey
	Items bought	Post-distribution household survey
Agency coordination-oriented	Who-what-where	Used as a dashboard by the clusters during the intervention
Market-oriented	Prices/affordability	Monitored during program
	Multiplier effect	Contribution to local economy

To summarize, it is clear that improved measures of market functioning are necessary in humanitarian cash programming.

EMMA and Triple-A Models

Looking at the three models introduced earlier, the Triple-A supply chain model and the EMMA model are very focused on describing *supply* aspects—the ability of the supply chain (market system) to respond to a changing or disrupted situation. On the other hand, the Access model makes more explicit factors related to the channel or consumer, such as affordability and adoption.

The EMMA approach differs from the Triple-A supply chain approach in that it seeks to quickly understand how the supply system is performing at a moment in time, and to envision how it will perform in support of potential cash interventions. Sometimes EMMA assessments lead to recommendations to use cash (when markets are functioning "well enough"); in other cases EMMA studies find that use of markets is not feasible. It would be useful to characterize the constraints on market functioning that EMMA studies have found across the wide range of emergency settings in which EMMA has been used. This focus on constraints suggests that measures that focus on constraints may be useful as status indicators to inform, for example, when to switch over from in kind to cash programming, or what impact bringing goods in "from outside" is having on constraints on local market functioning.

The Triple-A supply chain approach and the EMMA tool complement one another. The Triple-A supply chain approach focuses on the effectiveness of the coordination of supply actors to meet emerging demand, their ability to scale up quickly (agility), and the adaptability of the supply chain/market to new constraints/bottlenecks as events unfold in the post-disaster context and to adapt as agencies switch from goods to cash or vice versa. Last, the focus on alignment of incentives could help agencies assess what risks suppliers are willing to take and what limits their risk taking. These types of measures can capture how the supply system is working, and how the intervention is influencing the behavior and performance of the supply chain and thus, the market.

Table 5.6 details how the Triple-A Supply Chain model could be interpreted for the humanitarian context and suggests potential performance measures for measuring markets during or after cash-based interventions. Some of the suggested supply chain measures relate to suppliers' ability to respond and some to demand-supply mismatch. These can be useful, as they link to supplier risk, ability and willingness to participate, and can also suggest how programs might be designed differently and/or how coordination with suppliers might be improved, to better serve beneficiary needs and serve suppliers by reducing their risk as they restart their businesses. Furthermore, they could be used as indicators of the extent to which markets have been strengthened (e.g. through more visibility and coordination or through elimination of bottlenecks such as credit or warehousing) and are more resilient.

Table 5.6 Applying the Triple-A model to the humanitarian context

	Meaning in the Humanitarian Context	Potential Performance Measures
Agility: Can the supply chain respond to short-term changes in demand?	• Scalability • Responsiveness • Signaling/coordination	• Lumpiness of demand • Forecast accuracy (demand) • Lead times • Stock levels along the chain • Scalability of production rate and flow through supply chain • Shortages/stock outs • Excess product
Adaptability: Can the supply chain adapt when the situation changes?	• Switch from goods to cash or vice versa or to substitute commodities • Adjust to new realities; eg. infrastructure, limitations on credit, limitations on transport ➢ Resolve new bottlenecks ➢ Create new channels or capabilities; eg. put ATM machines in different types of places	• Coordination and rollover Measures that relate to constraints and bottlenecks – eg. cost and availability of credit • Measures that relate to adaption or innovation: rate of scale up of new channels or capabilities, performance of adapted supply chain
Alignment: Are incentives driving towards better performance?	• Incentives for ➢ Fair prices ➢ Availability of product ➢ Serving people in remote places	• Actual coverage of demand locations (sales) • Benefit to suppliers • Markup and profit along the supply chain • Measures of supplier risk • Measures of alignment through information sharing

The Access Model

The Access model can also suggest direction for measurement (See Table 5.7). Looking at the three "pillars:" availability, affordability and adoption, availability captures general supply chain considerations, as in the other two models. Affordability includes both affordability for beneficiaries (pricing of goods in the market) and cost effectiveness of the intervention. Adoption can be interpreted to include the extent to which beneficiaries use cash received to purchase what was intended (a programmatic issue) and the extent to which supply chain actors make available the intended products (as opposed to other products). In essence, the concept of adoption encompasses the signals that consumers send to the market regarding the goods they need to meet their needs, and the specific measures of adoption highlighted in the Access model can be helpful in monitoring the extent to which cash-based interventions stimulate market functioning.

Table 5.7 Applying the access model to the humanitarian context

	Meaning in the Humanitarian Context	Potential Performance Measures
Availability: Can the market system (supply chain) ensure provision of the desired supply of the commodity?	Similar to agility and adaptability in the Triple-A Supply Chain model.	Similar to those for agility and adaptability, plus • Distance beneficiaries travel to buy • Days/hours of operation of markets
Affordability: Can funders and purchasers afford the commodity?	• Ability of beneficiaries to afford the intended items/services • Ability of funders/agencies afford to fund the cash program	• Cost and availability of consumer credit • Costs to beneficiaries: transport, protection, "taxes", bribes • Items bought, place purchased, price paid, quality (household survey) • Administrative cost of cash disbursement • Program cost related to cash program • Value for Money of cash program
Adoption: Are market conditions and the design of the intervention such that market system (supply chain) entities are willing to participate as intended?	• Beneficiary choice (purchase choice) • Market description • Adoption of practices that strengthen markets and make them more resilient • Incentives for ➢ Fair prices ➢ Availability of product ➢ Serving people in remote places	Similar to those for alignment, plus • Items bought • Measures of adoption of desired practices • Availability of intended products • Measures of market strength/resilience • Market structure measures (level of competition, market size and growth, markup and profit along the chain)

Measures for the Case Studies

When we apply the principles of the Triple-A and Access Models across the five case studies, we see a wide array of constraints, positive changes/adaptations, and unaddressed opportunities with market systems to respond to the humanitarian intervention. The specific *expectation* or *objective* of the intervention related to markets is a key component of understanding how markets may respond for the use of cash or vouchers in a crisis response. Table 5.8 highlights the extent to which the key principles of the Triple-A and Access Models apply to the five case studies, taking into account the programming context, demands on the market system, and the type of the emergency response. The principle of 'Availability' from the Access model is effectively the same as the concepts of 'Agility' and 'Adaptability' in the Triple-A model, and has been omitted from this table to avoid redundancy.

Table 5.8 Relative importance of market system characteristics for the five case studies

	Jordan	Philippines	Lebanon	Haiti	Sahel
Anticipated change in the market system	Long-term scale-up	Short-term scale-up	Short-term scale-up	Reestablish market systems	Replace lost farm output
Market System Concerns: (rating reflects relative need for Agility, for Adaptability, etc.)					
Agility	High: immediate permanent scale-up	High: rapid scale-up of demand, unpredictable scale-down	High: rapid scale-up of demand, predictable (seasonal) scale-down	Low: little change in demand	Moderate: maintaining stock levels, with slight increase
Adaptability	Low: few alternative sources or delivery options	Low: at retail-level—sourcing from different/new suppliers	Low: increased demand but few changes in market system (new channels)	Low: constrained credit system	Moderate: substituting imported goods; market linkages in conflict areas
Alignment	Moderate: no significant incentive for municipal water sources to scale-up, few scale-up alternatives	High: high incentive to find new suppliers, expand availability if demand created through vouchers or cash	Moderate: slight need to change business to meet needs of population	High: re-starting neighborhood restaurants, grocers to match beneficiary needs for access to foods	High: continuation of markets in conflict areas to meet beneficiary needs
Affordability	Moderate: expanded access to water trucking, migration to cheaper distribution options over time	High: speed of delivery of shelter materials, minimal change in price	High: no change in market prices; 2.13:1 multiplier effect	Moderate: consumer price increase	Moderate: consumer price increase
Adoption	Moderate to high: use of different water delivery mechanism (trucking, filters)	Moderate: incentives to participate for both shops and beneficiaries; beneficiaries select items. Humanitarian agency acted as a 'channel leader'—requiring voucher shops to identify new sources, transportation routes	Moderate: broad use of cash for different items/commodities	Low: focus on re-starting previously existing businesses	Low: market functioning based on past behavior, consumption trends

Analyzing the five components of the Triple-A and Access Models across the case studies enables us to identify what metrics can be used in evaluating the performance of market systems in each case. Given the wide array of different programming contexts, the differing impact of the crises on markets, and the extent to which the humanitarian response relies on market systems, we would expect that different metrics may be needed in different contexts. The performance aspects from the Triple-A and Access Models do not neatly fit into each case study. Nevertheless, this exercise highlights some key learning points. Additionally, Table 5.8 highlights how the emergency response program (cash, vouchers, in-kind, etc.) can affect market performance by altering any of the market performance categories (agility, adaptability, alignment, affordability or adoption).

By looking at the five supply chain model concepts in 8, we can begin to understand specific metrics to measure market performance in response to cash and voucher programs. When considering Agility, in Jordan there is a high requirement to scale-up supply in response to the large-scale and long-term increase in water demand, which is quite different from Haiti, where the requirement is to re-build pre-existing market systems in order to respond to the same or lower level of demand as before the earthquake. Agility measures include coverage of supply (does everyone have adequate goods/services), shortages, and lead times to delivery.

For Adaptability, the Philippines case illustrates a high level of adaptability required in order to source and deliver appropriate construction goods in a timely manner, whereas the Sahel case presents a fairly moderate degree of adaptability where market actors switch sourcing/delivery of substitute foods, in this case rice, from previously-existing supply routes. Metrics for analyzing adaptability may focus on constraints in the market systems and the degree to which constraints are present or change over time.

The alignment principle emphasizes aligning incentives for both consumers and suppliers, and often cash and vouchers can be used to incentivize markets to continue functioning, or to function at a higher level. In the Sahel traders saw significant enough incentives to continue supplying markets in the conflict areas of northern Mali, whereas in the Philippines example the voucher program enabled stores to invest capital in sourcing and transporting greater volumes of goods to feed the market system for both voucher beneficiaries as well as non-beneficiaries.

Regarding affordability, the case from Lebanon demonstrates how a cash intervention enables beneficiaries to access essential winter items as well as achieving multiplier effect in the local economy, a tremendous value for the money of the intervention. In contrast, the Jordan example shows moderate affordability by enabling access to new channels of distribution (water trucking, filters) while awaiting migration to cheaper distribution options.

And lastly, the adoption principle varies across the cases from the Philippines where voucher programs lead businesses to adopt new approaches to sourcing their products, to the Sahel, where traders applies the same business practices and supplies the same commodities as previously in order to meet needs and preferences of the affected population. These cases illustrate that the Triple-A and Access Model concepts offer a valuable framework for analysis from which distinct metrics can be drawn to measure the performance of market systems under cash transfer programs.

5.5 Conclusion

While this chapter does not recommend specific measures, it does inform the discussion of how to measure markets that supply cash-based interventions, by analyzing case studies of cash-based interventions in individual emergency responses. The chapter also stresses the importance of establishing a model of what is to be evaluated and measured, and suggested three models—the Triple-A supply chain, the EMMA toolkit and the Access framework, as informative for considering how to represent markets and the supply chains behind them in order to develop metrics to assess performance of those markets. We are hopeful that the insights from this chapter will be useful to future researchers, and help them be grounded in the realities of real-world interventions and so lead to useful, relevant work.

References

Abidi H, de Leeuw S, Klumpp (2013, Aug) Measuring success in humanitarian supply chains. Int J Bus Manage Innov 2(8)
Andersen MB (1999) Do no harm: how aid can support peace or war. Lynne Rienner Publishers, Boulder
Austin L, Chessex S (2013) Minimum requirements for analysis of markets in emergencies. Cash Learning Partnership, Oxford
Beamon BM, Balcik B (2008) Performance measurement in humanitarian relief chains. Int J Public Sect Manage 21(1):4–25
Brady C (2012, Jan) An emergency market mapping analysis case study: changing responses to the Haiti earthquake. Oxfam. Oxford, UK.
Cash Learning Partnership (2015) Scoping study-emergency cash transfer programming in the WaSH and shelter sectors. www.cashlearning.org
Cuny FC, Hill RB (1999) Famine, conflict and response. Kumarian Press, West Hartford
De Leeuw S (2010) Towards a reference mission map for performance measurement in humanitarian supply chains. in Collaborative Networks for a Sustainable World, edited by L.M. Camarinha-Matos, X. Boucher and H. Afsarmanesh, 11th IFIP WG 5.5 Working Conference on Virtual Enterprises, PRO-VE 2010, St. Etienne, France, October 11-13, 2010. Proceedings. pp 181–188
DeToni A, Tonchia S (2001) Performance measurement systems. Int J Oper Prod Mgt 21, pp. 46–70.
DFID's Approach to Value for Money (VfM) (2011, July) Department of international development. https://www.gov.uk/government/uploads/system/uploads/attachment_data/file/67479/DFID-approach-value-money.pdf. Accessed 18 Nov 2014
Eisenhardt KM (1989) Building theories from case study research. Acad Manage Rev 14(4):532–550
EMMA Toolkit (2010) http://emma-toolkit.org/get/download/. Accessed 23 Nov 2014
Frost LJ, Reich MR (2008) Access: how do good health technologies get to poor people in poor countries, Harvard Center for Population and Development Studies, distributed by Harvard University Press. www.accessbook.org.
Gatignon A, Van Wassenhove LN, Charles A (2010) The Yogyakarta earthquake: humanitarian relief through IFRC's decentralized supply chain. Int J Product Econ 126:102–110
Global Logistics Cluster (2015) Unlocking cash and market-based programmes. http://logcluster.org

Gunasekaran A, Kobu B (2007) Performance measures and metrics in logistics and supply chain management: a review of recent literature (1995–2004) for research and applications. Int J Product Res 45(12):2819–2840

Holguin-Veras J et al (2013) On the appropriate objective function for post-disaster logistics models. J Oper Manage 31:262–280

Komrska J, Kopczak LR, Swaminathan J (2013) When supply chains save lives. Supply Chain Manage Rev 17:42–49

Lee HL (2004) The triple-A supply chain. Harv Bus Rev 82:1–10

Logistics Cluster (2014) The Supply Chain in Cash and Voucher Programmes, youtube video. https://www.youtube.com/watch?v=WXCCFt8QemQ. Accessed 23 Nov 2014

Oloruntoba R, Gray R (2006) Humanitarian aid: an agile supply chain. Int J Supply Chain Manage 11(2):115–120

Oxfam (2013) Water market system in Balqa, Zarqa & informal settlements of Amman and the Jordan Valley—August–September 2013

Oxfam (2014, Jan 14–18) Emergency mapping and analysis assessment: Corrugated Galvanised Iron (CGI) Market System

Schulz SF, Heigh I (2009) Logistics performance management in action within a humanitarian organization. Manage Res News 32(11):1038–1049

Sodhi MS, Tang CS (2014) Buttressing supply chains against floods in Asia for humanitarian relief and economic recovery. Production and Operations Management, Forthcoming. SSRN. http://ssrn.com/abstract=2234478

Tatham P, Hughes K (2011) Humanitarian logistics metrics: where we are and how we might improve. In: Christopher M, Tatham P (eds) Humanitarian logistics. Kogen Page Limited, pp 65–84

The IRC (2014, Aug) Emergency economies: the impact of cash assistance in Lebanon: an impact evaluation of the 2013–2014 Winter Cash Assistance Program for Syrian Refugees in Lebanon. http://www.rescue.org/sites/default/files/resource-file/Emergency%20Economies%20Evaluation%20Report%20FINAL%2009.09.14%20%282%29.pdf. Accessed 23 Nov 2014

Tomasini R, Van Wassenhove LN (2009) Humanitarian logistics. INSEAD Business Press, London

UN Office for Project Services (2013) Annual Statistical Report on United Nations Procurement. Marmorvej 51, 2100 Copenhagen. Denmark (2014)

United Nations (2013) Sahel regional strategy 2013. http://reliefweb.int/sites/reliefweb.int/files/resources/SahelStrategy2013_Dec2012.pdf. Accessed 23 Nov 2014

USAID (2014) Healthy markets for global health: a market shaping primer. http://www.usaid.gov/sites/default/files/documents/1864/healthymarkets_primer.pdf. Accessed 23 Nov 2014

World Food Program U.S.A. (2013) Famine prevented: a success story in the Sahel. http://wfpusa.org/sites/default/files/u-6876/Sahel%20success%20story_FINAL.pdf. Accessed 23 Nov 2014

Chapter 6
The Partner Proliferation Problem in Disaster Response Networks

Sara Hasani, Ramzi El-Haddadeh and Emel Aktas

Abstract The extraordinary conditions of a major disaster require mobilization of all available resources. This necessity, together with the stretch in the response budget in the public sector and the difficulty of raising funds in the private sector, draws various humanitarian actors with widely diverse capabilities into the affected area. This phenomenon is called the proliferation of actors, or the partner proliferation problem. This problem can have serious counterproductive effects on disaster operations, such as unmanaged independent efforts that lead to a duplication and confusion of effort. The disaster response phase generally lacks the contributions of a long-term outlook and pre-planning, which are adopted in existing long-term structures such as supply chains. The aim of this paper is to provide a structured review of the partner proliferation problem in the response phase and to suggest alternative courses of action for restructuring the disaster response network. Drawing on the concept of Virtual Organizations, the paper concludes that short-term collaboration is a suitable structure for the response phase. Short-term collaboration in the response phase is complementary to long-term collaborations such as supply chains in the recovery, mitigation, and preparedness phases of the disaster cycle. To that end, a conceptual framework is provided for re-structuring the disaster response network to align with the other phases of disaster management. Finally, further research is suggested to develop a decision making tool for partner configuration to meet the specific requirements of a disaster response network.

Keywords Disaster response · Partner proliferation · Short-term collaboration · Virtual organization · Disaster collaboration life cycle

---

S. Hasani (✉)
Business School, University of Bedfordshire, Luton, UK
e-mail: Sara.hasani@beds.ac.uk

R. El-Haddadeh
Brunel University, Uxbridge, UK

E. Aktas
School of Management, Cranfield University, Bedford, UK
e-mail: emel.aktas@cranfield.ac.uk

C. W. Zobel et al. (eds.), *Advances in Managing Humanitarian Operations*,
International Series in Operations Research & Management Science 235,
DOI 10.1007/978-3-319-24418-1_6

6.1 Introduction

Many cases of failure have been reported in disaster response operations due to the challenges faced by humanitarian partners (e.g. Haiti Earthquake in 2010 and the Indian Ocean Tsunami in 2004). One of these challenges is the partner proliferation problem that is the product of the extreme requirements of a disaster where all available sources are mobilized (Tierney and Trainor 2004) and any available partner is encouraged to participate in the response operations. This reaction suddenly stretches the response budget in the public sector (i.e. UN, Red Cross, governments) and multiplies the funds raised by the private sector (i.e. NGOs) (Rolland et al. 2010). Then, the overstretched humanitarian network struggles to channel the funding and resources in an effective manner. The result is the emergence of inexperienced actors in the response operation, such as companies joining in the activities outside their area of expertise with an attempt to improve their public image (Telford et al. 2006; Careem et al. 2006). It also brings in partners who range from competent to incompetent, reputable to disreputable, opportunistic to committed, and well established to just-formed, in addition to individuals and tourists who are eager to help. This oversupply of uncoordinated and inexperienced partners is referred to as the proliferation of actors (Inomata 2006). The actors enter the disaster-affected area in a chaotic pattern (Comfort 2007), which contributes to the proliferation problem.

The proliferation of actors is of special importance because the existence of various partners with different mandates, agendas, levels of professionalism, expertise and resources presents a major challenge to operational coordination. It also poses various threats to ethical and socio-economical issues in the affected area. A number of these actors have moved beyond life-saving activities and tried to make changes in the social environment and attempted conflict resolution. For example, many of them have created their own armed security forces or are in close relations with the military (Metcalfe 2012). This also brings about issues where a range of allegations of financial fraud and ethical misconduct of aid workers are reported along with sexual exploitation, abuse, and bullying (Odihpn.org 2014; Valburn 2012; Edition.CNN.com 2014).

These issues signal a necessity to deal with the proliferation problem with a wider approach than existing guidelines such as resource scheduling techniques (Rolland et al. 2010), reputation management systems for efficient selection of partners (Javaid et al. 2013), and diagnosis of the severity of the disaster (Hasani et al. 2014).

Although all types of dysfunctions in disaster situations have counterproductive effects on the entire disaster management operation and add negative value to the system (Telford et al. 2006), the proliferation problem in particular damages the quality of the response (Reineck 2010). This threatens the reputation of humanitarian aid organizations (Reineck 2010) and can destroy trust in the long-term. Entrance of inexperienced actors increases the load on the affected populations, local authorities, and coordination structures for information or services. It also increases the costs due to multiple offices tasked with similar activities and associated overheads, and leads to a counterproductive duplication and confusion of effort. In this situation the partners may compete over donations, funding, facilities, and publicity instead of taking advantage of each other's capabilities (Kent 2004, Telford et al. 2006, Balcik et al. 2010).

To help foster improved cooperation, the present study first articulates the challenges associated with the proliferation of partners and then synthesizes a series of solutions based on the existing literature, identified through a structured review in Scopus and analyzed in terms of the problems addressed, the methodologies used, and the key findings reached. To that end this paper elaborates the scope of the proliferation problem by defining the growth of disasters and the chaotic pattern of partners' entering the disaster area. Then the challenges facing the resolution of the proliferation problem are listed, followed by a few possible solutions. Investigating one of the possible solutions (restructuring) gives rise to a comparison between the existing short-term structures. The authors then build upon the existing literature to synthesize a conceptual model for restructuring the four phases of disaster management with emphasis on the similarities between the disaster response network and virtual organizations as a short-term manifestation of collaborative networks. Finally the article suggests directions for future research.

6.2 Problem Definition

Disasters are defined as intense forms of collective stress caused by a disaster agent (Britton 1986) and resulting in '*a disruption that physically affects a system as a whole and threatens its priorities and goals*' (van Wassenhove 2006, p. 476). The intense negative impacts of a disaster on people, goods, services, and the environment make the community incapable of coping (Kovács and Spens 2009) and in need of assistance from governments and international agencies.

A disaster can occur naturally or due to human activities, including 'slow onset' disasters such as famine and 'sudden onset' disasters such as earthquakes. Man-made disasters could be of an environmental nature (e.g. chemical leaks) or induced by political conflict (e.g. refugee crisis). Table 6.1 shows different types of disasters (van Wassenhove 2006).

This study focuses on natural 'sudden onset disasters' because they have rapidly increased in frequency and severity during the past decades (Fig. 6.1). Sudden onset disasters "occur with little or no warning and often cause excessive injuries and deaths far surpassing the national response capacities" (World Health Organization 2013, p. 10) while slow-onset disasters "*emerge gradually over time, often based on a confluence of different events*" (OCHA 2011, p. 3).

Figure 6.1 illustrates the number of natural sudden-onset disasters from 1983 to 2013. The total number of natural sudden onset disasters registered in the last decade (1449) is more than the disasters registered during 1983–2003 (1234). This growth is mainly associated with geographical and socio-economic reasons. For example climate change is likely to induce more weather-related disasters such as flash floods and landslides. In addition, the urban migration leads to poorly

Table 6.1 Categories of disasters. (Source: van Wassenhove 2006)

	Natural	Man-made
Sudden-onset	e.g. Earthquake, tsunami	e.g. Chemical leak, nuclear attack
Slow-onset	e.g. Famine, Drought	e.g. Refugee crisis, environmental crisis

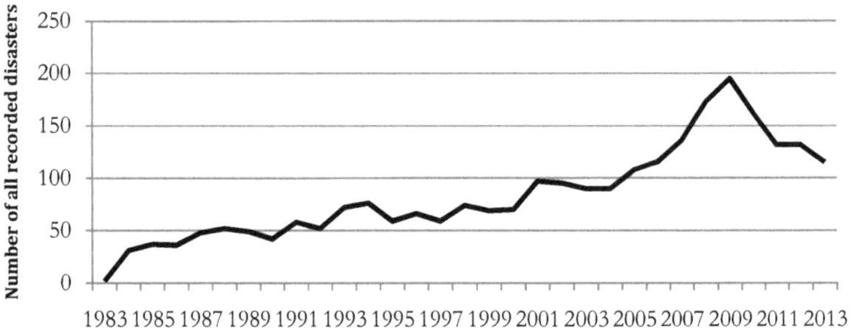

Fig. 6.1 The growth of worldwide natural sudden onset disasters recorded in OCHA database since 1983. (Compiled from relief web (2013))

structured settlements being built on exposed stretches of seismic faults, flooding plains or landslide-prone slopes by the vulnerable population (CRED, ISDR, cited by BBC 2004). Although the improved technical methods of recording and sharing information during the past decade could have contributed to the increase of available data, further research is required to identify the extant of this effect which is out of the scope of this paper.

However, the literature addressing negative impacts of natural onset disasters on humans and social, ecological, and economic environments appears to be insufficient in volume. Out of the 348,551 articles found since 1856, which contain either of the keywords "natural onset disaster(s), tsunami, flash flood, earthquake, volcano, eruption, and cyclone", 83.7 % are just related to the sciences such as earth sciences or engineering, which solely investigate the characteristics of the phenomena itself. The titles related to the areas of interest in this research such as decision sciences, management, economics and in part social sciences accumulates to 18,273 titles which is approximately 5 % of the total titles found in the literature search. However the interest of scholars has been exponentially increasing during the last decades as can be seen from a quick comparison in Table 6.2 that shows how the number of articles published on the related subjects has increased over time.

Table 6.2 shows that the articles published on the related subject areas have more than tripled from 545 titles in 2004 to 1785 titles in 2014. It also shows that the articles published in decision sciences containing the above keywords have increased almost 20 times from 14 articles in 2004 to 266 articles in 2014; an increase higher than any other subject area, suggesting a recognition of the need for research on the subject.

Disasters have negative impacts on humans and social, ecological, and economic environments. These impacts are impossible to eliminate but can be alleviated

Table 6.2 Comparison between articles published in 2004 and 2014

Subject	Year	
	2004	2014
Social sciences	430	1297
Business and management	78	117
Economics	23	105
Decision sciences	14	266
Total	545	1785

and minimized with effective disaster management (Moe and Pathranarakul 2006). However, experiences in the Haiti Earthquake in 2010 and the Indian Ocean Tsunami in 2004 reveal various failures. For example, 2500 people died of cholera in Haiti in the presence of 12,000 humanitarian organizations (Karunakara 2010). This was partly associated with the lack of safe drinking water and the fact that 7 months after the disaster 30% of camps did not have any kind of toilet (Heikkinen 2012). This occurred in the context of donations of USD 1482 per capita (Metcalf et al. 2011), which exceeded the GDP per capita of the country (USD 669 per capita, Worldbank 2014) in that year. During the Indian Ocean Tsunami, competition among aid workers to spend huge private donations led to a misallocation of resources and duplications of activities (Wright 2005). These negative impacts can be reduced if the partners are carefully selected according to the requirements of each particular disaster. That is why this research focuses on the partner proliferation problem and possible ways of addressing it.

What is more, response operations face various challenges such as mass scale effects in large geographical areas and population, and severe damages to people and property. In addition, the involvement of multiple parties, the time pressure for rescue operations and decision-making, severe resource shortages and vast unpredictability are amongst the biggest challenges facing humanitarian logistics operations (Jiang et al. 2012). Various scholars emphasize deficiencies in preparedness and planning and inadequacy of prepared rescuers (Benjamin et al. 2011; Kovács and Spens 2009). Others point out the proliferation of actors in the disaster situation (Reinecke 2010; Telford et al. 2006; Balcik et al. 2010). The majority of these challenges are faced due to the lack of relevant criteria including the standards and indicators, the weak collaboration and the inadequate infrastructure (Kovács and Spens 2009). Also, the uncertainty in demand and supply and the difficulty of inventory forecasting (Balcik et al. 2010) associated with the opportunistic behavior of partners (Pettit and Beresford 2009) together with high employee turn-over (Reinecke 2010; Telford et al. 2006) complicate the situation. The low recognition of the role of logistics (Kovács and Spens 2009) in humanitarian operation further aggravates the problem.

Despite the awareness of practitioners and scholars of the complications and issues related to the proliferation of actors in the disaster response phase, the extent of the negative effects of these challenges on the disaster operation is understudied. Also, the efforts to provide specific guidelines to tackle these challenges are limited to few studies. For example, Farazmand (2007, 2009) introduces the concept of surprise management as a solution to the challenges facing humanitarian operations. He basically draws upon the failure of the response operation during Hurricane Katrina and suggests engagement of citizens and adaptive collaboration. However, this solution lacks an adoptable guideline or framework to which the practitioners could refer and operationalize the surprise management. On the other hand, Rolland et al. (2010) provides a decision support system for resource scheduling in the response and recovery phases and Javaid et al. (2013) develops a reputation management system for the efficient selection of partners. Finally, the authors of this study proposed in an earlier paper a decision support system for diagnosing the severity of the disaster using the limited data available in the early hours (Hasani et al. 2014).

In part, this chapter is a response to the call by Moe and Pathranarakul (2006) who emphasized the necessity to minimize the negative effects of disasters and to Altay and Green (2006) who pointed out the lack of a network structure to facilitate the response phase of disasters. Although various papers are published especially in the past 2 years in response to these calls, the majority of them are concerned with the pre- and post-disaster phases including mitigation, recovery and preparedness (Crawford et al. 2014; Doocy et al. 2014; Hardy et al. 2013; Malhotra and Vetkatesh 2013; Karunasena 2011). A limited number of articles focusing on the response phase using collaborative partners investigate how the partners who actually responded to the disaster are different from the ones who were planned to participate (Guo and Kapucu 2015). This shows that emergency norms prevail over the bureaucratic norms (Schneider 2011), which is another confirmation of the fact that preplanning and a long-term outlook in the response phase is extremely problematic. Basically, the partners in an existing disaster response are a part of the collaborative structure and the performance of a disaster response is increasingly being assessed by comparisons between planned versus actual networks (Guo and Kapucu 2015; Hu et al. 2014; Hu and Kapucu 2014; Kapucu and Demiroz 2011; Choi and Kim 2007; Choi and Brower 2006). However, the above articles mostly focus on reporting the discrepancies between the planned and actual structures/partners without providing a guideline for dealing with these differences. A solution for controlling or reducing the discrepancies between the expected partners and the actual proliferated partners is missing. Although there are some studies which provide guidelines for scheduling and task allocation during the response phase (Fiedrich et al. 2000; Nourjou et al. 2014a, 2014b) or provide metrics for amending the partner coordination, the research focusing on the response phase and specifically addressing the proliferation problem is yet to be fully developed. To that end the present chapter addresses the partner proliferation problem in disaster response networks as one of the most recurring problems in humanitarian operations. In line with this problem, the research question of this chapter is:

> What collaborative structures are suitable to accommodate the partner proliferation problem within the special characteristics of a disaster?

To answer this question, an extensive literature review was undertaken in different search engines in the area of business, management and decision-making. The

Scopus collection retrieved 7094 results during 1947 to 2013 focusing on collaborative networks. The result shows a growth in the amount of literature focusing on collaborative networks since 1994 (when Kanter first coined the term), jumping from 22 articles in 1994 to 3033 during the next 20 years. The scholars in the areas of engineering, social sciences and business and management have produced more material on the collaboration subject than other areas. A quick comparison of these works shows the distinction between short-term and long-term collaboration during the past two decades.

Figure 6.2 shows that although both branches of literature have grown during the past years, long-term collaboration has attracted more interest. Long-term collaboration has been the subject of extensive literature reviews (Wu and Barnes 2011; Aissaoui et al. 2007, de Boer et al. 2001) because traditional collaboration networks such as supply chains, joint ventures, strategic alliances, and franchises usually have a long-term outlook (Gallear et al. 2012). However, the literature is missing a holistic literature review on collaboration with a short-term outlook such as virtual organizations, inter-organizational projects, product development projects, outsourcing projects, and temporary alliances.

The majority of the articles on long-term collaborations mainly focus on how to successfully manage the operation phase of an already formed collaboration. The experts display far less interest in the initiation phase. This finding is of special interest because the initiation phase is where the partners are selected and configured. This phase is very important in the configuration of a disaster response network (DRN) and dealing with the partner proliferation problem. That is why the authors' focus is on efficiently structuring the disaster response network to reduce the rush of available partners into the disaster area. The literature review reveals a difference between the approaches taken to tackle collaboration in the response phase using long-term structures such as supply chains and short-term structures such as virtual organizations or public projects. It also provides a basis for developing a conceptual framework applicable to restructuring the disaster response network (DRN). The

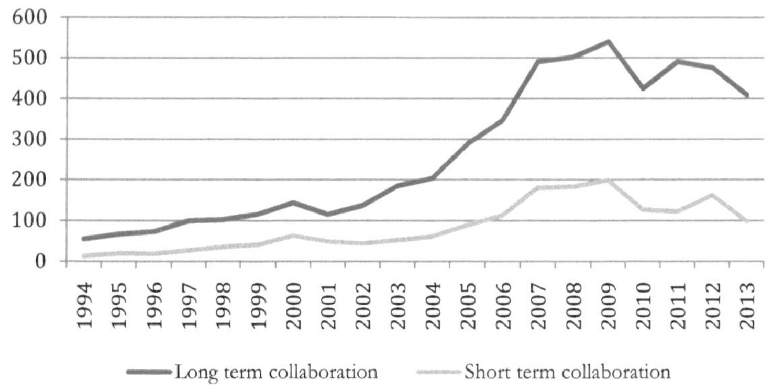

Fig. 6.2 The growth of collaboration studies with the focus on short/long term collaboration. (Source: Author)

authors argue that one of the reasons for failure in disaster response network configuration is the incompatibility of the disaster response situation with the existing collaborative structures used for managing the response operation.

The uncertainty and the lack of information (Tomasini and van Wassenhove 2009), together with the damaged infrastructure (Jiang et al. 2012), unequal and ineffective distribution of demand and supply and their respective fluctuations (Comfort et al. 2004, Tierney and Trainor 2004), and unsteady flow of the financial resources obtained by fund-raising from occasional donors (Oloruntoba and Gray 2006) all make the planning and the long-term outlook almost impossible. Also long-term approaches in business are usually profit-based whilst in disaster situations the non-financial factors such as the time value of commodities are much more critical than the associated costs (Oloruntoba and Gray 2006; Pettit and Beresford 2009), which makes the conventional profit-based values less appropriate. Therefore, due to the lack of control and information in disaster situations, the existing structures such as supply chains or project-based collaborations might fall short in practice because these structures necessitate a certain amount of knowledge about the required and available resources, the budget, and the time. These data are generally unknown in disaster situations due to the unstable nature of the disaster response network. This lack of data and planning opportunity in the response phase is of particular importance when dealing with the proliferation problem because the data required for shortlisting the partners are insufficient or even non-existent until the actual disaster strikes. To that end a method is required to enable the decision makers to select the partners in a short period of time. One of the solutions to this problem is restructuring the response network to accommodate the characteristics of the disaster situation with a non-financial short-term outlook, which has the capability to work with the minimum data available and without much pre-planning.

6.3 Challenges Facing the Resolutions of the Partner Proliferation Problem

Various scholars have proposed methods to reduce problems associated with the proliferation of partners. The majority of these methods emphasize restructuring the disaster response network either by adopting a collaborative structure (Telford et al. 2006; Balcik et al. 2010) or by shortlisting and selecting fewer beneficiaries (Telford et al. 2006), e.g. selecting certain organizations (Moore et al. 2003) which are able to effectively collaborate (Kovács and Spens 2009). However further guidance on the methods for selection, restructuring and configuration of effective partners are yet to be developed (Kovács and Spens 2009 Moore et al. 2003). The lack of guidance for how to restructure and select partners in order to reduce the proliferation problem can be explained by two main root causes: the lack of information sharing among all disaster response partners and the existing financial-based approach that requires all disaster response units to conduct operations within the boundaries of their allocated budget or raised funds.

First, the lack of control and sharing of information poses a challenge to humanitarian response operations due to the uncertainties associated with a disaster situation (Tomasini and van Wassenhove 2009). Without information, existing structures such as supply chains or project-based collaborations are likely to be inadequate in practice since they have been designed for a more structured and predictable environment. There are practices of allocating surge capacities and mobilization of all available resources in the prominent response organizations such as the UN and the Red Cross. Although these organizations are designed to respond to the uncertainty embedded in a disaster situation, the unsuccessful experiences of large scale disasters such as the Indian Ocean Tsunami and the Haiti Earthquake mentioned earlier signal the necessity for improved management of information resources.

In existing collaborative structures, the efficiency of the system is based on knowledge of where, when and how much goods and services should be delivered. These data are generally unknown in disaster situations due to the unstable nature of the disaster response network (van Wassenhove 2006). Also, the available data in the early hours after a disaster strike are generally incomplete due to the lack of time and the access to the affected area. This turbulence and unpredictability (Pettit and Beresford 2009) generates a challenging environment for planning and long-term management. For example, damaged infrastructure (Jiang et al. 2012) results in ineffective distribution of supply (Comfort et al. 2004; Tierney and Trainor 2004), making it even more difficult to meet the already fluctuated demand with the unsteady supply. Another challenge is the unpredictable flow of financial resources, which are obtained by fund-raising from occasional donors (Oloruntoba and Gray 2006). These issues are in contrast with a long-term planning approach where the budget of the project, the required tasks, and the available resources are either clearer or more predictable.

Second, the financial-based approach fails to address the special characteristics of the disaster response network. For example in a market environment, profit-based collaborative structures such as supply chains rely on the customers at the receiving end as the source of income and aim to maximize their profit while minimizing costs. However, in disaster situations non-profit factors such as the time value of commodities and critical ethical imperatives are much more important (Oloruntoba and Gray 2006; Pettit and Beresford 2009, IFRC Code of Conduct). In this situation, the donors are the source of income, not the actual receivers of the aid. In addition, the central goals of saving lives of the affected population or maintaining the dignity of people do not produce any monetary value and may bring about financially unjustifiable costs. These issues are in contrast with the principles of a financial-based approach and require a new method, which can address the efficiency of the system based on non-profit measures.

Due to the above challenges, there is a need to adopt a structure that is capable of dealing with the non-profit, ethical and short-term characteristics of the disaster situation. However, a network structure for dealing with partner proliferation (Telford et al. 2006; Balcik et al. 2010) , and in particular to facilitate the resolution of disasters, is missing (Altay and Green 2006). In the absence of a suitable partnering structure for disaster response networks, different forms of collaboration networks have been suggested. For example, following the popularity of long-term structures for disaster management in the literature (Maon et al. 2009; Eßig and Tandler 2010; Tatham and Spens 2011) the humanitarian community adopted a supply chain concept as their network structure. Examples include the International Federation of Red Cross and Red Crescent Societies and the World Food Program (Wassenhove et al. 2005). On the other hand, other scholars suggest temporary structures such as high reliability virtual organizations (Grabowski und Roberts 2011) and public projects (Moe and Pathranarakul 2006). This split signals a necessity to investigate the characteristics of each structure to identify the suitable structures for a disaster response network.

The following section argues that both short-term and long-term structures should be adopted in combination. To justify this argument, we draw upon the literature to support and further develop the idea that different phases of the disaster management life cycle require different network structures.

6.4 Disaster Response Networks (DRN)

Drabek (1986) was the first to analyze over 1000 disasters and articulate the four phases of the disaster cycle: (1) preparedness (planning and warning), (2) response (evacuation and emergency), (3) recovery (restoration and reconstruction), and (4) mitigation (perceptions and adjustment). A summary of the typical activities of the disaster life cycle (Altay and Green 2006) is presented in Fig. 6.3.

Many sources use this cycle for describing disaster management (UNICEF 2005; Altay and Green 2006; Benjamin et al. 2011). However, there is a difference between the life cycle used for long-term and short-term structures. For example, Howden (2009) uses this life cycle for addressing a supply chain of disaster management while Moe and Pathranarakul (2006) use it for addressing a disaster management project. Recently, Noran (2011) tries to distinguish between the two by suggesting a combination in which the short-term preparation, response and recovery stages are structured as a virtual organization with a short-term outlook, while mitigation and long-term preparation are managed as a long-term structure. He examines the subject through the enterprise architecture lens to build a "business model" and compares this structure with commercial peer structures where the collaboration forms a virtual organization to bid, win and combine the resources to complete each project promptly. However he does not show how the two structures and their life cycle could fit together and provide a continuous cycle of operations, nor does he address the ethical considerations. The current chapter complements this work by comparing the life cycles of both long- and short-term structures and provides a conceptual model to show where the disaster management life cycle fits within the combined and non-profit short-term/long term structure.

The life cycle of virtual organizations has been extensively modeled (Jägers et al. 1998; Jagdev and Thoben 2001; Sitek 2007, Sitek et al. 2010). It includes three phases: the initiation phase when partner selection takes place, the operation phase

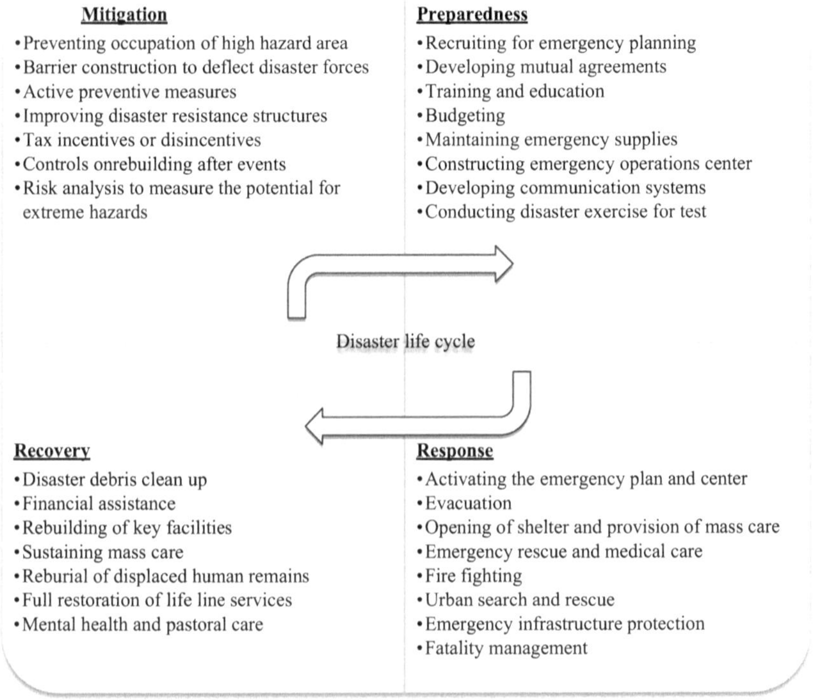

Fig. 6.3 Typical activities of a disaster life cycle. (Source: Altay and Green (2006))

when day-to-day activities take place and the dissolution phase when the objectives of the collaboration are accomplished or the need for collaboration disappears and the partnership dissolves (Fig. 6.4).

Figure 6.4 shows that in a long-term collaboration (for example supply chain) the initiation phase is followed by operations leading to a dissolution phase. However in this structure, the operation phase is much longer than the operation phase in short-term collaboration. Also the dissolution phase in the long-term collaboration may result in partners not working together for the same collaboration but staying in closer contact for future undertakings.

On the other hand, the short-term collaboration (for example virtual organization) starts from a Virtual Breeding Environment (VBE). "*A virtual breeding environment is a long-term pool of potential partners, which provides the environment for the establishment of cooperation agreements, common infrastructures, common ontologies, and mutual trust, which are the facilitating elements when building a new virtual enterprise*" (Camarinha-Matos and Afsarmanesh 2003, p. 157). Following demand creation, the initial phase including preparation and configuration tasks (Sitek 2007, Sitek et al. 2010; Ermilova and Afsarmanesh 2006, 2007) is conducted to select a subset of VBE (Camarinha-Matos and Afsarmanesh 2003). Initiation is followed by a short period of day-to-day operations (Sitek et al. 2010) until the market declines and virtual organization dissolves. The virtual organization will

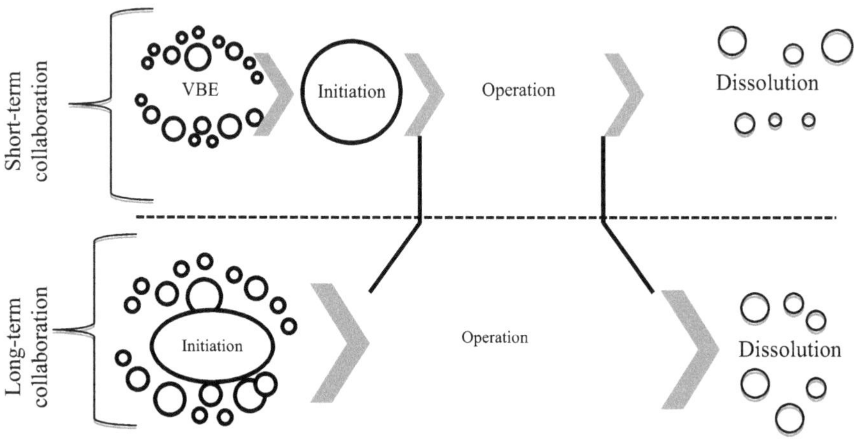

Fig. 6.4 Comparison between short-term and long-term collaboration life cycle. (Adopted from: Sitek 2007, Sitek et al. 2010; Thoben and Jagdev 2001)

assemble and update its breeding environment, which will facilitate the virtual organization partner selection in the next opportunity.

The difference between the two structures is mainly in the length of the operation phase. Also the VBE is a prominent theme in the emergence of a virtual organization, which gives rise to a quick initiation process. Due to these differences, the application of short-term and long-term structures should be different in a disaster response network. This is further explored below under a discussion of long-term and short-term structures.

Long-Term Structures

The subject of supply chains has attracted great attention from scholars focusing on partner selection in disaster management. A variety of approaches to long-term structure for partner configuration are exhibited in the literature review papers (Weber et al. 1991; Holt 1998; Degraeve et al. 2000; De Boer et al. 2001; Wu and Barnes 2011). Disaster operations in the customary structures for disaster response such as supply chains are performed before, during, and after a disaster with the goal of preventing the loss of human life, reducing the disaster's impact on the economy, and returning to a state of normality (Altay and Green 2006). However, due to a lack of control and insufficient information in a disaster situation, these approaches which consider supply chain as the suitable structure for disaster response may fall short in practice. For example the supply chain structure is put forward as a model based on the assumption that 80 % of disaster operation activities involve logistics (van Wassenhove 2006). Therefore a supply chain structure is suitable for managing disaster response operations (Balcik et al. 2010). However there are various arguments against the suitability of these long-term structures for all phases of disaster situations. A huge amount of a disaster's financial resources depend upon fund-raising from occasional donors and therefore cannot guarantee a steady flow of funds. Also, the demand and supply are rarely equal or distributed effectively (Comfort et al. 2004; Tierney and Trainor 2004) and they dramatically fluctuate during the course of disaster response operation (van Wassenhove 2006). Also the lack of standards and indicators, inadequate training, lack of collaboration, low recognition of logistics and inadequate infrastructure (Kovács and Spens 2009; Benjamin et al. 2011) necessitates a clear restructuring of the disaster response network in ways that all phases of a disaster's life cycle can be based on, including short-term. Following this mind-set, scholars have recently started to employ temporary organizational structures (Simpson and Hancock 2009) for some phases and long-term structures for other phases of the life cycle.

Short-Term Structures

Another suggestion, limited to few studies, is the use of short-term structures to address the temporary nature of response operations in addition to addressing the conflict of objectives associated with the proliferation of heterogeneous partners (Jiang et al. 2012). For example, some scholars see the response operations as a public project targeting the alleviation of poverty and elevating living conditions of people. While some consider this public project involves prediction, planning and execution (Moe and Pathranarakul 2006) in the long-term, others argue that the ad-hoc collaboration processes of disaster response cannot be planned ahead because activities such as individual citizen contributions and volunteering are not business-oriented (Camarinha-Matos and Afsarmanesh 2008; Nolte and Boegnick 2012).

Apart from the long-term/short-term segregation, some scholars also explore horizontal/vertical cooperation between entities. This cooperation can operate at the same level in the market, hence the horizontal cooperation (Schulz and Blecken 2010) or involve different actors along the value chain of one industry, which would be the vertical cooperation. Both horizontal and vertical cooperation can be short-term or long-term depending on the needs of the partners, but it is more likely that vertical cooperation has a long-term focus. Horizontal cooperation could be compared to virtual organizations and may have a short-term focus as well as a long-term focus when companies are in strong partnerships.

Working across organizations has been long recognized as a necessity in public management (Friend et al. 1974). However, in recent years more companies have adopted collaboration in response to a volatile and competitive business environment (Camarinha-Matos and Afsarmanesh 2008) to increase their survival chances and gain a competitive edge (Romero et al. 2009). An extreme case of a collaboration network is the virtual organization as a temporary alliance of independent enterprises (Camarinha-Matos and Afsarmanesh 2005) which is formed in response to a single market opportunity (Martinez et al. 2001) and dissolves with the market's decline (Brown and Zhang 1999). In fact, the virtual organization structure is a method used by traditional companies to access external resources (Jägers et al. 1998) and a niche market (Brown and Zhang 1999) by collaborating with others. According to Jägers et al. (1998), the idea of a virtual organization has emerged from lean and agile manufacturing with a shift of focus to inter-organizational relationships. Drucker (1988) heralded the first signs of organizations with characteristics similar to those of virtual organizations. He signaled the emergence of a new generation of organizations. However, the expression 'Virtual Corporation' (Davidow and Malone 1992; Byrne 1993) was coined later in the literature. Virtual organizations are agile supply chains with a shift in focus from intra-enterprise performance to inter-relationship between companies (Corvello and Migliarese 2007). We refer the reader to Thoben and Jagdev (2001) for a comparison of supply chains and virtual organizations.

Here we argue that to accommodate various phases of the disaster life cycle, different structures are required. Supporting this perspective, Noran (2011) not only suggests a short-term virtual structure for managing preparation, response, and recovery in disaster networks, but also a long-term structure to manage mitigation and long-term preparation. This suggestion of using virtual organizations as the short-term structure for disaster response phase has also been adopted by others. For example, Javaid et al. (2013) state that the nature of disasters requires short-term collaboration between the partners of a virtual organization. To that end, we explore below the suitability of a virtual organization structure for a DRN.

6.5 Suitability of Virtual Organizations for Disaster Response Networks

To provide evidence that a virtual organization is a suitable structure for the short-term stages of a disaster operation, we compare the characteristics of virtual organizations and disaster response networks to show that they are both temporary alliances of independent organizations. They both share resources and information to collectively access the damaged regions (analogous with "market") and provide for a one-time created demand. Using networks as their structure, they can decentralize and cover various sub-tasks in accordance with their heterogeneous nature, while dynamically adapting to the turbulent situation. When the demand declines due to the progress of disaster response operation, they can dissolve and become independent entities again. These common characteristics are highlighted in Table 6.3.

Table 6.3 illustrates that in both structures, time and cost effectiveness is crucial. The temporary network is created to cross boundaries of individual organizations and allow collective access to resources including donations. The network structure of individual yet mutually dependent and equally important partners emphasizes the lack of hierarchy. Instead, the uncertainty and highly unpredictable changes in the environment are addressed by spontaneous, complex, dynamic and flexible actions. In virtual organizations as well as disaster response networks the attempt

Table 6.3 Overlaps between characteristics of VO and DRN

Shared characteristics between VO and DRN	Literature on virtual organizations	Literature on disaster response networks
Cost/Time Effectiveness	Tan et al. (2008); Brown and Zhang (1999)	Nolte and Boenigk (2012); Comfort (2007); Tierney and Trainor (2004)
Collective access to resources/donation	Tan et al. (2008)	Nolte and Boenigk (2012)
Temporariness	Tan et al. (2008); Jägers et al. (1998); Brown and Zhang (1999)	Comfort et al.2004; Nolte and Boenigk (2012)
Lack of hierarchy	Tan et al. (2008); Brown and Zhang (1999); Jägers et al. (1998); Martinez et al. 2001)	Moe et al. 2007; Tierney and Trainor (2004)
Network structure	Jägers et al. (1998); Corvello and Migliarese (2007)	Nolte and Boenigk (2012); Tierney and Trainor (2004)
Independent participants	Tan et al. (2008); Jägers et al. (1998);	Nolte and Boenigk (2012); Tierney and Trainor (2004)
Spontaneity	Jägers et al. (1998); Brown and Zhang (1999)	Tierney and Trainor (2004)
Dynamism	Jägers et al. (1998);	Nolte and Boenigk 2012
Uncertainty	Jägers et al. (1998); Brown and Zhang (1999)	Tierney and Trainor (2004)
Participants equality	Jägers et al. (1998);	Tierney and Trainor (2004)
Boundary crossing	Jägers et al. (1998);	Nolte and Boenigk (2012)
Mutual dependency	Jägers et al. (1998); Brown and Zhang (1999)	Tierney and Trainor (2004)
Geographical dispersion of actors	Martinez et al. (2001)	Nolte and Boenigk (2012)
Niche market demand	Jägers et al. (1998); Brown and Zhang (1999); Martinez et al. (2001)	Tierney and Trainor (2004)
High partner turn over	Brown and Zhang (1999)	Tierney and Trainor (2004)
Unpredictable changes	Jägers et al. (1998); Corvello and Migliarese (2007)	Tierney and Trainor (2004)
Flexibility	Brown and Zhang (1999)	Nolte and Boenigk (2012)
Time Constrains	Brown and Zhang (1999)	Nolte and Boenigk (2012)
Complexity	Corvello and Migliarese (2007)	Nolte and Boenigk (2012)
Decomposable tasks	Martinez et al. (2001)	Tierney and Trainor (2004)

is to respond to the disaster-created demand (analogous to the niche market demand), by allocating decomposed tasks to geographically dispersed partners under time constraints to allow a collective access to resources/donations. VO here is the representative of a short-term structure with a temporary nature and so is a DRN. In other words, due to the shared characteristics between the two, this research argues that DRN can be well adopted into a temporary structure of a VO without the objective of making profit or achieving commercial gains. To better illustrate how the DRN with a VO structure could fit within the long-term structure of the disaster management life cycle, the disaster collaboration life cycle conformity model is put forward in Fig. 6.5.

Short-term collaboration in Fig. 6.5 shows where the short-term phases of disaster management fit within the collaboration. It illustrates that the disaster strike creates demand for humanitarian aid, necessitating preparation and configuration tasks (Sitek et al. 2010; Ermilova and Afsarmanesh 2006, 2007) in the initial phases. This includes selecting a subset of partners out of the virtual breeding environment (Camarinha-Matos and Afsarmanesh 2003), which is a pool of potential partners. The life cycle continues with the day-to-day operation (Sitek et al. 2010) of the disaster response network and dissolves when the need for humanitarian aid is diminished and the community is capable of coping with the situation on their own resources.

Long-term collaboration in Fig. 6.5 shows preparation, mitigation, and recovery phases between humanitarian partners. The operation of these three phases also follows a process of selecting suitable partners for long-term collaboration, similar to partner selection for a supply chain. The combination of the two life cycles gives rise to a conceptual model for disaster operation structure along the four phases of the disaster cycle.

To summarize, the disaster collaboration life cycle conformity model suggests that short-term and long-term collaboration in a DRN need to work in tandem. The long-term structures such as supply chains are suitable for structuring the majority of the disaster life cycle including recovery, mitigation, and preparation, whereas the response phase associated with the immediate aftermath of the disaster should adopt short-term structures such as virtual organizations. Unlike the immediate aftermath, there is time for gathering accurate data, planning, prediction, and implementation in recovery, mitigation, and preparation phases. Therefore in these phases the established networks such as supply chains can effectively form and operate. The response phase however requires a structure, which can form immediately and act upon uncertain data, because the community is under shock and in need of urgent help. For this phase, a virtual organization structure is more effective.

To further examine the suitability of virtual organizations to a DRN, we compare the life cycles of the two. The overlap between characteristics of virtual organiza-

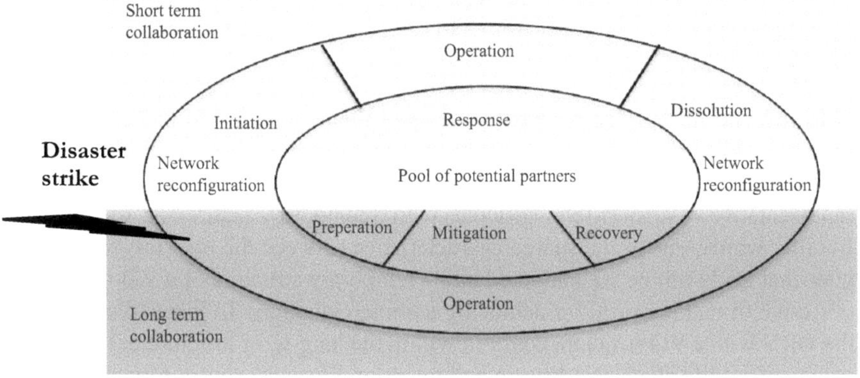

Fig. 6.5 Disaster collaboration life cycle conformity model. (Source: Authors)

tions and the disaster life cycle indicates that a virtual organization structure can accommodate the response phase requirements. To further clarify where the disaster response network and virtual organization overlap, Fig. 6.6 illustrates both life cycles.

The left side of Fig. 6.6 shows the life cycle of the VO and the right side shows the life cycle of the DRN. The middle column shows the period of the life cycle including before, during and after the business opportunity emerges (VO) or the disaster strikes (DRN). On the left side of Fig. 6.6, a sudden change in the environment such as a business opportunity creates demand for product and services. This kicks off the life cycle of a VO and it requires a network reconfiguration from a pool of potential partners. The result creates the response phase operated by a VO (short-term collaboration) within the three-phase life cycle (initiation, operation and dissolution). After dissolution of the VO, the network may be reconfigured into a long-term collaboration structure such as supply chains. The collaboration will initiate, assemble and update the pool of potential partners to facilitate partner selection the next time this is needed. This structure restarts when the market creates another demand.

On the right side of Fig. 6.6, the life cycle starts when a disaster strikes and creates demand for a humanitarian response. This is followed by the initiation and formation of the incident-specific network of partners who organize the response. When the demand for humanitarian response decreases, the specific network of humanitarian partners is dissolved and the recovery phase is started. This is followed by mitigation and preparedness, which keep the long-term network of partners up-to-date and better prepared for the next time they are called to respond to a disaster.

It is noteworthy to mention that these phases are seldom mutually exclusive or independent; rather, overlaps and interrelations enable them to operate concurrently (Shaluf 2008; Moan et al. 2009). For example, there is no crisp separation between

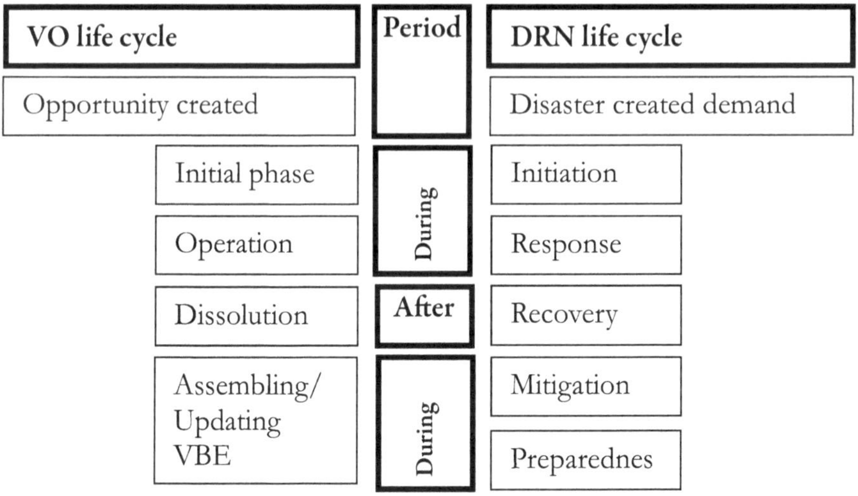

Fig. 6.6 The overlap between VO and DRN life cycle phases. (Source: Authors)

mitigation and recovery as the cycle progresses from recovery to mitigation, similar to preparedness and response because the disaster network must be built quickly given the situation before the disaster strike (Moan et al. 2009).

6.6 Discussion, Conclusions, and Limitations

This paper focuses on the partner proliferation problem in DRN as one of the most recurring problems in humanitarian operations. A literature review revealed growing interest in the possibilities of restructuring the DRN. To investigate this, we highlight the challenges facing the disaster response operation. These challenges are mainly associated with the lack of reliable data, which make long-term outlook and pre-planning extremely difficult. Instead, a short-term outlook toward collaboration is investigated and an example of a short-term structure, the virtual organization (VO) , is compared to DRN to outline its compatibility (Table 6.3). The result shows that their characteristics match in multiple aspects and suggests that a VO structure is a good candidate for restructuring the DRN.

For further investigation, long-term and short-term collaborations are compared (Fig. 6.4) to see where their different phases fit within each other's life cycle. The result is matched to the disaster life cycle, where a conformity model is created (Fig. 6.5). The latter shows that the three phases of the disaster life cycle (preparation, recovery, and mitigation) could be well supported within the framework of a long-term collaboration. On the other hand, the response phase of the disaster life cycle could be well supported within a short-term collaborative structure. Figure 6.6 shows how specific phases of the disaster life cycle could fit in a particular type of short-term collaboration, namely the virtual organization. Based on this analysis, we conclude that a VO is a suitable structure for addressing the proliferation problem in DRN. The main reasons are twofold: (1) the VO structure supports a short-term and temporary approach to operations that can be triggered by a sudden disaster-created demand, and (2) the potential to work with minimum planning and to respond to the sudden changes in the environment makes VO a suitable candidate for restructuring the DRN with its many uncertainties.

The analysis of literature and current practice reveals additional information on the severity and the consequences of the proliferation problem. It highlights the scope of this problem due to the growth of natural onset disasters in recent decades and shows that although scholars are aware of the extent of the negative consequences of the proliferation problem, there are no generally agreed upon frameworks to tackle this problem. Only a few suggestions are put forward to reduce the proliferation problem, including restructuring the disaster response network either by adopting a collaborative structure or by shortlisting and selecting fewer beneficiaries. In addition, while there are studies emphasizing the discrepancies between the planned partners and unexpected partners in the disaster response, a framework to address these discrepancies is yet to be developed.

Building upon these studies, the present paper argues that one of the reasons for the proliferation of partners is the inadequacy of existing collaborative structures to cope with the unpredictable, non-profit based, ethically constrained situation of a disaster aftermath where basically few data are available and many principles of business efficiency do not hold. To address this, the present study revisits the characteristics and the life cycle of existing structures including long-term (e.g. supply chain) and short-term (e.g. virtual organization) collaboration. A comparison between the two structures and the characteristics and the life cycle of the DRN gives rise to a conceptual model for re-structuring the DRN within the disaster life cycle. The model suggests that the virtual organization as a short-term manifestation of collaboration is a suitable structure for the response phase, while long-term collaboration structures such as supply chains can be best used for the management of mitigation, recovery and preparedness phases of a disaster response operation.

There are important limitations to these conclusions. The model presented here is a conceptual model and its practicality is yet to be examined. Also, the restructuring suggested here is only one of many possible solutions to the proliferation problem and further investigation is required to identify and analyze other potential solutions. It is also noteworthy to point out that while this research can lead to motivations and techniques for the selection of partners, the political task and ethical constraints of dealing with the non-shortlisted beneficiaries and the stakeholders they represent is out of the scope of the present research.

Another key future challenge is to provide specific guidance for restructuring the network, including (1) how and on what basis should we configure the VBE or the pool of partners, (2) how and on what basis should we select the humanitarian partners required for virtual organization under the time pressure of the disaster aftermath, (3) how to dissolve the virtual organization when the humanitarian aid demand declines after the disaster, and (4) how to deal with the convergence problem with the entrance of unexpected partners. The above challenges are the subject of ongoing research by the authors intending to develop a decision support tool. This study is part of an extensive research on the configuration of humanitarian partners in a disaster response network, which includes prediction of the human impact of disasters and scenario-based decision-making models.

References

Aissaoui N, Haouari M, Hassini E (2007) Supplier selection and order lot sizing modeling: a review. Computers & Operations Res 34 (12), 3516–3540.

Altay N, Green W (2006) OR/MS research in disaster operations management. Eur J Oper Res 175(1):475–493

Balcik B, Beamon BM, Krejci CC, Muramatsu KM, Ramirez M (2010) Coordination in humanitarian response chains: practices, challenges and opportunities. Int J Prod Econ 126(1):22–34

Benjamin E, Bassily-Marcus A, Babu E, Silver L, Martin M (2011) Principles and practice of disaster response: lessons from Haiti. Mt Sinai J Med A J Transl Pers Med 78(3):306–318

Britton NR (1986) Developing an understanding of disaster. J Sociol 22(2):254–271

Browne J, Zhang J (1999) Extended and virtual enterprises—similarities and differences. Int J Agile Manage Syst 1(1):30–36

Byrne A (1993) The virtual corporation. Bus J Week 3304:98–103
Camarinha-Matos LM, Afsarmanesh H (2003) Elements of a base VE infrastructure. Comput Ind 51(2):139–163
Camarinha-Matos LM, Afsamarnesh H (2005) Collaborative networks: a new scientific discipline. J Intell Manuf 16(4/5):439–452
Camarinha-Matos LM, Afsarmanesh H (2008) On reference models for collaborative networked organizations. Int J Prod Res 46(9):2453–2469
Careem M, Silva C De, Silva R De, Raschidt L, Weerawarana S (2006) Sahana: overview of a disaster management system, Proceedings of the International Conference on Information and Automation. 361–366
Choi SO, Brower RS (2006) When practice matters more than government plans a network analysis of local emergency management. Administration & Society 37(6):651–678
Choi SO, Kim BT (2007) Power and cognitive accuracy in local emergency management networks. Public Administration Rev 67 (S1)198–209
Comfort LK, Ko K, Zagorecki A (2004) Coordination in rapidly evolving disaster response systems the role of information. Am Behav Sci 48(3):295–313
Comfort, LK (2007) Crisis management in hindsight: cognition, communication, coordination, and control. Public Adm Rev 67(May):189–197
Corvello V, Migliarese P (2007) Virtual forms for the organization of production: a comparative analysis. Int J Prod Econ 110(1–2):5–15
Crawford L, Langston C, Bajrachraya B (2014) Participatory project management for improved disaster resilience. Int J Disaster Resil Built Environ 4(3)
Data.worldbank.org (2014) GDP (current US$) | Data | Table. http://data.worldbank.org/indicator/NY.GDP.MKTP.CD. Accessed 27 Oct 2014
Davidow W, Malone M (1992) The virtual corporation. Harper Collins, New York
De Boer L, Labro E, Morlacchi P (2001) A review of methods supporting supplier selection. Eur J Purch Supply Manage 7(2):75–89
Degraeve Z, Labro E, Roodhooft F (2000) An evaluation of vendor selection models from a total cost of ownership perspective. Eur J Oper Res 125(1):34–58
Doocy S, Russell E, Gorokhovich Y, Kirsch T (2014) Disaster preparedness and humanitarian response in flood and landslide-affected communities in Eastern Uganda. Disaster Prev Manage An Int J 22(4):326–339
Drabek T (1986) Human system responses to disaster: an inventory of sociological findings. Springer-Verlag, New York
Drucker P (1988) The coming of the new organization. Harv Bus Rev 66(1):45–53
Edition.cnn.com (2014) Charity: aid workers raping, abusing children—CNN.com. http://edition.cnn.com/2008/WORLD/europe/05/27/charity.aidworkers/index.html?eref. Accessed 27 Oct 2014
Ermilova E, Afsarmanesh H (2006) Competency and profiling management in virtual organization breeding environments. In: Camarinha-Matos L, Afsarmanesh H, Ollus M (eds) IFIP international federation of information processing, Volume 224, network- centric collaboration and supporting frameworks. Springer, Boston, pp 131–142
Ermilova E, Afsarmanesh H (2007) Modeling and management of profiles and competencies in VBEs. J Intell Manuf 18(5):561–586
Eßig M, Tandler S (2010) Disaster Response Supply Chain Management (SCM): integration of humanitarian and defense logistics by means of SCM. In: Feichtinger W, Gauster M, Tanner F (eds) Economic impacts of crisis response operations—an underestimated factor in external engagement. Bundesministerium für Landesverteidigung und Sport, Wien, pp 283–310
Farazmand A (2007) Learning from the Katrina crisis: a global and international perspective with implications for future crisis management. Public Adm Rev 67(s1):149–159
Farazmand A (2009) Hurricane Katrina, the crisis of Leadership, and chaos management: time for trying the "surprise management theory in action." Public Organ Rev 9(4)
Fiedrich F, Gehbauer F, Rickers U (2000) Optimized resource allocation for emergency response after earthquake disasters. Saf Sci 35(1–3):41–57

Friend J, Power J, Yewlett C (1974) Public planning: the inter-corporate dimension. Tavistock, London

Gallear D, Ghobadian A, Chen W (2012) Corporate responsibility, supply chain partnership and performance: an empirical examination. Intl J of Production Economics 140(1):83–91

Grabowski M, Roberts K (2011) High reliability virtual organizations. ACM Trans Comput Hum Interact 18(4):1–23

Guo X, Kapucu N (2015) Network performance assessment for collaborative disaster response. Disaster Prev Manage 24(2):201–220

Hardy V, Roper K, Kennedy S (2013) Emergency preparedness and disaster recovery in the US post 9/11. J Facil Manage 7(3):212–223

Hasani S, El-Haddadeh R, Aktas E (2014) A disaster severity assessment decision support tool for reducing the risk of failure in the response operations. In: Brebbia C (ed) Risk analysis IX. Wessex Institute of Technology, UK

Heikkinen H (2012) Corporate-humanitarian partnership building: the case of peepoople's humanitarian emergency response partnerships. Dissertation. Hanken school of economics

Holt GD (1998) Which contractor selection methodology? Int J Proj Manage 16(3):153–164

Howden M (2009) How humanitarian logistics information systems can improve humanitarian supply chains: a view from the field. Proceedings of the 6th International ISCRAM Conference—Gothenburg, Sweden

Hu Q, Knox CC, Kapucu N (2014) What have we learned since September 11, 2001? Anetwork study of the Boston marathon bombings response. Public Adm Rev 76(4):698–712

Hu Q, Kapucu N (2014) Information communication technology utilization for effective emergency management networks. Public Manage Rev doi:10.1080/ 14719037.2014.969762, (ahead of print), pp 1–26

IFRC Code of Conduct http://www.ifrc.org/en/publications-and-reports/code-of-conduct/

Inomata T (2006) Towards a United Nations Humanitarian Assistance Programme for Disaster Response and Reduction: Lessons Learned from the Indian Ocean Tsunami Disaster. Join Inspection unit, Geneva. ftp://ftp.fao.org/docrep/fao/meeting/012/jiu-rep-2006-5-e.pdf. Accessed 2 Aug 2014

Jägers H, Jansen G, Steenbakkers C (1998) Characteristics of virtual organizations. PrimaVera Working Paper Series 98(02). http://dare.uva.nl/en/record/153412. Accessed 3 Aug 2014

Jagdev HS, Thoben K (2001) Production planning & control: the management of operations anatomy of enterprise collaborations, pp 37–41

Javaid S, Majeed A, Afzal H (2013) A reputation management system for efficient selection of disaster management team, 15th International Conference on Advanced Communications Technology (ICACT), ISSN 1738-9445, 829–834

Jiang Y, Yuan Y, Huang K, Zhao L (2012) Logistics for large-scale disaster response: achievements and challenges. 45th Hawaii International Conference on System Sciences, 1277–1285

Kapucu N, Demiroz F (2011) Measuring performance for collaborative public management using network analysis methods and tools. Public Perform & Manage Rev 34(4):549–579

Karunakara U (2010) Haiti: where aid failed. [online] the Guardian. http://www.theguardian.com/commentisfree/2010/dec/28/haiti-cholera-earthquake-aid-agencies-failure. Accessed 27 Oct 2014

Karunasena G (2011) Sustainable post-disaster waste management: construction and demolition debris. Post-disaster reconstruction of the built environment

Kent A (2004) A critical look at risk assessments for global catastrophes. Risk Analysis 24(1):-157–168

Kovács G, Spens K (2009) Identifying challenges in humanitarian logistics. Int J Phys Distrib Logist Manage 39(6):506–528

Malhotra R, Venkatesh U (2013) Pre-crisis period planning: lessons for hospitality and tourism. Worldw Hosp Tour Themes 1(1):66–74

Maon F, Lindgreen A, Vanhamme J (2009) Developing supply chains in disaster response operations through cross-sector socially oriented collaborations: a theoretical model. Supply Chain Manage An Int J 14(2):149–164

Martinez M, Fouletier P, Park K, Favrel J (2001) Virtual enterprise—organisation, evolution and control. Int J Prod Econ 74(1–3):225–238
Metcalfe V, Martin E, Pantuliano S (2011) Risk in humanitarian action: towards a common approach. London: Humanitarian Policy Group working paper, 1–6
Metcalfe V, Haysom S, Gordon S (2012) Trends and challenges in humanitarian civil—military coordination a review of the literature. Humanitarian Policy Group working paper, London, pp 1–34
Moan F, Lindgreen A, Vanhamme J (2009) Developing supply chains in disaster response operations through cross-sector socially oriented collaborations: a theoretical model. Supply Chain Manage An Int J 14(2):149–164
Moe TL, Pathranarakul P (2006) An integrated approach to natural disaster management: public project management and its critical success factors. Disaster Prev Manage 15(3):396–413
Moe TL, Gehbauer F, Senitz S, Mueller M (2007) Balanced scorecard for natural disaster management projects. Disaster Prev Manage 16(5):785–806
Moore S, Eng E, Daniel M (2003) International NGOs and the role of network centrality in humanitarian aid operations: a case study of coordination during the 2000 Mozambique floods. Disasters 27(4):305–318
News.bbc.co.uk (2014) BBC NEWS | In Depth | Natural disasters 'on the rise'. http://news.bbc.co.uk/1/hi/3666474.stm. Accessed 27 Oct 2014
Nolte IM, Boenigk S (2012) A study of ad hoc network performance in disaster response. Nonprofit Volunt Sect Q 42(1):148–173
Noran O, (2011) Towards a collaborative network paradigm for emergency services. Adapt Value Creating Collab Netw (IFIP Advances in Information and Communication Technology) 362:477–485
Nourjou R, Szekely P, Hatayama M, Ghafory-Ashtiany M, Smith S (2014a) Data model of the strategic action planning and scheduling problem in a disaster response team. J Disaster Res 9(3):381–399
Nourjou R, Smith S, Hatayama M, Okada N, Szekely P, (2014b) Dynamic assignment of geospatial-temporal macro tasks to agents under human strategic decisions for centralized scheduling in multi-agent systems international. J Mach Learn Comput 4(1):39–46
Odihpn.org (2014) Sexual exploitation and abuse by UN, NGO and INGO personnel a self-assessment—Issue 52—Humanitarian Exchange Magazine—Humanitarian Practice Network. http://www.odihpn.org/humanitarian-exchange-magazine/issue-52/sexual-exploitation-and-abuse-by-un-ngo-and-ingo-personnel-a-self-assessment. Accessed 27 Oct 2014
Oloruntoba R, Gray R (2006) Humanitarian aid: an agile supply chain? Supply Chain Manage an Int J 11(2):115–120
Pettit S, Beresford A (2009) Critical success factors in the context of humanitarian aid supply chains. Int J Phys Dist Logist Manage 39(6):450–468
Reinecke I (2010) International Disaster Response Law and the co-ordination of international organisations. ANU Undergrad Res J 2:143–163
Responseweb.int (2011) OCHA and slow-onset emergencies, Occasional Policy Briefing Series, Brief, 6. http://responseweb.int/sites/responseweb.int/files/resources/report_36.pdf. Accessed 2 Aug 2014
Rolland E, Patterson RA, Ward K, Dodin B (2010) Decision support for disaster management. Oper Manage Res 3(1–2):68–79
Romero D, Galeano N, Molina A (2009) Mechanisms for assessing and enhancing organisations' readiness for collaboration in collaborative networks. Int J Prod Res. 47(17):4691–4710
Schneider SK (2011) Dealing with disaster: public management in crisis situzations, 2nd edn., M. E. Sharpe, Armonk, NY
Schulz SF, Blecken A (2010) Horizontal cooperation in disaster response logistics: benefits and impediments. Int J Phys Dist Logist Manage 40(8/9):636–656
Shaluf I (2008) Technological disaster stages and management. Disaster Prev Manage 17(1):-114–126

Simpson N, Hancock P (2009) Fifty years of operational research and emergency response. J Oper Res Society 60(May):126–139

Sitek M (2007) Collaboration Formation In Virtual Organisations By Applying Prospective Performance Measurement. PhD University of Bremen

Sitek P, Seifert M, Graser F (2007) Partner profiling to support the initiation of collaborative networks. In: Proceedings of 13th international conference on concurrent enterprising. Concurrent innovation: an emerging paradigm for collaboration competitiveness in the extended enterprise, 4–6 June, Sophia-Antipolis, France. 213–220

Sitek P, Seifert M, Thoben K-D (2010) Towards an inter-organisational perspective for managing quality in virtual organisations. Int J Qual Reliab Manage 27(2):231–246

Tan W, Zhao X, Jiang C, Chen X, Tang A, & Tan W (2008) The differences and conjunctions on performance management between entity enterprise and virtual enterprise, 65–69

Tatham P, Spens K (2011) Towards a humanitarian logistics knowledge management system. Disaster Prev Manage 20(1):6–26

Telford J, Cosgrave J, Houghton R (2006) Joint evaluation of the international response to the Indian Ocean tsunami. Tsunami Evaluation Coalition, London

Thoben K, Jagdev H (2001) Typological issues in enterprise networks. Prod Plan Control Manage Oper 12(5):421–436

Tierney K, Trainor J (2004) Networks and resilience in the World Trade Center disaster, 1st edn. MCEER, Buffalo, NY

Tomasini R, Van Wassenhove L (2009) From preparedness to partnerships: case study research on humanitarian logistics. Int Trans Operat Res 16(5):549–559

UNICEF (2005) Emergency field handbook. UNICEF, New York, pp 1–417

Valburn M (2012) After the quake, praise becomes resentment in Haiti. Center for Public Integrity. http://www.publicintegrity.org/2012/01/10/7838/after-quake-praise-becomes-resentment-haiti. Accessed 12 Oct 2015

Van Wassenhove L (2006) Humanitarian aid logistics: supply chain management in high gear. In: Journal of the Operational Research Study, 57(5):475–489

Weber CA, Current JR, Benton WC (1991) Vendor selection criteria and methods. Eur J Oper Res 50(1):2–18

WHO. (2013) Classification and minimum standards for foreign medical teams in sudden onset. World Health Organisation. http://www.who.int/hac/global_health_cluster/fmt_guidelines_september2013.pdf. Accessed 3 Aug 2014

Wright T (2005) UN failed to coordinate tsunami response, Red Cross report says. NY Times, 6 Oct 2005. http://www.nytimes.com/2005/10/05/world/asia/05iht-response.html. Accessed 27 Oct 2014

Wu C, Barnes D (2011) A literature review of decision-making models and approaches for partner selection in agile supply chains. J Purch Supply Manage 17(4):256–274

Chapter 7
How Humanitarian Culture Informs Change Adoption: A Case Study of Humanitarian Logistics

Rebecca Walton, Robin Mays and Mark Haselkorn

Abstract This chapter presents a longitudinal case study of change management within a geographically distributed emergency logistics team at a large international humanitarian organization. We found four categories of factors affecting change adoption: (1) individual barriers to change, (2) individual drivers of change, (3) collective barriers to change, and (4) collective drivers of change. These factors reflected specific instances of humanitarian culture at play in influencing whether change is adopted. Our findings identified not only singular factors but, more importantly, the interplay among factors that contributed to the ultimate outcome—whether change was adopted. Key findings include that (1) individual barriers have a stronger influence for advancing or inhibiting change in humanitarian teams and (2) collective factors establish an environment that can either catalyze or prohibit individual change factors. These findings have practical implications for facilitating change adoption in humanitarian organizations.

Keywords Change management · change adoption · humanitarian culture · logistics · humanitarian organizations · longitudinal research

7.1 Introduction

Although operational change management in business organizations has been long studied, there is significantly less research exploring change management within the operational support systems of humanitarian organizations in general and in logistics operations in particular. The rare analyses of change management in the

R. Walton (✉)
Technical and Professional Communication, Utah State University, Logan, UT, USA
e-mail: rebecca.walton@usu.edu

R. Mays · M. Haselkorn
Human Centered Design & Engineering, University of Washington, Seattle, WA, USA
e-mail: rmays@uw.edu

M. Haselkorn
e-mail: markh@uw.edu

C. W. Zobel et al. (eds.), *Advances in Managing Humanitarian Operations*,
International Series in Operations Research & Management Science 235,
DOI 10.1007/978-3-319-24418-1_7

humanitarian organizational support context highlight the need for a better understanding of and approach to facilitating change to improve the performance of humanitarian organizations' support systems (Ontko et al. 2007). This dearth of understanding of and within humanitarian organizations indicates a potential problem: humanitarian logistics operations may miss opportunities for change that would improve their practices and, thus, more effectively and efficiently support the missions of their organizations. An important first step in remedying this problem is identifying how change operates and the factors that promote and inhibit change within the humanitarian organizations in which these logistics teams operate.

Humanitarian culture plays a central role in the adoption (and rejection) of change. Because humanitarian organizations are designed to support lowest level decision-making, field-level practitioners within these organizations hold the most influential roles in the adoption of change (Chambers 1997; Lindenberg and Bryant 2001). Thus, human perceptions and behavior at both the team and individual levels are highly significant for affecting change management. Taking into account these aspects of humanitarian culture, we address the research question, "What individual and collective factors impede and promote change adoption in the operational processes of a humanitarian logistics team?"

In this chapter, we explore whether and to what extent research-based recommendations for change that were articulated by the team itself affected the operational practices of the emergency logistics team of a large non-governmental organization (NGO). From 2007 to 2011, we conducted a longitudinal case study of change management within the organization's geographically distributed emergency logistics team. Our initial research identified strengths and weaknesses of the team's operational practices, leading to recommendations for improvement based on their team-defined goals. Our follow-up study, the focus of this chapter, identified barriers to change and drivers of change that emerged in the operationalizing of those recommendations.

7.2 Literature Review

Our research is framed by literature on the culture of international humanitarian organizations and organizational change management.

Culture of Humanitarian Organizations

Humanitarian logistics support systems differ from for-profit business logistics models in specific ways. Many of the differences between these systems emerge from the fundamentally different cultures of humanitarian organizations versus for-profit businesses. These cultural differences—e.g., in values, norms, organizational structures, processes, and priorities—are rooted in differing mission objectives and lead us to question claims that "many of the tools and techniques used in commercial supply chains, despite the challenges posed by the differences [with humanitarian logistics], can be applied in the humanitarian arena" (Torabi et al. 2013a, p. 2). In contrast, we align with scholars arguing that international humanitarian operations are fundamentally incongruent with for-profit business models due to widely differing organizational cultures, driven by differently aligned core objectives and system constraints (Mays et al. 2012; Wassenhove 2006).

First, while traditional for-profit business organizations operate very effectively with a top-down model of decision-making, humanitarian organizations operate in the reverse way to be most effective. Humanitarian emergency logistics systems are characterized by a need to accommodate significantly higher ranges of rapid, unpredictable and dynamic uncertainty (Ponnaiyan 2013). Stakeholders, power structures and processes also emerge and evolve with each disaster response. With rapidly changing situations on the ground, humanitarian logistics are necessarily grounded within field-led organizational and operational decision-making as opposed to top-down organizational hierarchies (Chambers 1997; Walton et al. 2013). In a typical for-profit business organization, the logistics system is designed to inform and support a small group of decision-makers positioned at the top of the triangle who have the best perspective to make decisions and direct support of what is often a dispersed network of workers (whose perspective encompasses only a piece of the whole picture) at the bottom. Guided by top-level decision makers, the work of the network then supports the big picture. In humanitarian organizations the triangle is inverted. There exists a small group of practitioners with the best perspective of the situation positioned at the bottom of the organization, practitioners who are making decisions with affected communities and sending requests for support to headquarters and donors (a dispersed network of supporting workers whose perspective encompasses only pieces of the full picture) at the top. In contrast to top-down models, headquarters' efforts support (rather than direct) the decisions made by field-level practitioners with affected communities.

Second, humanitarian culture is informed by the humanitarian imperative values of *humanity, impartiality, neutrality*, and in particular, *independence* (ICRC 2004). In concert with the broader humanitarian mission, humanitarian logistics systems are designed to prioritize the humanitarian imperative above cost optimization and support locally driven efforts for the most successful operation (Berenguer-Falguera 2012, p. 66; Holguin-Veras et al. 2012; Tomasini and Wassenhove 2009; Torabi et al. 2013b). Humanitarian logistics systems' ability to accommodate a wider, impact-oriented definition of effectiveness is enabled by a means-centered (as opposed to task-centered) approach to success (Mays et al. 2012). Such impact-oriented systems often operate in tension with efforts to improve resource efficiency (i.e., whereas efficient practice in a for-profit context optimizes for cost factors, efficient practice in humanitarian contexts optimizes to meet the humanitarian imperative—which may be more expensive in terms of financial cost) (Maiers et al. 2005; Walton et al. 2013).

These key characteristics of humanitarian organizations have important implications for evaluating current practice in humanitarian logistics. The need to accommodate maximum uncertainty and lowest level empowerment creates organizational characteristics such as minimal standard operational procedures in order to accommodate the need for maximum adaptation, little operational transparency beyond lowest-level decision-makers, and limited funding for assessing organizational support functions. These characteristics afford less value for evaluating practice by traditional for-profit business or quantitative measures, measures which may be less relevant to NGOs' overall impact-oriented views of successful practice (Mays et al. 2012, Tomasini and Wassenhove 2009; Walton et al. 2013). In addition, as humanitarian logistics systems address a more complicated mix of interactions and factors—many of which are not easily quantified—quantitative data may not be the most helpful tool for decision-making in these conditions (Holguin-Veras et al. 2013; Mays et al. 2012). For example, in terms of optimal speed, Walton et al. observed that "single-faceted, quantitative measures of time are inadequate to represent how stakeholders within humanitarian organizations experience speed" (2013, p. 226).

Research in the field shows that quantitative measures have not proven helpful thus far in supporting decision making for optimal performance in humanitarian work. Benini et al. noted in their Pakistan case study (2006) and Benini and Conley in their Iraq case study (2007) that basing decisions on quantitative measures (such as stock level) did not lead to optimal performance in humanitarian operations—e.g., matching relief goods to survivors' needs. Further, humanitarian decision-makers and logistics managers are less likely to value optimal or effective performance in terms of task-centered achievements measured by numeric goals but rather in terms of relational gains and wider impact to community well-being (Lindenberg and Bryant 2001; Mays et al. 2014; Tomasini and Wassenhove 2009). Although humanitarian effectiveness is driven by a different definition of success, few have developed performance measurement mechanisms to inform an organizationally appropriate evaluation of work (Sheehan 1996). This gap highlights the need for additional research based in an understanding of humanitarian culture that can inform logistics practices and measures.

Research related to effective humanitarian logistics is a growing body of work, but it remains minimally accessed and understood (Oloruntoba and Gray 2006; Walton et al. 2013). Much research on humanitarian organizations is conducted as case studies due to highly varied geographies, missions, circumstances, and ways of operating that challenge simple generalizations and call for localized perspectives. Insider epistemology (i.e., making meaning that is informed by a deep insider perspective) is characteristic of the richer and more comprehensive contributions to this literature (e.g., Benini et al. 2006; Long and Wood 1995; Oloruntoba and Gray 2006; Walton et al. 2013). Humanitarian culture is characterized by highly autonomous ways of operating in tandem with their impact-oriented mission. Specifically regarding change adoption, Lindenberg and Bryant (2001) claimed that "one of the most important barriers to change in NGOs is the strong and individualistic and independent style of staff" (p. 56). Thus, research that can appropriately inform the management of humanitarian organizations in general and humanitarian logistics in particular must be rooted in a deep understanding of humanitarian culture—particularly the values of autonomy and independence.

Managing Change in Humanitarian Organizations

The differences between businesses and humanitarian organizations that demand different logistics processes—e.g., different organizational cultures, different organizational priorities, different employee motivations—also demand different change management strategies. Managing change in a way that is culturally appropriate for humanitarian organizations requires a keen focus on organizational mission and strategic support of lowest-level empowerment. With for-profit companies operating in the social services sector, some nonprofit organizations find themselves competing in what has continued to move toward a performance-based market. The pressure of competing against for-profit companies can lead some humanitarian organizations to compromise their missions in efforts to adopt change that could increase a particular view of "productivity" (Ryan 2002). However, rather than compromise or downplay their mission, many nonprofit management strategies instead place impact at the center of organizationally appropriate performance measurement and subsequent change adoption (Kaplan 2001). For example, the Balanced Scorecard (Kaplan and Norton 1996), a widely used tool in nonprofit management, places the overarching non-profit mission at the top of the card with financial measures conveying constraints and affordances relevant to the primary objective—that of meeting the needs of those they serve (Kaplan 2001). Similarly, Behn's (1995) work on innovative organizations links the importance of a clear organizational mission in motivating members of the organization to embrace change in non-profits.

Buy-in at an individual level is vital for successful and sustainable change adoption in the non-profit sector, and specifically within international relief and development non-profits (Behn 1995; Lindenberg and Bryant 2001). Organizational change is often an uncomfortable premise that brings losses as well as gains (Bridges 2009), and individual change adoption is rarely motivated by numbers-based arguments (Kim and Mauborgne 2003) nor by scientific facts alone (Rogers 2003). Instead, change adoption in these types of organizations is heavily influenced by human factors, as described in Rogers' (2003) theory of social change: "… most people depend mainly upon subjective evaluations of an innovation that is conveyed to them from other individuals like themselves who have previously adopted the innovation" (p. 18). Rogers noted that later adopters of change tend to depend less on data and more on the experiences and opinions of people with whom they identify. We found this dynamic to be true in our case study of a humanitarian logistics team, particularly when practitioners observed their colleagues adopting changes that enabled them to better support the mission of the organization. Our study suggests that, outweighing the discomfort of transitions, a commitment to organizational mission is what often motivates humanitarian practitioners to adopt change.

People working for humanitarian organizations, as well as those in non-profits more generally, do not work primarily for compensation but rather because they personally value the organizational mission and want to support it (Collins 2001; Kaplan 2001, Mays et al. 2014). That individual passion can create an organizational strength, but it can also create organizational challenges for shared focus and strategic alignment when individuals have personal, sometimes competing or contradictory, visions for achieving mission (Kaplan 2001). To force strategic alignment and facilitate faster change adoption in nonprofit organizations, Peter Drucker, a best-selling author on nonprofit management, has advocated for traditional business principles and hierarchical restructuring (2009). However, although top-down approaches may offer the promise of faster diffusion of change in for-profit business environments (where organizational values are congruent with top-down directives and hierarchical structures), such top-down approaches do not offer the same promise of quick, sustainable change in organizational cultures that value lowest-level empowerment (Hoffman 2010; Lindenberg and Bryant 2001; Singhal et al. 2010). Lowest-level empowerment is supported by flat or reversed hierarchies (Behn 1995) which intentionally afford greater autonomy to decision-makers on the ground. This brings into view our earlier description of the reversed hierarchy found within humanitarian cultures, including their emergency logistics teams, where on-the-ground decision-makers are practitioners in the field who have the closest relationships with local communities, the deepest insight into navigating governmental constraints, and the most up-to-date knowledge of disaster status. These "frontline" practitioners are the most important and most challenging members of the organization to individually motivate to adopt changes (Behn 1995). Their cautious attitude toward change makes sense when considering their central position of impact in the reversed hierarchy. Thus, in the humanitarian context, where autonomy and independence are central to effective work, individual motivation and mindset plays a central role in influencing change adoption (Lindenberg and Bryant 2001).

7.3 Methods

This research was conducted in two phases: (1) the initial research to understand and recommend improvements to humanitarian logistics processes and (2) the follow-up study of organizational change resulting from these recommendations. The sections below describe data collection and analysis for both phases of research with an emphasis on the second phase.

Phase I: Initial Research

Phase I research methods were highly participatory and focused on work processes, as we sought actionable recommendations for the research participants and other stakeholders. We began by mapping current work processes of the emergency logistics team, gathering data through participatory ethnomethodology and conversational analysis of process information and task flows (Button 1991). The global team consisted of ten full-time members, three with greater than 15 years of direct field experience, two with more than 15 years of experience coordinating support logistics at the central level, three with 5–10 years of field experience, and two with less than 5 years of humanitarian logistics experience. In 2007 and 2008, Mays worked with a business analyst to map existing processes and data elements from the onset of disasters to the point of delivery. Walton and Mays met with stakeholders individually and in cross-functional groups to document the overall logistics process. From late 2008 to mid-2009, Walton interviewed approximately 40 employees of the humanitarian organization, including field requestors (recipients and other stakeholders of the logistics team's service) and members of the emergency logistics team itself. Data collection also included onsite research by Walton and a business process analyst at two national offices in regions that had recently experienced emergencies. The data collected in internal customer interviews, focus groups, and onsite observations were inductively analyzed for themes and incorporated into process model flows which were iteratively analyzed by participants and researchers for process improvements.

The research findings were the focus of the humanitarian logistics team's annual strategy meetings in November 2009 and were conveyed in a written report presented by Haselkorn via video conference. An external change management organization facilitated three days of discussion with the humanitarian logistics team regarding the research findings. During the annual strategy meeting, the team generated six high-priority action items for its future work, based on the findings of the initial research (see Table 7.1).

Phase II: Follow-Up Study

The primary focus of this chapter is on outcomes of our follow-up study conducted approximately a year and a half after the November 2009 annual strategy meeting, exploring the team's progress in pursuing the six changes (Table 7.1) that it had prioritized. This follow-up study was conducted (1) to identify changes in emergency logistics processes relevant to the findings of the report and (2) to gauge the impact, if any, of the original research. Follow-up data were collected through semi-structured interviews with the external facilitator who coordinated the 2009 annual meeting, one emergency logistics team member who did not attend the annual meeting, and six of the nine emergency logistics team members who attended the annual meeting. This included Mays, an insider-researcher as a member of the team and one of the attendees at the annual meeting. The response rate to interview requests was affected by availability of the emergency logistics team, which is frequently deployed to disaster locations worldwide. At the time of the follow-up interviews, six of the nine attending members of the team were available and agreed to participate.

The June 2011 follow-up interviews focused on the six action items, stemming from the initial research findings (Table 7.1). The semi-structured interviews

Table 7.1 Action items[a] generated by the emergency logistics team based on initial research findings

Action Item 1 (AI-1)	*Add more structure to information and communications tools*. This action item stemmed from findings that some information used by stakeholders in logistics responses was distributed (e.g., across disparate email messages, often with unrelated subject lines) or implicit (e.g., existing as expert knowledge in the minds of stakeholders but not available in written form)
Action Item 2 (AI-2)	*Focus on relationships that are vital to create*. This action item stemmed from findings that the logistics team could better meet stakeholder needs by building relationships with both internal and external partners: for example, field requestors in national offices (internal) and industry partners that might donate operations-related goods (external)
Action Item 3 (AI-3)	*Expand the range of logistical solutions offered*. The core mission of the emergency logistics team is to meet the needs of field requestors (to optimally equip requestors to support local beneficiaries during disasters). This action item stemmed from findings that requesters' needs were not optimally met by a limited selection of one-size-fits-all services but instead that they need flexible, customized solutions based on the type of emergency, available in-country resources, and other factors
Action Item 4 (AI-4)	*Improve communications among stakeholders involved in the movement process*, including within the emergency logistics team itself, between the logistics team and other employees of the organization (such as field requestors at national and regional levels), and industry partners. This action item stemmed from findings that highlighted confusion and a lack of understanding among some stakeholders regarding the emergency logistics team and its services
Action Item 5 (AI-5)	*Increase standardization of processes* (e.g., the request process and vendor management). This action item stemmed from findings that individual members of the emergency logistics team were overworked and that duties could be better shared across team members if processes were established
Action Item 6 (AI-6)	*Provide the field with information that goes beyond core service*. This action item stemmed from findings that the field requestors comprising the emergency logistics team's internal customers desired more preparedness and planning-related services

[a] Action items listed in order of priority from highest to lowest as ranked by the logistics team

followed a protocol in which interviewees were asked to reflect upon each of the six action items, describing what changes (if any) had occurred, the motivations for changes, the form of changes, how changes had affected their own work, and any resistance to change. In this way, the follow-up study focused on changes that the emergency logistics team had committed to making that were relevant to the initial research findings. Walton conducted an initial analysis of the interview notes and structured reflections, identifying patterns (i.e., emergent themes) which were then verified by Mays. Walton and Mays then iteratively reviewed notes and transcripts, jointly developing a formal coding scheme to describe, label, and organize the emergent themes. This coding structure (Table 7.2) and the resulting findings are presented in the Results & Discussion section.

Table 7.2 Research themes: factors operating as barriers to and drivers of change

	DRIVERS OF CHANGE	BARRIERS TO CHANGE
INDIVIDUAL	**Openness**: Having a willingness to try new things	**Inflexible independence**: Operating with a rigidly independent working style characterized by resisting collaboration and authority
	Personal accountability: Acknowledging that the way I worked (e.g., practices, processes, communication level) contributed to poor operational performance	**Culpability blindness**: Believing that the way I worked (e.g., practices, processes, communication level) did not contribute to poor operational performance
	Vision: Possessing a clear vision of the desired future, whether for the team or for one's own role	**Insecurity**: Fearing that flux in operational processes could harm my role or value
	Positive experience: Observing or engaging in successful practice	**Gaps**: Lacking skills, experience, or internal capacity
COLLECTIVE	**Meeting needs**: Offering services that are desired by field requestors (Note: This is the core mission of the logistics team)	**Inertia**: Disinterest in or resistance to new ways of operating because of implicitly agreed upon ways of operating
	Mission enforcement: Making visible operationally poor performance and requiring accountability for the way that performance affects the organization's ability to successfully achieve its mission	**Lack of enforcement**: Failure of leadership to address poor performance or adherence to explicitly agreed upon practice
	Empowerment: Being given the authority and autonomy to make decisions and take action	**Peer permission**: Requiring the approval of team members to make decisions and take action

7.4 Results & Discussion

Several themes emerged regarding individual and group factors that affected change adoption. These factors are grouped into four categories shown in Table 7.2.

The majority of factors affecting change adoption were human-centered and resided with individual team members, though collective conditions also affected change adoption. The first category, *individual drivers of change*, includes individuals' mindsets and behaviors that facilitated change adoption, particularly in their own work practices. The second category, *individual barriers to change*, is comprised of individuals' attitudes and behaviors that impeded change adoption in their own practices and those of their teammates. Several themes under these categories mirror each other in pairs: e.g., "openness" and "inflexible independence" represent opposite positions on trying new ways of operating; "personal accountability" and "culpability blindness" are opposite perceptions of an individual's role in poor operational performance. The third category is *collective drivers of change*, and the final category of themes is *collective barriers to change*. While the collective change drivers and barriers affected individual team members, collective drivers and barriers emerged from conditions set by group behavior, whether formalized or ad-hoc: e.g., "empowerment" and "peer permission" are conditions surrounding team decision making. These conditions affect individuals' decision-making and actions, but team members do not individually establish these conditions. Some collective barriers and drivers are mirrored pairs as well: "empowerment" and "peer permission" are opposite team approaches to decision making; "lack of enforcement" and "mission enforcement" are opposite responses to operational performance and adherence to previously agreed upon ways of working.

These factors are not the same as personality traits. Rather, the same personality trait could be expressed in ways that act as barriers or drivers of change. For example, independence under one set of conditions can be expressed as "openness," which enables change, but expressed in another way, this trait can become "inflexible independence," which impedes change initiatives. This finding has implications for organizations seeking to effect change, implications detailed in the Conclusions section.

The six action items generated by the humanitarian emergency logistics team at the 2009 annual meeting (Table 7.1) had varying levels of impact on the team's practices by June 2011. Relevant data emerged in connection with five of the six action items, all but AI-2, which seemed to relate to work that was already beginning to take place by the time of the annual meeting and thus had less relevance to change management. Regarding the other five action items, some resulted in readily adopted change, while other action items met resistance, and still other action items initially met resistance but later led to change. Thus, the factors in Table 7.2 interrelated in ways that led to three types of change outcomes: (1) readily adopted change, (2) ongoing resistance to change, and (3) initial resistance shifting into change adoption. Our follow-up study identified not only factors impeding and promoting change but, more importantly, the interplay and relationships among the themes (i.e., how these factors interacted to contribute to the ultimate outcome: whether change was adopted). Discussed in detail in the three sections that follow, these interplays occurred in complex contexts in which many factors were present, but Table 7.3 presents the most influential factors contributing to the ultimate change outcomes. Six interplays emerged from our case study data. From left to right, Table 7.3 identifies the key relevant factors, the action produced by this combination of factors, and the resulting change outcome.

Below we describe how factors interacted in each type of change outcome.

Readily Adopted Change

The action items that most readily led to change adoption were those calling for new practices, as opposed to calling for changes to existing practices. These changes

Table 7.3 Interplays among change factors and resulting change outcomes. (Dark gray shading indicates barriers, and light gray shading indicates drivers)

	Combination of Relevant Factors: Individual Factor(s)		Combination of Relevant Factors: Collective Factor(s)	Action			Outcome
Readily Adopted Change	Vision & Openness	+	Empowerment	circumvented	Peer Permission	=	Change
	[any]	+	Meeting Needs	produced	Positive Experience	=	Change
Resistance to Change	Insecurity	+	Inertia	activated	Peer Permission	=	No Change
	Inflexible Independence	+	Lack of Enforcement	activated and enabled	Peer Permission	=	No Change
	Gaps	+	Lack of Enforcement	amplified	Inertia	=	No Change
Shift Toward Change Adoption	Culpability Blindness	+	Mission Enforcement	transformed barriers to	Personal Accountability & Openness	=	Change

were less susceptible to collective barriers such as "inertia" because they did not counter established organizational or collective practice. Change that called for new practice also highlighted the power of individual drivers to lead to change in humanitarian organizations. Action Item 6 (AI-6: Provide the field with information beyond core services) led to the most readily adopted change. Traditionally, the humanitarian emergency logistics team offered two services to the field: (1) providing relief goods from globally distributed warehouses and (2) sending a logistician to an emergency site. One of the key findings of the initial research was the desire of field requestors for preparedness and planning services. The initial research revealed that field offices found preparedness activities (such as localized pre-positioning of relief goods) to be extremely valuable in the initial days after a disaster, significantly decreasing the time for goods to reach beneficiaries and increasing the number of beneficiaries served. Based on those findings, the emergency logistics team generated AI-6.

Three individual drivers of change—"vision," "openness," and "positive experience"—emerged in the team's response to this action item. Although individual barriers were present, two collective drivers of change—"empowerment" and "meeting needs"—were essential to overcoming resistance and adopting change. As conveyed in the interview quote below, the individual driver "vision" was initially present because AI-6 was congruent with at least one team member's vision:

> When the early research from the field reflected they really want [preparedness and planning], that gave me the ability to go forward with a lot of my ideas, as well as to learn additional things they wanted that I didn't know. That gave me momentum to proceed, particularly from [the team leader].

The above quote shows the drivers "openness" and "vision" present in a team member who was well positioned to advocate for change adoption. These individual drivers led to a key collective change driver: empowerment. Because research supported the team member's vision, the team leader helped to cultivate an environment with the empowerment necessary to initiate preparedness and planning services, eliminating the need for peer permission:

> We did workshops on preparedness and response checklists in [the Middle East] and did one project in Nepal to have a preparedness plan to understand the country's risks and resources. We helped the country office set up systems like prepositioning, vendor agreements, putting people on a surge capacity list—all local specific things according to the region. I didn't need any kind of buy-in or consensus to proceed with that. There were times it sounded like [logistics team members] wanted to put the brakes on it, but because it was under me and [the team leader] gave me authority to proceed, it didn't affect me.

The above quote shows the power of the collective driver "empowerment" in mitigating effects of the collective barrier "peer permission," which may have otherwise halted change adoption.

Two other drivers of change—"meeting needs" and "positive experience"—also combined to support readily adopted change. When some team members began offering preparedness and planning services and those services met the needs of the field, it increased the field's awareness of the emergency logistics team and the value it offered. Preparedness and planning initiatives were overwhelmingly successful, leading to a volume of service requests exceeding team capacity. Thus, this change adoption produced "positive experience" (an individual change driver) for both field requestors and logistics team members engaged in these changes:

> I do internal logistics assessments and training for the national office level… It's been like this for three years, but now it's more active because regions are recognizing the value of [our team] participating in [their regional events], and I'm being invited to participate. I don't need to chase their invite. It works well.
>
> Country offices are realizing they need to build capacity… The country offices see the need to have preparedness in their countries. And countries are also seeing the support we give.

These quotes show the relationship between the collective driver "meeting needs" and the individual driver "positive experience." As the logistics team expanded its services in response to the initial research findings, the team better met the needs of its internal customers and developed additional positive experiences. The collective driver "meeting needs" triggered and then reinforced the individual driver "positive experience," strengthening change adoption and helping it spread among team members, removing previously existing barriers such as "inertia" and "peer permission."

Resistance to Change

The second context in which themes emerged was when change met resistance. In general, the action items meeting the strongest resistance were those addressing long-standing, established practices. Two action items meeting strong resistance were AI-3 (Expand the range of logistical solutions) and AI-4 (Improve communication with stakeholders in the movement process).

The individual barriers to change "insecurity," "independence," and "gaps" and the collective barriers "inertia," "lack of enforcement" and "peer permission" emerged in this context, and the amplifying effect of the collective barrier "lack of enforcement" became clear. With amplification from "lack of enforcement," the individual barriers to change were so powerful that they impeded change even after the team had expressed concerns of becoming irrelevant to the organization if they did not improve their operational performance. The quote below relates how the team responded during an envisioning exercise at the 2009 annual meeting, after discussing the findings of the research report:

> Towards the end of the session, we did an exercise where we talked about, 'Okay, we've been here and had this long, hard talk. What would happen if we didn't continue? What would that look like?' People said it would be, they were negative: 'We'd not innovate, we'd die out, etc.'

Yet even after expressing fears of "dying out" as a team, some members resisted implementing changes to long-established processes. Some of this resistance stemmed from "inertia":

> Change is something we all don't want. It is uncomfortable… People are comfortable doing what they think is best for them and what they know.
> Change is traumatic, and resistance would be there because people want their comfort zone and don't want it disrupted if something doesn't look like the way that people have been doing it.
> Most of them have been doing this for 20 years, and they don't want to change.
> It's worked this way. Why would we do anything else? It's always worked this way.

The four quotes above show the clear connection between change and discomfort. When people become experts with ingrained practices, changing those practices is uncomfortable. Established organizational practice can thus create the collective barrier "inertia." "Inertia," being based in a past practice that was successful, also speaks to the power of the individual driver "positive experience" in influencing practice over the long term. As Rogers (2003) suggested, for people to embrace innovation (i.e., engage in change adoption), the relative advantage must be great enough to warrant the discomfort, i.e., to overcome the collective barrier. This follow-up study showed an interesting relationship between the barriers of "inertia" and "insecurity" (what Rogers called "uncertainty"). Team members expressed insecurities regarding the risk of changing established practices, especially whether their efforts to change would be supported by others in the organization and whether the change would "stick" or be successful at meeting needs. In other words, "insecurity" increased the power of "inertia" by decreasing the perceived relative advantage of innovation to be successful.

In some cases, "insecurity" posed such a strong barrier that change appeared to have actively negative consequences. For example, one team member conveyed reluctance to freely express concerns, even after the team received an email encouraging them to do so. Fear of repercussions from expressing dissent to one's teammates or leadership suggests one reason for resistance to AI-4. Another barrier to change was insecurity regarding whether changes would reduce one's value. This insecurity activated the barrier “peer permission” to produce a competitive, rather than collaborative, environment in which individuals resisted changing their work practices or accommodating peers’ changes in an effort to preserve their own value:

> There is a threat also that maybe I am not useful or if I don’t have a role, I can be worried about how change will affect me. It results in having frantic people trying to find credit and a competitive attitude, which is not healthy on a team. Change brings up those kinds of things.
>
> People are protective of themselves and defensive and insecure about [their] jobs even if they appear to have all the answers.

Other insecurities stemmed from a history of unenforced requirements and a lack of consequences for team members resisting previously agreed upon operational changes. Thus, the collective barrier “lack of enforcement” amplified the individual barrier “insecurity”:

> What I thought was interesting [was] the process was mentioned but not the enforcement of the process. People would not do it and would wait until the team leader came after them. No one held them accountable.
>
> They didn’t think change would stick. So why bother.
>
> People only do what they want to do, and the team leader lets them.

The collective barrier to change “lack of enforcement” also amplified the individual barrier “inflexible independence.” This dynamic emerged as one of the most prominent barriers in the team:

> No one likes to be told what to do, especially old farts. Outsiders who don’t know, who’ve never been in disaster response, even though [in the research report] there were comments and quotes from [our own organization], that was shrugged off as, ‘They are unaware; they don’t really know who we are and what we do. There is an outsider who doesn’t know us telling us what our problems are.’
>
> Some of the resistance is a power play: you’re not going to make me change.
>
> People will never agree with everything you say. People will push back by nature… We still struggle as a team to get everyone on same page and do the same thing.

Thus, as indicated in the quotes above, if no effective mechanisms were in place to hold team members accountable for following previously agreed-upon practices, they could employ a rigidly independent working style that actively resisted change.

The quote below shows “inflexible independence” expressed through a resistance to communicate with other team members, activating the barrier “peer permission,” a practice which impeded change and in this case also impeded relief efforts:

> In the field, there were communication breakdowns all over, especially within [the emergency response team]. [Team member] and [team member] didn’t talk to people outside the field. [One of the team members in the field] would talk to only [another team member outside the field] and occasionally me, but he wouldn’t do a daily call to allow us to fully support him.

In a flat or inverted hierarchy, “inflexible independence” in the lowest-level decision-maker blocked change not only in the practices of that particular individual but also in the practices of his/her teammates. With lowest-level empowerment, implementing change requires “peer permission,” and this factor is used by the decision-maker (informed person) on the ground to enforce his/her decision-making.

Where the individual barrier "inflexible independence" was present and the collective driver "empowerment" was missing, the collective barrier "peer permission" posed a highly effective barrier to change adoption. This was true when a resistant peer held a stakeholder role in the new behavior. For example, when individuals' ability to adopt change is dependent upon the cooperation of their peers, those peers present a barrier to change adoption by withholding cooperation:

> These things could go forward because I was directly in charge and [the team leader] supported it. Anything that required buy in from the team, if I had to partner with my counterpart [team member], it was difficult to move forward. I don't know why. We tried to find out those reasons, but they weren't explicit, and we kept running into people not returning calls or participating in meetings.

A final barrier to change was "gaps," particularly gaps in skills or personal capacity to adopt change. Where "lack of enforcement" was also present, change adoption rarely stuck. For example, the initial research identified a need for more structured communication tools, leading the team to create AI-1. When team members discussed their communication tools in the follow-up interviews, they said that a Sharepoint-based communication tool was developed with participation from the team over a 2-year period, but once developed, it was not widely used, and usage was not required or enforced. Only a few weeks before the follow-up interviews, the team was introduced to another tool, Yammer, which was welcomed by some members but rejected by others:

> [We] have SharePoint. It works but is clunky. Doesn't have broad buy in.
> It will be interesting to see how fast people will get tired [of Yammer]… Age matters. Younger [people] pick it up faster, and some on our team are not really getting it. They say, 'Why should I be on Yammer? [It] is a waste.'
> We've got Sharepoint, but everybody hates it because it's not as user friendly as advertised. They just started hammering us on, now we're using Yammer as Facebook for our organization… I was overwhelmed, so I just shut it off.

The above quotes suggest that technology design also plays a role change adoption, indicating the relevance of a range of related factors, including some beyond the scope of this chapter. These examples described in this section show how barriers reinforced each other and created an environment that encouraged additional barriers. For example, "insecurity" strengthened "inertia" by decreasing the perceived value of change. Also, where there was "lack of enforcement," individual barriers were amplified. "Peer permission" was activated by "inflexible independence" to block changes in not only individuals' work practices but also those of their teammates. The next section shows that when certain change drivers were present, resistance to change was overcome.

Shift Toward Change Adoption

The final context in which themes emerged was when change initially met resistance which was then overcome. In this context, the collective driver of change "mission enforcement" was central to dissipating barriers such as "culpability blindness" and "inertia." Further, this collective driver created an environment in which individual drivers could flourish, supporting the value of lowest-level empowerment which is so congruent with the culture of humanitarian organizations.

Almost every member of the team mentioned the January 2010 Haiti earthquake as a major turning point for the team in change adoption—particularly in motivating increased standardization of processes (AI-5). By resisting change and continuing to operate in old ways, individuals did not encounter the driver "positive experience." In other words, their work did not optimally meet needs in the field, which began to break down the barrier "culpability blindness":

> There is a huge gap identified in Haiti and Pakistan… One thing that came out of Haiti is a gap in procurement, including vendor management. On the ground, there is a gap when it comes to effective procurement services.
>
> Now two big needs came out of Haiti and Pakistan: procurement and [relief items] tracking. In Haiti, we realized we need to do things differently.
>
> Haiti was kind of a turning point. We're trying to standardize emergency procurement, the way that people [do it] in a disaster, there is a standard way to procure based on a kit being put together by [the emergency logistics team] with forms, et cetera. [We are] also going to do transport and warehouse. That has never been done. Never have people tried to standardize [these] processes. This happened because of Haiti.

The initial research highlighted bottlenecks due to limited human capacity and recommended that standard processes be developed to allow responsibilities to be shared among stakeholders while preserving quality of service. However, this finding was not initially valued: it received only two votes in the team's prioritization exercise during the annual meeting, ranking as the next-to-the-last priority. As Rogers (2003) described, people in the early and late majorities categories need greater motivation to adopt change than those in the innovator category. For several members of the humanitarian emergency logistics team, that motivation to change stemmed from experiences during the Haiti earthquake (i.e., the failure of old ways of operating to support organizational mission by meeting needs in the field) and aftereffects within the humanitarian organization (i.e., the increase of the collective driver "mission enforcement"). These conditions moved the team past what Kim and Mauborgne (2003) described as the most difficult aspect of change management: generating agreement about the cause of problems and the need for change.

This outcome—initial resistance followed by a major turning point—was influenced by interplay among themes. While poor operational performance (i.e., not meeting needs) was a factor in creating an environment conducive to change, it was primarily the collective driver "mission enforcement" that served as a change catalyst. Because of the overall poor operational performance in Haiti, the logistics performance received unusually high scrutiny from wider organizational stakeholders. Across the organization, there was a call for accountability, explanations of problematic operational performance, and changes to prevent recurrence of those problems:

> It [the Haitian disaster] was unique in that it was an urban setting and being near US and was very visible to donors and was an extra spin on things, making it harder.
>
> What happened was they are blaming us [the emergency logistics team] for procurement problems in Haiti. Certain people, well, they are blaming [a larger organizational team] for

> problems in procurement, IT, and maybe HR. What hit us was procurement… We had no lessons learned after that. We have all these big muckety-mucks blaming us and now [the team leader] says we should have a lessons learned, and now we spent 11 hours writing a report on Sunday for [the team leader] to take to a meeting.

The increased accountability for supporting the organization's mission triggered an internal shift in some team members, from the barrier of "culpability blindness" to the change driver "personal accountability":

> Haiti was such an ordeal, and we're still getting the same critiques, some valid and some invalid, coming out of that. Some people on the team [are] saying to the team, 'We need to do better; we need to improve our own process.' Even [one team member] is saying, 'I need to map out my processes.' In Italy when we presented we needed to do better [based on the initial research report], there was a lot of pushback saying, 'No, we're doing great.' But since Haiti, there is a big feeling [that] we need to do better.
>
> Some existing functions are changing… There is more of an openness to things in the Italy paper [the research report] like standardizing processes and relationships. There is more of an openness, attributed to what happened in and after Haiti, ongoing critiquing of work in Haiti… There is a shift to 'we need to get our own house in order.'
>
> Haiti is an example of the struggle [to adopt change in standardizing processes]. Procurement was done without the standardization. It was a failure on our part. People were not doing things they need to do.

By making visible mission failures and requiring team members to account for poor operational performance, the collective driver "mission enforcement" prompted a shift from the collective barrier "inertia" to the individual driver "openness":

> In humanitarian crisis, things are changing. What we did in the 1990s, we can't do today. We did it and saved lives, but change is difficult because we have all these standards. I feel like I went through a personal struggle. You look at it and all the developments around you and realize there needs to be a change; you have experience, but you need to change.

Several team members described their efforts to change the way that relief items are prepositioned before emergencies and tracked during emergencies. The individual change driver "openness" is apparent in these efforts to develop a new way of operating:

> In Haiti tracking was a big issue… I think we still remain with that issue, but we are doing some things to change the system. We are developing a database which [one team member] will be using to track inventory of [relief items]. That should help to ensure that all information is in the system and all information is in one place, data is extracted and is in an adequate form for management.
>
> I'm leading the [relief items] management bit to build [relief items] management capability. We are making changes in terms of logistical solutions we offer. Now we will not only make sure supplies are prepositioned in the five warehouses, but we launched an initiative to encourage national offices to preposition items and make them available to the wider [emergency logistics team] network.

These quotes illustrate Rogers's social change theory at work: for some stakeholders, research must intermingle with personal experience to prompt change adoption (2003). For change to "stick," individuals in the organization must want to change their practice (Kim and Mauborgne 2003). And where change drivers mitigate existing barriers, change adoption is attainable. Where "mission enforcement" increased, it not only counteracted "lack of enforcement" but also created an environment conducive to individual change drivers like "openness" and "personal accountability." Because humanitarian organizations typically have flat hierarchies in which individuals wield considerable power and autonomy, these individual drivers are key components in leading to and sustaining change adoption. This finding has significant implications for managers seeking to promote change adoption within humanitarian organizations, which are discussed in the Conclusions section.

In summary, while collective barriers and drivers did not appear able to affect change independently, those factors had powerful catalytic effects on individual factors, thus highly influencing the outcome of change adoption. For example, the collective driver "meeting needs" triggered the individual driver "positive experience," creating a stronger environment for change adoption. Additionally, two of the most influential change factors were the opposite factors "lack of enforcement" and "mission enforcement." Their presence, or lack thereof, had significant effects on a range of individual change factors: e.g., "lack of enforcement" amplified individual barriers such as "inflexible independence," "gaps," and "insecurity," creating an environment hostile to change adoption. In contrast, "mission enforcement" had a profound impact in mitigating barriers perceived as power plays such as "culpability blindness," "inflexible independence," and "inertia," creating an environment more conducive to individual drivers needed for change, like "personal accountability" and "openness."

7.5 Conclusions

This chapter addresses a major challenge for humanitarian practice—facilitating change adoption—by identifying human perceptions and behaviors that promote and impede change adoption. These outcomes augment the more usual focus on the predicted quality of the design of the proposed changes (e.g. the introduction of new technology). Congruent with the values of autonomy and independence common to humanitarian organizational culture, individual barriers and drivers were key influencers of change, with collective factors playing catalytic or amplifying roles. Specifically, we found that (1) individual barriers afford a stronger overall influence for advancing or inhibiting change in the humanitarian logistics team's practices and that (2) collective factors establish an environment that can either catalyze or prohibit individual change drivers.

Further, in reflecting upon the interplay of factors as a whole, it was instructive to observe the subtle but powerful roles that "peer permission" and "positive experience" played in influencing change within the reversed structure of a humanitarian organization. Peer permission acts as the primary inhibitor of change for on-the-ground decision-makers. Even if one individual has the drivers "openness" and "vision," for example, that person cannot change his or her practice if that practice is collaborative and requires the buy-in of teammates: individual stakeholders hold the power to block change in collaborative tasks by withholding permission. To resolve this barrier, "peer permission" can be circumvented by "empowerment," eliminating the collaborative requirement. Or resistance to change can be dissolved by the powerful driver "positive experience," which occurs when individuals support the mission of the organization by meeting the needs of the field. When the driver "positive experience" is present, those individuals adopt changes in both autonomous and collaborative work to continue logistics work that meets the needs of the field. Thus, when "peer permission" is granted, it is invisible. Only when "peer permission" operates as a barrier does its effect come to the forefront. We want to be clear that we do not believe this suggests a need for more top-down hierarchy in humanitarian logistics teams, nor in nonprofit organizations more broadly. As discussed in the literature review, strict hierarchical structures are incongruent with core values of humanitarian organizational culture such as lowest-level empowerment (Chambers 1997; Lindenberg and Bryant 2001). The power of the driver "positive experience" is congruent with mission being the highest priority and measure of success for humanitarian work, including logistics work. This dynamic highlights "peer permission" and "positive experience" as important ways that humanitarian organizations operate to maintain empowerment and influence at the field level—a dynamic with valuable implications for managing change in humanitarian logistics.

In this case study, participants consistently linked effective practice with relationship building and mission support. When the team failed to support the mission, individual team members saw a need to change. When some participants could see mission success after adopting change, those changes caught on:

> People are very dedicated, so they'd change to do more good. If they felt the change would do more good, they'd do it. They are extraordinarily dedicated.

One of the key positive influences of "mission enforcement" is directly challenging "culpability blindness." The participants were highly committed to mission, and when the barrier "culpability blindness" began to break down and be replaced with the. driver "personal responsibility," people were more likely to develop further drivers like "openness," and "peer permission" did not operate as a barrier. Peer permission became effectively invisible because team members were collaborating to adopt change where they believed it held promise for supporting mission. This is not to say that "mission enforcement" alone can enforce change adoption in humanitarian organizations. The logistics team we studied had conditions that allowed the barriers "culpability blindness" and "inflexible independence" to flourish, even when those barriers impeded the ability of the logistics team to meet needs of the field:

> The team was suffering from that… No one wanted to say that that was a problem, that enforcement was an issue.

When the driver "mission enforcement" entered the picture, it was immediately after a problematic response to the Haiti earthquake: a context when team members had refused to adopt change, did not collectively "meet needs," and had a lack of "positive experience." "Mission enforcement" served as the tipping point factor that provided an impetus to change both individually and collaboratively.

There are several practical implications of these findings that could be relevant to managers seeking to support change adoption within the unique organizational structure and culture surrounding humanitarian logistics. First, these findings are congruent with Bratton's strategies for motivating organizational change (Kim and Malborgne 2003), suggesting that organizations can cultivate drivers within individual team members through training, for example to instill a shared vision of the desired future as an important early step in creating an environment conducive to change. In this case study, we saw that when individuals had a clear "vision," those individuals sought additional drivers of change, such as "empowerment," and moved forward with change adoption. Training, such as simulations, offers a place to create a clear shared "vision" and "positive experience" (individual drivers of change), as well as to mitigate "insecurities" regarding whether change will stick (insecurity being an individual barrier). Another implication of this research for team leaders is to consider their endogenous tools for effecting change—particularly in considering how their influence can create collective change drivers affecting individual team members. For example, in one area we saw the team manager's provision of "empowerment" serve as an effective way to move change forward by allowing a team member with the individual drivers "openness" and "vision" to transform the barrier of "peer permission." When this change adoption subsequently created the driver "positive experience," change adoption spread throughout the team. Finally, when endogenous approaches to creating change drivers in the logistics team are exhausted, the larger organization may be able to help dissipate individual barriers and encourage change via the drivers "meeting needs" and "mission enforcement"—i.e., by making more visible the specific mission successes and shortfalls.

Another practical implication of this research is how to apply accountability in a way that is culturally appropriate for humanitarian organizations. Autonomy and independence are important values in humanitarian organizations, so directive, top-down approaches to enforcing accountability could be antithetical to those values and unlikely to cultivate change drivers within team members. One important aspect of culturally appropriate accountability is participatory practices. In the organization we studied, logistics team members had significant input into the tools, standards, and processes that structured their work practices. This participatory approach is congruent with the organizational culture and was well received among team members. However, as this case study showed, participatory development of logistics standards and processes alone is insufficient to ensure that stakeholders will follow those standards and processes. Therefore, a second culturally appropriate practice is necessary: applying creative accountability strategies to cultivate individual change drivers by *making more visible the linkages between people's practices and mission outcomes*. These creative strategies may include self-driven approaches: e.g., having team members develop self-identified goals for their operational performance and tracking outcomes relevant to those goals. By extending existing literature on humanitarian logistics and contextualizing humanitarian logistics within the broader organizational culture, our research illustrates the importance of balancing cultural values such as autonomy with the operational accountability necessary to promote change adoption in ways that help the organization and the individuals comprising the logistics team to meet their humanitarian mission.

Stepping back, we see the important general conclusion that our usual focus on the changes themselves–the logistics intervention and the evidence for improved performance and results–must be augmented with an understanding of the individual, organizational, and cultural change management factors. If these are ignored, we will end up with carefully designed logistics "improvements" that never are adopted and therefore have no impact.

The implications for future research are significant. There is a complex interplay of change drivers and change barriers, both within and across the individual and collective humans at the center of change, and this complexity is increased by the multi-faceted and dynamic nature of the priority action items of change themselves. This argues that our long-term research goal is not the compilation of change management generalizations that will be applied "one size fits all" across all humanitarian logistics teams—a science of change management for humanitarian logistics. Rather, we must acknowledge and focus on the central role of humans operating in highly decentralized and dynamic situations and not let virtual, highly distributed work mask the complexity of the case of interest. As Nardi et al. (2002) put it,

> The rhetoric of virtuality involves a process of 'deletion' in which real people are 'deleted' as work is described as invisibly distributed in a 'system,' and intelligence is no where in particular, certainly not in the minds and hands of specific workers. (p. 209)

But of course, that is exactly where the intelligence behind change resides, both to advance change and to hinder it. Our research stance in cases like these should be more that of a designer than a scientist. We must learn how best to analyze the realities of change, through ethnographic approaches and other methods, and how to make useful sense of that analysis–useful in that it increases our ability to design systems and environments that enable humans to identify, create, and manage desired change among a wider variety of effective work practice models.

References

Behn RD (1995) Creating an innovative organization: ten hints for involving frontline workers. State Local Gov Rev 27(3):221–234

Benini A, Conley C (2007) Rapid humanitarian assessments and rationality: a value-of- information study from Iraq, 2003–2004. Disasters 31(1):29–48

Benini A, Conley C, Dittemore B, Waksman Z (2006) Survivor needs or logistical convenience? Factors shaping decisions to deliver relief to earthquake-affected communities, Pakistan 2005–2006. Version 30. Vietnam Veterans of America Foundation, Information Management and Mine Action Programs, Washington DC

Berenguer-Falguera G (2012) A new approach in supply chain design: Studies in reverse logistics and nonprofit settings (Doctoral dissertation). Retrieved from ProQuest Dissertations and Theses. (Accession Order No. AAT 3526515)

Bridges W (2009) Managing transitions: making the most of change. Perseus Books, Philadelphia

Button G (ed) (1991) Studies of work in human-computer interaction. Cambridge University Press, New York

Chambers R (1997) Whose reality counts? Putting the first last. Intermediate Technology Publications, London
Collins J (2001) Good to great. HarperCollins, New York
Drucker P (2009) Managing the nonprofit organizations: practices and principles. Collins Business, New York
Hoffman AJ (2010) Climate change as a cultural and behavioral issue. Organ Dyn 39(4):295–305
Holguín-Veras J, Jaller M, Wachtendorf T (2012) Comparative performance of alternative humanitarian logistic structures after the Port-au-Prince earthquake: ACEs, PIEs, and CANs. Transp Res Part A: Policy Pract 46(10):1623–1640
Holguín-Veras J, Perez N, Jaller M, Wassenhove L, Aros-Vera F (2013) On the appropriate objective function for post-disaster humanitarian logistics models. J Oper Manage 31(5):262–280
Kaplan RS (2001) Strategic performance measurement and management in nonprofit organizations. Nonprofit Manage Leadersh 11(3):353–370
Kaplan RS, Norton DP (1996) The balanced scorecard: translating strategy into action. Harvard Business School Press, Boston
Kim WC, Mauborgne R (2003, April) Tipping point leadership. Harv Bus Rev. http://inet.katz.pitt.edu/studentnet/projects/ceestaging/upmc2012/Tipping%20Point%20Leadership.pdf
Lindenberg M, Bryant C (2001) Going global: transforming relief and development NGOs. Kumarian Press, Bloomfield
Long D, Wood D (1995) The logistics of famine relief. J Bus Logist 16(1):213–229
Maiers C, Reynolds M, Haselkorn M (2005) Challenges to effective information and communication systems in humanitarian relief organizations. Proceedings of the 2005 IEEE International Professional Communication Conference
Mays R, Racadio R, Gugerty M (2012) Competing constraints: the operational mismatch between business logistics and humanitarian effectiveness. Proceedings of the 2012 IEEE Global Humanitarian Technology Conference
Mays R, Walton R, Lemos M, Haselkorn M (2014) Valuing what works: success factors in disaster preparedness. An independent analysis of Red Cross/Red Crescent practitioner needs. http://preparecenter.org/resources/valuing-what-works-success-factors-disaster-preparedness
Nardi BA, Whittaker S, Schwarz H (2002) NetWORKers and their activity in intensional networks. Comput Support Coop Work 11:205–242
Oloruntoba R, Gray R (2006) Humanitarian aid, an agile supply chain? Supply Chain Manage 11(2):115–120
Ontko M, Williamson S, Haselkorn M, Kemp R (2007) An examination of the effectiveness of lessons-learned reporting within the humanitarian sector. J Inform Technol Soc Change, Spring 28–48
Ponnaiyan S (2013) Supply chain network planning for humanitarian operations during seasonal disasters (Doctoral dissertation). Retrieved from ProQuest Dissertations and Theses. (Accession Order No. AAT 3579232)
Rogers EM (2003) Diffusion of innovations, 5th edn. The Free Press, New York
Ryan WP (2002) The new landscape for nonprofits. In: Futter V (ed) Nonprofit governance and management. American Bar Association, Chicago
Sheehan R (1996) Mission accomplishment as philanthropic organization effectiveness: key findings from the excellence in philanthropy project. Nonprofit Volunt Sect Q 25:110–123
Singhal A, Shirley S, Marston EH (2010) Turning diffusion of innovation paradigm on its head: the positive deviance approach to social change. In: Vishwanath A, Barnett G (eds) Advances in the study of the diffusion of innovation. Peter Lang, New York
The ICRC code of conduct: Humanitarian principles in practice (2004, Sept 20) http://www.icrc.org/eng/resources/documents/misc/64zahh.htm
Tomasini RM, Van Wassenhove LN (2009) From preparedness to partnerships: case study research on humanitarian logistics. Int Trans Oper Res 16(5):549–559
Torabi SA, Aghabegloo M, Meisami A (2013a, May) Performance measurement in humanitarian relief chains: a combined Fuzzy DEMATEL-ANP approach. In Proceedings of the 24th annual conference of the Production and Operations Management Society, pp 1–10

Torabi SA, Baghersad M, Meisami A (2013b, May) Emergency relief routing and temporary depots location Problem with considering roads restoration. In Proceedings of the 24th annual conference of the Production and Operations Management Society, pp 1–10

Van Wassenhove LN (2006) Humanitarian aid logistics: supply chain management in high gear. J Oper Res Soc 57(5):475–489

Walton R, Mays R, Haselkorn M (2013) What makes response rapid? Humanitarian practitioners' views on speed in dynamic and uncertain logistics environments. In: Hellingrath B, Link D, Widera A (eds) Managing humanitarian supply chains: strategies, practices and research. BVL International/DVV Media Group, Bremen, pp 222–243

Chapter 8
Information Visibility in Humanitarian Operations: Current State-of-the-Art

Natalie Privett

Abstract Humanitarian operations can be greatly improved through increased supply chain visibility, that is, availability of information throughout the supply chain. This is broadly true for all types of humanitarian operations, whether disaster relief, global health efforts, capacity building, or community development. This chapter establishes basic supply chain visibility needs in humanitarian contexts and explores current state-of-the-art technologies and applications employed to address these. Identified visibility needs for humanitarian operations include tracking of location, tracking of inventory levels, temperature monitoring, tracing product information, information sharing, and decision making support. Through review of current state-of-the-art technologies and applications aimed at increasing visibility in humanitarian operations, this chapter demonstrates that they are indeed varied in approach and complexity; while each tackle different needs, no one solution satisfies all. Thus, multifaceted and disjoined efforts must be employed to achieve even low levels of visibility today. This chapter concludes with identification of persisting gaps in current state-of-the-art technologies and applications, which can direct future efforts of researchers, developers and users.

Keywords Technology · Visibility · Cold chain · Information · Mobile phone projects · Satellite technology · Tracking · Tracing

8.1 Introduction

Humanitarian supply chains function in the most challenging environments, and, consequently, information and visibility are obscured more than typical supply chains. Supply chain visibility is the clarity and accessibility of information and data throughout the supply chain, from manufacturer to final recipient. For example, information regarding products in transit, such as location and movement, is

---

N. Privett (✉)
Robert F. Wagner Graduate School of Public Service, New York University, New York, NY, USA
e-mail: natalie.privett@nyu.edu

C. W. Zobel et al. (eds.), *Advances in Managing Humanitarian Operations*,
International Series in Operations Research & Management Science 235,
DOI 10.1007/978-3-319-24418-1_8

part of supply chain visibility, as is data on availability and stock levels of products at storage and dispensation points. Presumably, supply chain and operational decisions– from ordering to distribution –depend on this information presented by supply chain visibility. However, in humanitarian settings, such data and information is often scarce or inadequate. Consequently, increased visibility has the potential to greatly improve humanitarian operations. Supply chain visibility can provide actual data to fuel more effective and accurate decisions, enable evidence-based interventions, expose issues for effective remedy, and increase accountability. This chapter establishes fundamental humanitarian supply chain visibility needs, explores current state-of-the-art technologies employed to improve visibility in humanitarian supply chains, and draws conclusions about the remaining gaps through comparison of the presented technologies. As such, this research assembles and evaluates current needs and efforts in humanitarian supply chain visibility, which can inform current research and practice of obtainable information and directs developers to existing gaps.

8.2 Methodology

Humanitarian supply chain visibility needs were collected through 22 semi-structured interviews and a review of relevant literature (including white papers, case studies, organizational newsletters, etc.). Interviewees were selected based on their expertise, experience, and role in the supply chain. Together the interviewees span the chain from manufacturer to nongovernmental organizations to the UN and country governments. Literature was reviewed from a variety of sources including academic articles, white papers, trade publications, technical reports, and news reports. This review not only informed the humanitarian supply chain visibility needs, but also provided details on pilots and implementations of various technologies.

In this chapter, humanitarian supply chain visibility needs are briefly discussed to provide context for the reader to understand the impact and potential impact of described technologies and projects. The state-of-the-art of humanitarian visibility technology is then described by reviewing technologies and pilot projects currently and recently applied in the humanitarian sector. Finally, a comparison of the technologies, specifically a comparison of strengths, weaknesses, and addressed needs, informs conclusions.

8.3 Humanitarian Supply Chain Visibility Needs

Humanitarian supply chain visibility needs are what the technologies described in this chapter attempt to address, at least in part. As such, outlining and understanding these needs provide context by which we can compare and understand the impacts of the described technologies and projects. In fact, it is by mapping the technologies back onto these needs that we identify the key remaining weaknesses and gaps in humanitarian supply chain visibility technology (See Conclusion).

Fig. 8.1 Humanitarian visibility needs

HUMANITARIAN VISIBILITY NEEDS

- Robust to Contextual Challenges
- Tracking of Location
- Tracking of Inventory Levels
- Temperature Monitoring
- Tracing of Product Information
- Information Sharing
- Decision Making Support

Humanitarian visibility needs encompass needs faced in accelerated disaster response settings as well as prolonged crises and development situations. Any solution that attempts to address these needs must first be robust to the unique contextual challenges of humanitarian settings. Beyond this, humanitarian visibility needs can be generally described as tracking of location and inventory, temperature monitoring, tracing of product information, information sharing, and decision making support. (See Methodology section on how these were determined.) These are shown in Fig. 8.1 and each is further expounded upon in the following subsections. These supply chain visibility needs are unique compared to commercial supply chain needs not only because humanitarian organizations often work in more physically challenging environments. Humanitarian supply chains most often involve many stakeholders with distinct objectives, systems and operations, and a decentralized system that complicates operations, communication, and decision making. Responsiveness and speed are also much more of a priority for humanitarian supply chains.

Robust to Contextual Challenges

The lack of visibility in humanitarian settings is partially due to the unique contextual challenges of these settings, including damaged, weak, or non-existing infrastructure (e.g., roads, electricity, and mobile networks), rurality or remoteness of operations, lack of human resources, and environmental conditions. These contextual challenges are especially relevant considering that the developing world experiences the brunt of slow onset disasters both natural (e.g., famine and drought) and man-made (e.g., political crises and refugee situations). While natural disasters do not strike the developing world more frequently, they do distress more victims in such places. These unique features and challenges must be accommodated by humanitarian supply chain operations. As such, solutions must limit their dependency on infrastructure and human resources while remaining robust even in the most remote of locations. As Sandwell (2011) quotes an interviewee, "'I'd like to go fully electronic but until electricity is available around the world 24/7 [disaster often bring down any available power] and people are fully trained this isn't going to happen any time soon.'"

Tracking of Location

Tracking of location is a broad category encompassing both products and vehicles. Product can be tracked at different levels, by container, pallet, box, package, or unit of use (i.e., individual item level). Such tracking can provide shipment visibility to the system, greatly aiding in supply chain management and ordering decisions while helping the front-lines anticipate and prepare for receiving goods. A survey by the Fritz Institute following the 2004 tsumani found that only 26 % of surveyed logisticians had access to track-and-trace software. "As a consequence, supply lines were clogged, huge amounts of goods piled up at airports and transport and warehousing were inadequate" (Thomas 2005).

Furthermore, tracking of product and vehicle location can help to increase shipment security, improve personnel safety, bolster accountability and reduce loss and theft. Lack of fleet visibility has been cited as a key challenge by international humanitarian organizations (Besiou et al. 2011), particularly considering that altogether the collective global humanitarian fleet includes an estimated 70,000 units (Brophy-Williams et al. 2012). Furthermore, "transportation is the greatest single occupational health and safety risk faced by field staff," which only further emphasizes this need (Brophy-Williams et al. 2012).

Tracking of Inventory Levels

Tracking of stock levels can greatly improve decision making in humanitarian supply chains, especially as inventory discrepancies are a persistent problem in these supply chains. Inventory information can be used to better manage donations, procurement, warehousing and allocation resulting in increased efficiency, effectiveness, transparency and donor accountability. However, this level of control is often in tension with the responsiveness and speed required of these organizations. Many (though not all) solutions aimed at tracking item and/or vehicle location, though, have the added potential benefit of monitoring inventory levels.

Temperature Monitoring

While many types of products move along humanitarian supply chains, specific types of pharmaceutical products present unique challenges. Pharmaceutical products can be classified by their sensitivity to storage and transport conditions as follows: cold chain (e.g., vaccines), ambient temperature, moisture sensitive (e.g., tablets), and regulated active ingredient products (WHO et al. 2007). Wastage of pharmaceutical products typically occurs because of heat exposure, exposure to freezing temperatures, or expiration. Such exposure results in wastage and consequential monetary loss as well as high risk to patients as temperature exposure can reduce or destroy pharmaceutical and vaccine efficacy (Department of Vaccines and Biologicals of the WHO 2005). The root causes of such exposure and consequential wastage can include improper cold chain procedure, inadequate cold chain storage, mishandling, climate, or a combination of these. Climate in many developing and disaster prone regions is often so harsh that it can affect both the quality and transportation of goods. For example, harsh heat can melt packaging and crack tablets, rendering goods useless. Severe cold can freeze vaccines, significantly damaging their efficacy. Condensation caused by extreme temperature changes can compromise tablets and paper packaging forcing the products to be discarded.

Tracing of Product Information

Tracing product information is a crucial and pressing need. Information that may be desirable to trace includes product ingredients, origin, manufacturing site, history, batch number, specification, expiration date, and associated program or organization. Currently, predominant uses of tracing product information are in fighting counterfeit pharmaceuticals and distributing regulated drugs.

The rise of counterfeit pharmaceuticals on the market poses increasing risks to both consumers and company reputations, and has become a significant global problem for health and pharmaceutical products. Tracing various points of product information has been the main strategy against ensuring that deliveries are not counterfeit. Tracing specific information can essentially create and track product serialization as an anti-counterfeit measure. As such, there is wide interest and need for tracing product information even in humanitarian supply chains.

Another need for tracing product information is to facilitate distribution of regulated products and associated legal compliance. Narcotics and other regulated active ingredient products, such as codeine, require strict control of distribution. Furthermore, regulations and control of these substances vary between countries, creating transportation and logistics headaches when crossing borders. As such, tracking the active ingredients of a shipment can help comply with legal requirements and provide necessary control.

Information Sharing

Lack of information availability and sharing lends to a fragmented and uncoordinated system. In fact, lack of coordination of the current aid delivery system was cited as critical in a recent survey of global health and humanitarian stakeholders (Privett and Gonsalvez 2014). Moreover, an uncoordinated system is a complex system, encumbering and hindering the management and distribution of aid. Yet common, shared information and knowledge are necessary to develop coordination and "enable rapid adaptation to what is, by definition, a turbulent and uncertain environment (Stephenson 2005)."

Information sharing facilitates collaboration and partnership, such as that cited by Rietjens et al. (2009) between the military and humanitarian organizations in Afghanistan. It also enables system-wide situational assessment and appropriate response organization. Furthermore, information sharing leads to knowledge creation, learning, and adaptation (Kapucu 2008; Celik and Corbacioglu 2010). In fact, in a comparative analysis of response to the 1999 Turkey earthquakes, Celik and Corbacioglu (2010) find that "difficulties in accessing and exchanging timely and accurate … information inhibited coordination during the Marmara [earthquake] response while increased communication … improved coordination and search-and-rescue activities during the Duzce response."

Decision Making Support

As Besiou et al. (2011) state, "decision making in humanitarian operations is usually based on intuition and experience which are not always sufficient to fully understand the global impact of these decisions." Visibility equips such organizations to make informed decisions. However, information alone is often insufficient support for decision making. According to Thomas and Kopczak (2005), the main challenges that impede effective decision making in the humanitarian sector include inadequate technology use and lack of trained staff, issues that are intertwined and incredibly complex in the humanitarian sector. As Maiers et al. (2008) explore in detail, such information and communication systems involve "not only IT but also people, practices policies and organizational environments." In accordance, Blecken and Hellingrath (2008) offer information technology and software as a possible solution to improve visibility, responsiveness, monitoring, and reporting.

8.4 Current Visibility Projects and Technologies

This section explores the current state-of-the-art projects and technologies being used to improve visibility in humanitarian supply chains as organized by their technology. Though many of these have implications and functionality beyond supply chain visibility, these were chosen and reviewed based on their impact particularly on visibility. In cases where several technologies overlap significantly in functionality, use, and impact, only one is described in detail though others may be referenced. They are summarized in Table 8.1 in terms of the visibility needs discussed

Table 8.1 Visibility needs and technology summary table

		Tracking of location	Tracking of fleet vehicles	Tracking of inventory levels	Temperature monitoring	Tracing of product information	Information sharing	Decision making support
Mature	Barcode	✓		✓		✓		
	RFID	✓		✓		✓		
	GPS	✓	✓					
Mobile phone	SMS for Life			✓			✓	✓
	Humanitarian Data Toolkit							✓
	Open Data Kit (ODK)						✓	✓
	UNICEF Rapid SMS			✓			✓	✓
	Save the Children "McAID"			✓			✓	
	INFOSAID						✓	✓
	Mapping with Mobiles						✓	✓
Satellite-based	Satellite Imagery							✓
	ARGOS Early Warning Sys	✓		✓			✓	✓
	HumaNav		✓				✓	✓
	BAT-340		✓				✓	✓
Temperature	Vaccine Vial Monitors				✓			
	Freeze Indicators				✓			
	Data Logger Tags				✓			
	USB Data Loggers				✓			
	Optimize Albania				✓		✓	
Software	HELIOS						✓	✓
	LLS/SUMA						✓	✓
	Sahana Eden						✓	✓
	Resource Map						✓	✓

Table 8.2 Website references for technologies and projects

	Technology or project	Website
Mobile Phone	SMS for Life	http://malaria.novartis.com/innovation/sms-for-life
	Humanitarian Data Toolkit	https://innovation.internews.org/pilots/introducing-humanitarian-data-toolkit-components-and-workflows
	Open Data Kit (ODK)	opendatakit.org
	UNICEF Rapid SMS	www.rapidsms.org
	Save the Children "McAID"	*McAID.info*
	INFOSAID	http://www.cdacnetwork.org/i/20140728102420-genh0
	Nokia Data Gathering	www.indt.org/projects/nokia-data-gathering/?lang=en
Satellite	ARGOS Early Warning Sys	http://www.argos-project.eu/
	HumaNav	http://www.novacom-services.com/en/humanav
	BAT-340	http://www.globaltrack.com/bat340.htm
Temperature	Vaccine Vial Monitors	http://www.path.org/projects/vaccine_vial_monitor.php
	Freeze Indicators	http://solutions.3m.com/wps/portal/3M/en_US/Microbiology/FoodSafety/product-information/product-catalog/?PC_Z7_RJH9U523003DC-023S7P92O3O87000000_nid=71PHMRZCJCbeX2JZNTZSLTgl
	LogTag Recorders Data Logger Tag	http://www.logtagrecorders.com/
	VaxAlert™ Data Logger Tag	http://www.sensitech.com/products/vaxalert
	Q-Tag®2 plus Data Logger Tag	http://www.berlinger.ch/en/berlinger/main/ambient-tag/temperature-monitoring/q-tag-2-plus.html
	Optimize Albania	http://www.who.int/immunization/programmes_systems/supply_chain/optimize/albania/en/
Software	HELIOS	http://www.helios-foundation.org/
	LLS/SUMA	http://www.lssweb.net/
	Sahana Eden	http://eden.sahanafoundation.org/
	Resource Map	http://resourcemap.instedd.org/en

in the previous section. Table 8.2 provides website references for technologies and projects.

Mature Technologies

All companies have limited visibility into their supply chains, especially as these chains spread globally. Several mature technologies exist that continue to dominate tracking and tracing in these chains, namely *barcodes, radio frequency identification* (*RFID*) and *global positioning systems* (*GPS*). These mature technologies are often already implemented by manufacturers and suppliers. Thus, humanitarian supply chains can take advantage of existing tags prepared by manufacturers and thus limit investment. This may involve collaboration efforts with suppliers to leverage the technology and traced information to the greatest extent possible.

Barcode (1D and 2D)

Both 1D and 2D *barcodes* are optical machine-readable representations of data about the item to which they are attached. The difference between the two is that 2D *barcodes* have a notably higher data capacity. However, information cannot be updated or changed. The fact that barcodes physically attach to products prevents separation and enables individual product level tracking. However, if the *barcode* is attached to packaging, separation may still occur and individual item tracking is not truly achieved. Readers are necessary to read the data. Software can also enable reading capabilities on other devices (e.g., PDAs, smartphones), making barcodes a quite flexible solution.

Radio Frequency Identification (RFID)

Like barcodes, *RFID* tags attach to products. The tags consist of a microchip and antenna. In-place readers are also required, which emit, receive, and convert electromagnetic waves to communicate with the tag. Thus, *RFID* tags do not have to pass by a physical reader like barcodes. Instead, they can be read via a broadcast signal whenever they are within a reader's coverage area. There are two main types of RFID tags: passive tags draw power from a field created by a reader, while active tags can use a battery and also broadcast a signal. Information on active tags can be updated and changed, an important difference. Depending on tag cost, individual products can be tagged like barcodes. While the uses of *RFID* technology to track and trace products are many, one application area is in fighting pharmaceutical counterfeits, a considerable and growing problem. In fact, Nigeria, with a current counterfeit drug rate of approximately 35 %, is adopting an *RFID* system for just this purpose (Swedberg 2010).

Global Positioning System (GPS)

The *GPS* receiver is composed of an antenna, receivers, processors, and a clock or oscillator. It uses in-place standard satellite positioning service, that is, already in-place capital. *GPS* typically reports time, location, and relative movement. Strengths of *GPS* in the humanitarian context are that it is relatively independent of infrastructure and able to operate even in the most remote and affected regions.

GPS technology is mainly used for vehicle tracking, though other uses are clearly possibly. For instance, Kapucu (2008) provides an example of *GPS* assisting with evacuation assessment in hurricanes. *GPS* can also be used in collaboration with geographic information systems (GIS), which manage, analyze, manipulate, and visually display location data. Though the addition of GIS is a great enhancement to *GPS* data, it creates more dependency on expertise and human resources. For this reason, only approximately 23 % of respondents had both GIS and *GPS* during the Turkey earthquakes (Celik and Corbacioglu 2010).

Mobile Phone Projects and Technologies

Mobile technology and connectivity are booming. Worldwide, over 93 out of every 100 people—more than 5 billion—are mobile phone subscribers (International Telecommunication Union 2014). In developing countries, mobile phone subscribers are nearly 88 per 100 or more than 6 billion (International Telecommunication Union 2014). Furthermore, a 2011 WHO survey found that 83 % of 112 responding countries are using mobile phones for healthcare (World Health Organization 2011). Leveraging this trend, near real-time data collection and decision support can be significantly boosted, even in remote regions, with limited dependence on infrastructure. However, stakeholders—providers, developers, and donors—"must work together to build integrated, interoperable systems" (Morris 2009). Several have emerged in response. The following are example approaches piloted in developing countries using the common, everyday technology of mobile phones to improve humanitarian supply chains. They all rely on reporting and require training.

SMS for Life

SMS for Life, a public-private collaboration between Roll Back Malaria, Novartis, Google, IBM, Vodaphone, and African cell phone providers, was a 21-week pilot project from 2009 to 2010 involving 129 health facilities servicing 1.2 million people in Tanzania (Barrington et al. 2010a). In this project, health workers send messages containing stock level information. A central collection website allows stock level reports in on-line, paper, and map display formats. Secure access to the website is controlled by user IDs and passwords allocated to pre-specified groups (Barrington et al. 2010b).

This project was intended to collect accurate stockout and inventory information at health facilities and it did provide reliable weekly data. Data accuracy was improved, achieving 94 % in the pilot (Barrington et al. 2010a). Aims also included increasing availability and reducing stockouts. Approximately 300,000 additional people had access by the end of pilot (Barrington et al. 2010b). This scalable solution requires training, personal cell phones, and a collaboration structure between multiple public and private partners. Furthermore, reliance on self-reporting can affect data accuracy.

Humanitarian Data Toolkit

Piloted in 2013, the *Humanitarian Data Toolkit* is a project of the Internews Center for Innovation and Learning in collaboration with Modi Research Labs, Columbia University and Captricity that aims to enable local research capacity for data collection in order to facilitate timely needs assessments during disasters (European Journalism Centre 2014) The toolkit is a portable, rugged box that includes "all the necessary equipment and software to conduct a needs assessment:" a laptop that acts as the server and host, a portable scanner, mobile smartphones, a foldable solar panel, power strips, USB ports, Ethernet cables, paper surveys, mobile phone survey software, survey data aggregation software, paper survey digitizing software, data analysis software, and training and support materials (Susman-Pena 2013).

The pilot projects in the Dadaab, Kenya refugee camps in 2013 explored feasibility and impact with very positive results. Overall, the toolkit worked well with only minor technical issues. For example, the phone keyboards were small, making them difficult to use, and the GPS capabilities were found to be prohibitively slow. Furthermore, real-time data transmission from the field was found to be not worthwhile, mainly due to its drain on field mobile phone batteries and the fact that data analysis can only occur when the data set is complete (Susman-Pena 2013). A main objective of the *Humanitarian Data Toolkit* is sustainability and scalability. Iterative design continues and Internews is looking to NGOs to test and improve the toolkit while helping with adoption.

Open Data Kit

Beginning in 2008 as a google.org project, *Open Data Kit* (*ODK*) is an open-source, modular toolkit that enables both collection and delivery of data and information (Hartung et al. 2010). It is now run out of the University of Washington with the goal of empowering users to build information services for developing regions (OpenDataKit 2015A). The now seven tools include (1) Collect, a mobile data collection platform, (2) Aggregate, a data storage support tool, (3) Build, a drag-and-drop form generator, (4) Form Uploader, (5) Briefcase, a data transfer module, (6) Validate, a tool to ensure compliance and functionality of created forms, and (7) XLS2XForm, a tool allowing Excel designed forms (Hartung et al. 2010; OpenDataKit 2015B). One innovation of *ODK* over other listed data collection technologies is that the Collect module supports a variety of data types, including text, location, images, audio, video and barcodes (Hartung et al. 2010).

The International Federation of Red Cross and Red Crescent Societies has been working with local societies to deploy *ODK* for mobile data collection to facilitate aid distribution in countries including Panama, Haiti, Peru, Guatemala and the Philippines. As IFRC innovation delegate Alberto Cabrera stated in an interview in Panama, "Without a doubt these tools will help the PRC further support local and national authorities by facilitating data collection and ensuring a better selection of beneficiaries (Guevara 2014)." In another example, New York City with InSTEDD is using *ODK* to gather field data to aid New York in emergency preparedness and response efforts (Anokwa 2011).

UNICEF Rapid SMS

Rapid SMS was developed by UNICEF's Innovation division with Dimagi and the Open Mobile Consortium (Chester 2010). It is an open-sourced, customizable framework that manages real-time data collection, presentation, communication, and coordination using mobile phones. More specifically, the project was intended to facilitate data collection, monitor reporting, present real-time data, analyze data, enable exportation, generate custom reports, visualize, communicate and coordinate (RapidSMS 2008).

Customizable SMS forms allow collection of both quantitative and qualitative data and the open-source code is the customizable base for a suite of different products. The Web Interface allows multiple users around the world to simultaneously access the data and SMS users. This is where information can be edited, exported to Excel, and graphically displayed. Rapid SMS has been piloted and/or implemented in Malawi, Ethiopia, Senegal, Nigeria, Somalia, and Kenya (RapidSMS 2011). Initial results reveal a weakness of this solution in that it is not necessarily coupled with reactive operations (Conesa and Kopczak 2009).

Save the Children "McAID"

Save the Children USA implemented its development assistance program, Jibon o Jibika ("Life and Livelihood"), in collaboration with international and local NGO partners. The program's aim is to reduce the high levels of food insecurity and malnutrition in southwest Bangladesh (USAID et al. 2009). The objectives also included facilitating quick and easy look-up of beneficiary information, improving data accuracy, real-time information availability, nutritional status tracking, and facilitating donor reporting.

As part of this program, an integrated information management system, "*McAid*," was developed for the Maternal and Child Aid Program and became fully functional in January 2009. In this, distribution workers carry portable, long life personal digital assistants (PDAs) containing a beneficiary information database and distribution plans. When a distribution is made and a child weighed or vaccinated, the patient's individual bar code ration card is scanned and information is entered into the database (Conesa and Kopczak 2009). This barcode scanning reduces data entry and reporting burdens and achieves high data accuracy. However, PDA devices must be rugged. An integrated PDA application facilitates data collection, tracking, tracing, and warehouse inventory reconciliation. Extensive training is required for PDA users. Finally, the *McAid* software auto-synchronizes to transfer data from the field to a central repository using Wi-Fi, USB, and broadband connectivity (Danton and Stevenson 2010).

Infosaid

INFOSAID was a set of 2010–2012 pilot projects aimed to enhance communication between responders and affected populations, thus, maximizing timely and accurate information. A centralized hub allowed for both dissemination and collection of information between community focal points and aid agencies. It was enabled via FrontlineSMS and FrontlineForms, which allowed for customized data collection and increased data accuracy and which can be exported into Excel for analysis (FrontlineSMS Web Administrator 2009). The usefulness and impact of *INFOSAID*, though, still largely depended on human resources and technical capacities to collect, verify and analyze the incoming data (Chapelier and Shah 2013).

Evidence from the pilot projects based in Kenya and Somalia during drought suggests that information dissemination increased in both speed and frequency and food aid operations were noticeably improved in Kenya. One example highlighted the use of mobile phones to inform staff of village raids and consequent displacement. Staff was then able to respond by redirecting food aid to a safe distribution point (Chapelier and Shah 2013). The main challenges of the pilot were technical in nature, such as issues with mobile phone charging and solar chargers, as well as lack of measurement and evaluation support, a common issue in humanitarian settings.

Nokia Data Gathering

The *Nokia Data Gathering* (*NDG*) system was originally piloted in 2007 for use during Brazil's dengue epidemic. The Humanitarian Information Facilitation Center (HIFC) more recently piloted the use of mobile phones utilizing *Nokia Data Gathering* software in Zimbabwe, during the country's early recovery and rebuilding phase following several emergencies. The *NDG* software not only enables mobile phones for the collection of survey data traditionally collected by paper-based surveys, but it also enables the near real-time transmission and analysis of this data (International Media Support 2012a). The intention is to establish a communication channel between disaster-affected people and humanitarian decision makers, improving needs assessment and humanitarian aid overall (International Media Support 2012b).

Electrical and network infrastructure remain the main technical challenges of the pilot. According to International Media Support (2012a), less than half of the deployed data gatherers were able to send data from the field due to lack of network coverage. However, the phones can store data for transmission once connectivity is restored. Alternatively data can be extracted directly from the memory card (Humanitarian Information Facilitation Centre and International Media Support 2012). Mobile phone power and charging was another technological issue with mobile phones requiring charging after 3 days.

Satellite-based Technologies

A few approaches have employed satellite-based technologies to improve humanitarian supply chains. Though all of these technologies have potential to be effective despite remoteness and have limited dependency on in-country infrastructure, they also may involve satellite usage costs.

Satellite Imagery

The most basic use of satellite technologies for humanitarian relief and emergency response is satellite imagery and maps, which are often combined with GIS. In the event of a major disaster, the United Nations can invoke the International Charter on Space and Major Disasters to induce space agencies to release satellite data, which is made freely available through mechanisms such as Respond, Humanitarian Information Centres, Reliefweb, AlertNet and the UNOSAT web portal (Pisano 2005). Otherwise, acquiring satellite resources and paying satellite usage charges would be prohibitively costly. The European Space Agency continues to work to increase satellite capabilities, such as all-weather, day and night, and high spatial resolution imagery (European Space Agency 2013). These images are largely used for coordination and information gathering, and, for example were used widely as early as the 2004 tsunami and the 2005 Kashmir earthquake (Pisano 2005).

The *Respond Project*, an alliance begun in 2004 as part of the European Global Monitoring for Environment and Security (GMES) initiative, aims to improve access to maps, satellite images and geo-information in humanitarian crisis events. The Respond Atlas is their published portfolio of images that demonstrate the range of services provided (European Space Agency 2011). Respond, however, is not limited to rapid onset disasters, but also supports slow onset and refugee crises. In such cases, once an emergency is recognized, both archived and newly acquired satellite imagery are made available for comparison. Comparison of satellite images over time can highlight both short-term and long-term changes and damage to landscape, infrastructure, villages and movements of people.

There are many additional projects that aim to use satellite technology and imaging to track man-made humanitarian crises, such as the *Satellite Sentinel* project (2010–2012) with its mission to deter civil war in Sudan by combining satellite imagery, field reporting and mapping into a single platform (Wang et al. 2013) and the *Standby Task Force Satellite Imagery Project*, which combines satellite imagery and crowd sourcing (Standby Task Force 2014).

ARGOS Early Warning System

The *ARGOS Early Warning System* is the product of a joint collaboration between The Mentor Initiative and Comic Relief. This system was built to identify and respond to malaria epidemics quickly—"days and weeks instead of weeks and months

(The MENTOR Initiative 2009)" as well as collect real-time stock information. As part of this system, workers travel to a data collection station, which can be labor and time intensive. The station runs on 2 year batteries and is the price of a laptop computer. It uses graphic icons to collect information (Conesa and Kopczak 2009), namely the number of diagnosed and treated patients and stocks of essential drugs at the individual health facility level. Due to its visual nature it does not rely on literacy or language, which is a strength. The ARGOS platform, an in-place system, relays this to a satellite to a central computer, which automatically analyses and displays fully analyzed data on a secure website (Conesa and Kopczak 2009). Ministries of Health and partners are alerted during peaks of positive malaria diagnosis (The MENTOR Initiative 2009).

HumaNav

HumaNav is an initiative begun in 2008 by UNITAR's Operational Satellite Applications Programme (UNOSAT) in collaboration with Novacom Services. The basic intent was to use satellite vehicle tracking to ensure security and safety of fleet vehicles and personnel, increase utilization, efficiency, communication, road safety and vehicle recovery, and reduce costs and environmental impact. *HumaNav* consists of a reusable in-vehicle monitoring system on each vehicle that can communicate location, mileage, speed, and performance real-time via satellite. Related alerts can be sent via SMS or voice to both drivers and bases. *HumaNav* also offers analysis capabilities around vehicle utilization, fuel consumption, costs and CO2 emissions. A secure website application accessible by multiple parties is intended to provide real-time monitoring of fleet locations while also creating a community of humanitarian users. Though functionally similar to Globaltrack's BAT-340 solution (see below), HumamNav has differentiated itself by fostering a community of user organizations that meet regularly to share experiences and best practices (Novacom Services 2013).

HumaNav was piloted in collaboration with UNHCR and ICRC in Nepal, Zimbabwe, Somalia, and Sudan. During the pilot, East Africa Oxfam calculated more than $1.5 million in savings over 1 year through recovered vehicles, reduced fuel consumption, reduced maintenance costs, and increased utilization (UNHCR et al. 2010). In April 2011, HumaNav officially launched and, by November of the same year, 1000 vehicles were equipped (UNITAR 2011). Then in 2012, the Working Group on Emergency Telecommunication publicly endorsed HumaNav as a solution designed to increase fleet security and safety and adapted to the humanitarian context (Brophy-Williams et al. 2012).

BAT-340

Globaltrack and Chevin Fleet Solutions have collaborated to offer a vehicle tracking and management solution specifically designed for humanitarian applications.

Like *HumaNav*, the primary purpose is to increase security and safety of transport and personnel. Additional objectives include accurate data collection, cost and fleet reduction, fleet management, maintenance and usage monitoring, geofencing, and communication.

The reusable device, Globaltrack's *BAT-340*, is an integrated satellite and GPS receiver solution that reports real-time position, speed, and direction. It has been optimized for harsh African terrain, as evidenced by its winning of the Frost & Sullivan 2010 Award for most innovative tracking product for Africa (Globaltrack and Chevin Fleet Solutions 2011), and operates on a 3-year replaceable lithium battery. Rapid Deployment Kits are also available, which are a portable unit powered by the vehicle cigarette lighter. A message terminal allows direct 2-way communication with the driver. It also has an emergency panic button, audible and visible notifications, and real-time GPS information. On the software side, Chevin's web-based fleet management system presents real-time information (location, speed, direction, distance driven, and messages), history, and analysis (charts, maps, reports, etc.) and also enables fleet management, inventory functions, SMS or email alerts, and driver communication.

Temperature Sensor and Monitor Technologies

Condition monitoring technologies are often as simple as possible to maximize their applicability in humanitarian settings. Current sensor technologies can be divided into two categories: sensors and monitors. Temperature sensors (*Vaccine Vial Monitors* and *Freeze Indicators*) sense and indicate whether a temperature breach has occurred; that is, they are binary indicators. Sensors are typically visual, simple, independent of infrastructure, and physically attached to products or boxes. Temperature monitors (data logger tags and USB data loggers) observe and keep a continuous record of temperature. Unlike sensors, monitors are unattached and accompany product, such that they cannot track individual items. They usually alert temperature failure in an effort to enable intervention and prevention of product wastage.

All temperature sensor and monitor technologies have more or less the same intended impacts, namely to ensure efficacy and safety of administered vaccines, prevent administration of compromised vaccines, reduce unnecessary wastage of vaccines, and improve access in remote regions. All listed technologies in this section achieve these impacts and often achieve even additional supply chain visibility as discussed below. Note that all of the described technologies are WHO pre-certified.

Note that none of these technologies can reliably enable intervention and prevention of temperature failure mainly because they provide only visual alerts requiring inspection to detect failure, with the exception of the SMS-based notifications in the Optimize Albania project (see below). Also, when temperature detection is not at the individual item level, there is still no guarantee that the results hold for all items. For example, even if applied at the box level, temperatures can vary significantly throughout a box. This potential variation is only more significant when devices are applied at higher levels.

Vaccine Vial Monitors

Vaccine Vial Monitors (*VVMs*) are actually sensors that were developed by TEMP-TIME and PATH under the HEATmarker™ technology, originally used in the food industry. They were first employed for the oral polio vaccine and today are used for all vaccines in developing countries (PATH 2011).

What PATH now calls "the world's smartest sticker," is made to either adhere to the vaccine vial or be directly printed on the label, enabling item level detection. It is a mean kinetic temperature indicator that progressively darkens with heat exposure over time showing cumulative heat exposure by following the Arrhenius equation, such that the speed of the color development is faster at higher temperatures and slower at lower temperatures. Thus, *VVMs* can be used to further reduce unnecessary wastage of vaccines through heat exposure inventory management (most exposed out first); however, education is required to achieve this. More than 20 different monitors are available with thresholds from days at refrigerated temperature to 6 years at 80 °F. However, all *VVMs* only sense and communicate heat exposure.

Freeze Indicators

Freezing is an overlooked but serious problem throughout the cold chain. Vaccines are some of the most sensitive products, and account for US$ 136 million of UNICEF's spending in 2005 (Matthias et al. 2007). A review by Matthias et al. (2007) finds that vaccines were exposed to freezing temperatures in 14–35 % of refrigerators and shipments. Beyond just vaccines, freezing can also rupture and consequently waste liquid products. *FreezeWatch* was developed by 3M for use with any product that is sensitive to freezing, but is mainly targeted towards pharmaceuticals. *FreezeWatch*, also a visual indicator, is a vial of sensitive liquid backed on indicator paper. If temperatures drop below the threshold (available in −4 and 0 °C) for more than 1 h, the vial bursts and the liquid stains an indicator paper (WHO 1999). It is placed directly on boxes and in vaccine refrigerators. The complement to *VVMs*, *FreezeWatch* only senses and indicates cold exposure. The *Monitor Mark*, also by 3M, is another similar technology. Electronic versions, such as the *Freeze-tag* by Berlinger are similar in ultimate function, but utilize electronic as opposed to chemical temperature sensing.

Data Logger Tags

Data logger tags are small battery-powered devices that monitor and record temperatures of shipments. These tags display visual alerts of temperatures outside of the preset threshold. They are typically included in shipments at the box, carton, or pallet level, and, therefore, do not monitor the temperature of individual items. Two examples are the *LogTag Recorders*, used by Médecins Sans Frontières (MSF), *VaxAlert™*, and the *Q-Tag®2 plus*

LogTag Recorders Ltd. manufactures a business card sized temperature monitor, which is currently used by Médecins Sans Frontières (MSF) to monitor their cold chain transport. The tag consists of a low cost temperature sensor with visual indicators. It records time and temperature and is powered by a replaceable 2 to 3 year battery. It can store up to 8000 readings with a measurement range of −40–85 °C (±0.5–±0.8 °C accuracy) and adjustable sampling frequency of 30 s to 18 h. LogTag Analyser is a freely available software companion facilitates set-up, download of readings, analysis, and export to Excel. By providing a real-time visual alert of temperature failure, this device intends to enable intervention and prevention of wastage.

VaxAlert™, manufactured by Sensitech, is an example of a 10-day electronic temperature monitoring device that WHO recommends included in every international vaccine shipment. The reusable tag consists of a low cost temperature sensor and display that presents alarm status, mode, elapsed time, last recorded temperature, and alarm history. Several alarms can be configured (three for *VaxAlert*), including cumulative exposure alarms. The alarm status and history are recorded with time stamps. *VaxAlert* is powered by a 6 month battery and has a measurement range of −30–55 °C (±0.5–±1.0 °C accuracy) and sampling frequency of 1 min. *Q-Tag®2* plus by Berlinger is another similar device also approved by the WHO.

USB Data Loggers

Reusable USB data loggers are similar to the data logger tags described above. USB data loggers are typically a USB drive incorporating a temperature sensor and LCD display. Some loggers also incorporate other sensors, such as relative humidity. They usually display most recent, maximum and minimum logged values as well as status lights and visual alerts. They often allow unlimited logging by rewriting the oldest data and are typically powered by a 1 year lithium battery. Sampling frequency is generally adjustable from seconds to several hours. Control software that allows setup and data download typically accompanies the loggers. While the simple USB data loggers can be included in a shipment similar to a data logger, USB data loggers can be read at an intermediate supply chain stage or at the customer site. Temperature information can be downloaded for analysis and communication of cold chain performance throughout the supply chain.

Optimize Albania

In the monitoring solutions above, alarms are ineffective when they are not directly observed by health workers and stored temperature information cannot prevent temperature deviation. In response, the Albanian Institute of Public Health implemented a SMS-based gateway in 24 of its storage facilities in 2011 (Bino 2012). In this enhanced system, an activated alarm triggers an SMS text message that is sent to a central server, which forwards the notification on to the appropriate supervisor.

The central server continues to store the alarm status until the problem has been addressed. This pilot project exhibited many benefits, such as increased procedural compliance, a centralized repository of equipment performance, follow-up measures and additional benefits to nurses and supervisors. However, the pilot study was not able to identify a case where the SMS text message alerts saved a vaccine from temperature exposure.

Software Technologies

Even when data exists, lack of decision making support continues to be a point of frustration at both the global and field levels. Information is characterized as fragmented, inconsistent, inaccurate, and inadequately represented graphically (Mock and Garfield 2007). Information management solutions, nevertheless, can improve visibility and transparency in humanitarian supply chains (Tomasini and Van Wassenhove 2009). Humanitarian-specific software technologies seek to provide decision making through open-source, customizable solutions as independent of infrastructure as possible. All require reporting, delaying data, and typically do not fully capture material or information flows.

HELIOS

HELIOS is supply chain management software designed specifically for humanitarian organizations by the Fritz Institute, Blue Star Infotech, and Iris Logic as a response to manual processes and lack of infrastructure to support sizeable software implementations (Fritz Institute 2008). *HELIOS* aims to create a community of *HELIOS* users—humanitarian organizations—and thereby improve supply chain coordination, efficiency, optimization, and visibility. It further intends to improve donor communications and accountability by tracing donations.

HELIOS consists of a database, five optional modules, and an interface. The database can be hosted externally by a data center (for cost), where installation, backup, and administration outsourced to the host, or hosted internally by an organization's server. Infrastructure requirements are controlled through the ability to host externally. To circumvent internet connectivity issues, *HELIOS* can be deployed on servers in the field where connectivity is limited. Its five optional modules include Request Processing, Procurement (ordering, purchasing, and administering), Mobilization (assignment and distribution of donations), Warehousing, and Reports (Blecken and Hellingrath 2008). Several modules automate routine tasks. Interface and data transfer are enabled through Internet Explorer or Software as a Service (SAAS).

Logistics Support System (LSS/SUMA)

The Humanitarian Supply Management System (*SUMA*) is a joint effort of the Pan American Health Organization and the World Health Organization with financing provided by a variety of multilateral donors. SUMA launched in 1990 and the *Logistics Support System* (*LSS*) was added in 2005 to help manage the distribution of relief supplies (FMS Professional Solutions Group 2010). Specifically, *LSS* manages supplies from their pledge through port arrival to warehousing and finally distribution to beneficiaries. *LSS* is promoted as a complement to agency-specific tracking systems. It is itself not a commodity tracking system, but a coordination platform that aggregates both inventory and pipeline data from any variety of agencies and organizations to provide a high-level overview for supply coordination and management during an emergency. Most recently, *LSS/SUMA* was used in the 2010 Haiti earthquake and subsequent cholera outbreak (Pan American Health Organization 2010a, b).

Sahana Eden

Eden is an open source information management solution by the Sahana Software Foundation for use in all phases of disaster management—mitigation, preparation, response, and recovery. *Eden* has been used by the IFRC to create their Resource Mapping System and by the Disaster Risk Reduction Project Portal to create an information sharing platform for the Asia Pacific region (Kohli et al. 2011). Eden intends to improve coordination, management, and operations of organizations, people, projects, inventory, and assets throughout the disaster management phases. Data collection and facilitation of assessments are also aims.

The *Eden* solution consists of a free and open source software platform with optional modules. The software can be run from a local flash drive, server, or hosted server (for cost). Like *HELIOS*, infrastructure requirements are controlled through the ability to host externally and *Eden* can be run independent of internet connectivity. Sahana *Eden* is designed to allow individuals to go into the field with just a snapshot of data and later synchronize to the system (Currion et al. 2007). It supports databases, messaging, and data export. Modules include Organization Registry (tracks active organizations), Project Tracking, Disaster Victim Registry (traces individuals), Camp Registry (reports, tracks and helps manage individuals; maps capacities and locations of shelters), and Request Management System (tracks and monitors supply and demand). Optional Modules include Inventory Management (tracks inventory and facilitates transshipment and transactions), Messaging (alert generation), Situation Mapping (map annotation), and Synchronization Module (synchronizes data disconnected from system) (Currion et al. 2007) (Kohli et al. 2011). Functionally, it aims to be modified and augmented easily through the creation of new modules. In fact, its XForms data collection capabilities have now been expanded for support and integration with *Open Data Kit* (discussed in previous section) (Hartung et al. 2010).

Resource Map

InStedd's Cambodia innovation lab developed the open source *Resource Map* technology. The Resource Mapping Tool allows map creation, sharing, online information updates, exportation, alerts, and queries. Such creation and management can be labor intensive. Integration of the mobile platform allows for updates and queries via SMS, making it accessible anywhere and enabling real-time data collection (InStedd 2012a). This approach also allows a variety of methods of data collection and query, such as SMS and internet.

The main objective of the tool is to enable visualization of information geospatially (InStedd 2012a) to improve decision making, monitoring, evaluation, and reaction. It also aspires to create a collaborative environment and enable information sharing. Operationally, it intends to track resources and report to funders. It is currently being piloted in a beta version. Initial feedback seems positive, as the Clinton Health Foundation has used it to provide stock situations down to the facility level for the National Center for Malaria Control (InStedd 2012b).

8.5 Conclusions

Increased visibility has the potential to greatly improve humanitarian operations. This review demonstrates that current efforts to increase visibility in humanitarian supply chains are diverse and unique to their challenging settings. While they each make strides in visibility by resolving different needs, as summarized in Table 8.1, none satisfies all. Though each mechanism described here improves visibility, more complete visibility must be achieved through patchwork solutions of multifaceted and disjoined mechanisms.

Over all of these technologies and projects, a set of key outstanding weaknesses has been identified and are illustrated in Fig. 8.2. Humanitarian supply chains are often characterized by notably limited infrastructure, including roads, electricity, and internet; technologies must be adapted to these unique environments. While great success has been achieved, dependency continues to be a weakness of these technologies and projects. Such dependency includes reliance on electricity, computers, internet connectivity, cellular networks, and existing systems (e.g., RFID reader network). Typically, as a solution moves away from being infrastructure dependent, it becomes more dependent on human resources. In line with the latter, others of these solutions are labor intensive, dependent on reporting, and require extensive training. While human resources constraints and dependency themselves are serious issues facing humanitarian supply chains, particularly in more long-term crises and development interventions, exactly how humanitarian organizations accomplish their work is rooted in humanitarian principles, which may value local human resource use despite its disadvantages. For example, the Red Cross Code of Conduct states the intention to build disaster response on local capacities and beneficiary involvement (Mays et al. 2012). Additional outstanding weaknesses include

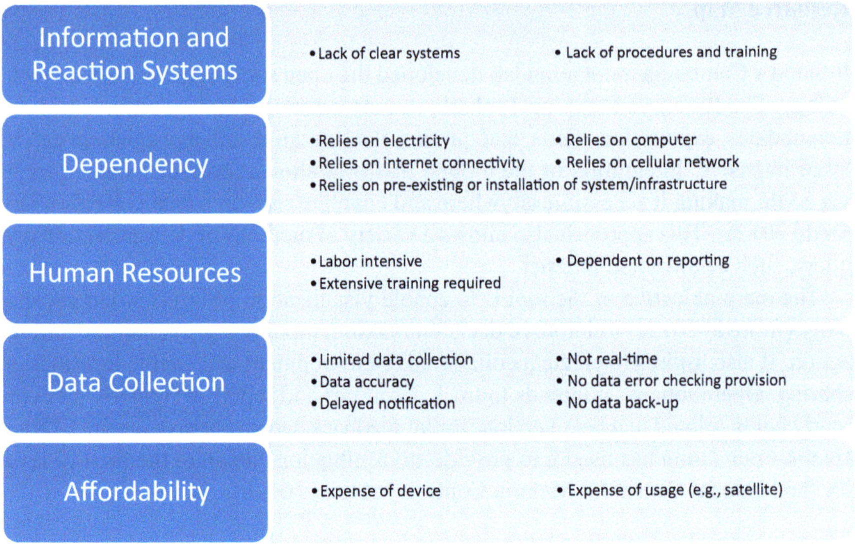

Fig. 8.2 Key outstanding weaknesses

that data collection is clearly limited and existing methods continue to be plagued by inaccuracy, deficient error checking provisions, and lack of back-up. Some data is still not real-time and delayed notification does not enable prevention. Affordability (cost) is another key weakness in the form of the devices, systems, and/or usage.

Most importantly, for this information to be beneficial, information and reaction systems must be synchronized. Only actionable information is valuable. Thus, the greatest limitations are the lack of systems, procedures, and training that enable meaningful and appropriate reaction to the information provided.

References

Anokwa Y (2011, September 21) New York city using ODK for health operations. OpenDataKit. https://opendatakit.org/2011/09/new-york-city-using-odk-for-health-operations/. Accessed 30 April 2015

Barrington J, Wereko-Brobby O, Ziegler R (2010a) SMS for life: Tanzania pilot project report. Basal: Roll Back Malaraia

Barrington J, Wereko-Brobby O, Ward P, Mwafongo W, Kungulwe S (2010b) SMS for life: a pilot project to improve anti-malarial drug supply management in rural Tanzania using standard technology. Malar J 9:298

Besiou M, Stapleton O, Van Wassenhove LN (2011) System dynamics for humanitarian operations. J Humanit Logist Supply Chain Manag 1(1):78–103

Bino S (2012) Albania: demonstrating the benefits of an online immunization registry. PATH, Geneva (World Health Organization)

Blecken A, Hellingrath B (2008) Supply chain management software for humanitarian operations: review and assessment of current tools. In: Fiedrich F, Walle BV (eds) Proceedings of 5th International ISCRAM Conference. 5th International ISCRAM Conference, Washington, D.C., pp 342–351

Brophy-Williams S, Hardman J, Leaning J, Meier P, Olafsson G, Pham PN et al (2012) World Disasters report: focus on technology and the future of humanitarian action. International Federation of Red Cross and Red Crescent Societies, Geneva

Celik S, Corbacioglu S (2010) Role of information in collective action in dynamic disaster environments. Disasters 34(1):137–154

Chapelier C, Shah A (2013) Improving communciation between humanitarian aid agencies and crisis-affected people: lessons from the infoasaid project. Humanitarian Practice Network, London (Overseas Development Institute)

Chester P (2010, May 25). Malaria kills: distributing 63 million bednets in Nigeria with RapidSMS. MobileActive.Org. http://mobileactive.org/. Accessed 28 Oct 2011

Conesa S, Kopczak L (2009) Positioning paper and lessons learned: coordination & development of SMS project. UNICEF supply division innovation initiative. MIT-Zaragoza International Logistics Program, Zaragoza

Currion P, de Silva C, Van de Walle B (2007) Open source software for disaster management. Commun ACM 50(3):61–65

Danton H, Stevenson K (2010) FY 2010 annual results report save the children/Bangladesh. USAID, Office of Food for Peace. USAID, Dhaka

Department of Vaccines and Biologicals of the WHO (2005) Monitoring vaccine wastage at country level: guidelines for programme managers. WHO, Geneva (World Health Organization, Department of Vaccines and Biologicals)

European Journalism Centre (2014, February 4) Featured tool: humanitarian data toolkit. Emergency Journalism. http://emergencyjournalism.net/featured-tool-humanitarian-data-toolkit/. Accessed 20 Oct 2014

European Space Agency (2011, November 22) Satellites respond to humanitarian needs. France. http://www.esa.int/Our_Activities/Observing_the_Earth/Envisat/Satellites_respond_to_humanitarian_needs

European Space Agency (2013, September 1) How satellites help manage humanitarian crises. Copernicus (37)

FMS Professional Solutions Group (2010) Logistics Support System (LSS) providing international humanitarian relief logistics for the pan American health organization and six UN agencies. FMS Professional Solutions Group. https://www.fmsinc.com/consulting/portfolio/logistics-support-system.aspx. Accessed 20 Oct 2014

Fritz Institute (2008) HELIOS OVERVIEW. http://www.fritzinstitute.org/prgTech-HELIOS_Overview.htm. Accessed 5 Nov 2011

FrontlineSMS Web Administrator (2009, March 2) FrontlineSMS: now with forms. FrontlineSMS. http://www.frontlinesms.com/2009/03/02/frontlinesms-now-with-forms/. Accessed 29 April 2015

Globaltrack and Chevin Fleet Solutions (2011) Press release: humanitarian fleet industry heavyweights join forces. Globaltrack and Chevin Fleet Solutions, Johannesburg

Guevara E (2014, July 18) Panamanian Red Cross uses new technology to streamline distribution of humanitarian aid. International Federation of Red Cross and Red Crescent Societies. http://www.ifrc.org/en/news-and-media/news-stories/americas/panama/panama-red-cross-uses-new-technology-to-streamline-distribution-of-humanitarian-aid-66450/. Accessed 30 April 2015

Hartung C, Anokwa Y, Brunette W, Lerer A, Tseng C, Borriello G (2010) Open data kit: tools to build information services for developing regions. Proceedings of the 4th ACM/IEEE international conference on information and communication technologies and development (p 18). ACM

Humanitarian Information Facilitation Centre and International Media Support (2012) Mapping with mobiles: lessons learnt from the use of smartphones for fast humanitarian surveys in Zimbabwe. International Media Support, Zimbabwe

InStedd (2012a) Resource map. http://instedd.org/technologies/resource-map/. Accessed 24 Jan 2012a
InStedd (2012b) Resource map: live data, anywhere, anytime. http://resourcemap.instedd.org/. Accessed 24 Jan 2012b
International Media Support (2012a) Mapping with mobiles. International Media Support, Copenhagen
International Media Support (2012b, May 2) Report shows smartphones make surveys easier in Zimbabwe. Digital opportunity. http://www.digitalopportunity.org/feature/report-shows-smartphones-make-surveys-easier-in-zimbabwe. Accessed 6 Oct 2014
International Telecommunication Union (2014, June 20) ITU world telecommunication/ICT database. Geneva. http://www.itu.int/en/ITU-D/Statistics/Pages/default.aspx
Kapucu N (2008) Collaborative emergency management: better community organising, better public preparedness and response. Disasters 32(2):239–262
Kohli S, Tressel P, Howden M, Boon F, Lev E, Koenig D et al (2011) Sahana Eden. The Sahana Foundation, Mountain View
Maiers C, Reynolds M, Haselkorn M (2008) Challenges to effective information and communication systems in humanitarian relief organizations. In: Hayhoe GF, Grady H, Connecting people with technology: issues in professional communication. Baywood, Amityville, pp. 119–131
Matthias D, Robertson J, Garrison M, Newland S, Nelson C (2007) Freezing temperatures in the vaccine cold chain: a systematic literature review. Vaccine 25:3980–3986
Mays R, Racadio R, Gugerty M (2012) Ompeting constraints: the operational mismatch between business logistics and humanitarian effectiveness. IEEE, Seattle, pp 132–137 (Global Humanitarian Technology Conference (GHTC))
Mock N, Garfield R (2007) Health tracking for improved humanitarian performance. Prehosp Disaster Med 22:377–383
Morris K (2009) Mobile phones connecting efforts to tackle infectious disease. Lancet Infect Dis 9:274
Novacom Services (2013) HumaNav. Novicom Services. http://www.novacom-services.com/en/humanav. Accessed 30 April 2015
OpenDataKit (2015A) About. OpenDataKit. https://opendatakit.org/about/. Accessed 30 April 2015
OpenDataKit (2015B) Tools. OpenDataKit. https://opendatakit.org/about/tools/. Accessed 30 April 2015
Pan American Health Organization (2010a, February 15) Earthquake in Haiti—January 2010. LSS. http://www.lssweb.net/haiti-january-2010.html. Accessed 20 Oct 2014
Pan American Health Organization (2010b, December 1) Cholera outbreak in Haiti—October 2010. LSS. http://www.lssweb.net/haiti-november-2010.html. Accessed 20 Oct 2014
PATH (2011) World's smartest sticker. PATH. http://www.path.org/projects/vaccine_vial_monitor.php. Accessed 3 Aug 2011
Pisano F (2005, December) Using satellite imagery to improve emergency relief. Humanitarian Exchange, pp 36–40
Privett N, Gonsalvez D (2014) The top ten global health supply chain issues: perspectives from the field. Op Res Health Care 3:226–230
Rapid SMS (2008, October) Ethiopia—Supply chain management—Case study. RapidSMS. http://www.rapidsms.org/case-studies/supply-chain-management-during-food-crises/. Accessed 28 Oct 2011
Rapid SMS (2011) Overview. RapidSMS. http://www.rapidsms.org/overview/. Accessed 28 Oct 2011
Rietjens S, Verlaan K, Zaalberg T, De Boer S (2009) Inter-organisational communication in civil–military cooperation during complex emergencies: a case study in Afghanistan. Disasters 33(3):412–435
Sandwell C (2011) A qualitative study exploring the challenges of humanitarian organisations. J Humanit Logist Supply Chain Manag 1(2):132–150

Standby Task Force (2014, October) Our model. Standby Task Force. http://blog.standbytaskforce.com/our-model/. Accessed 20 Oct 2014
Stephenson M (2005) Making humanitarian relief networks more effective: operational coordination, trust and sense making. Disasters 29(4):337–350
Susman-Pena T (2013) What we learned testing the humanitarian data toolkit. Internews Center for Data and Learning, Washington, D.C.
Swedberg C (2010) Nigerian drug agency opts for RFID anticounterfeiting technology. RFID J. http://www.rfidjournal.com/articles/view?7856
The MENTOR Initiative (2009) The MENTOR initiative. The MENTOR Initiative. www.mentorinitiative.org. Accessed 10 November 2011
Thomas A (2005, December 18) Linking preparedness and performance: the tsunami experience. Humanitarian Exchange, 32
Thomas A, Kopczak L (2005) From logistics to supply chain management: the path forward in the humanitarian sector. Fritz Institute, San Francisco
Tomasini R, Van Wassenhove L (2009) From preparedness to partnerships: case study research on humanitarian logistics. Int Trans Op Res 16:549–559
UNHCR, ICRC, UNITAR, NOVACOM (2010) Humanav operational benefits. UNOSAT, Geneva
UNITAR (2011) HumaNav community workshop convenes as service celebrates 1000 vehicles target. UNITAR, Geneva
USAID, TANGO, Save the Children (2009) Jibon O Jibika (Life and Livelihoods): final evaluation report. USAID, Bangladesh
Wang B, Raymond N, Gould G, Baker I (2013) Problems from hell, solution in the heavens?: identifying obstacles and opportunities for employing geospatial technologies to document and mitigate mass atrocities. Stab: Int J Secur Dev 2(3):1–18
WHO (1999) Temperature monitors for vaccines and the cold chain. WHO, Geneva (WHO, Vaccines and Other Biologicals)
WHO, Fund U. N., Programme U. N., Fund U. N., Bank W (2007) A model quality assurance system for procurement agencies. WHO, Geneva
World Health Organization (2011) mHealth: new horizons for health through mobile technologies. World Health Organization, Geneva

Chapter 9
Efficient Inventorying and Distribution of Blood Product During Disasters

Cigdem Gonul Kochan, Shailesh S. Kulkarni and David R. Nowicki

Abstract Poor supply management of blood products poses a critical challenge for healthcare supply chains in support of humanitarian operations. A mismatch between blood product supply and blood product demand can have catastrophic consequences that can lead to loss of life, especially when humanitarian efforts are deployed as a result of a disaster. This study is motivated by the Doctors Without Borders/Médecins Sans Frontières (MSF) (http://www.doctorswithoutborders.org/news-stories/special-report/syria-two-years-failure-international-aid) report. The MSF report describes the complexity, enormity, and critical importance of meeting the medical care and blood product needs for civilians under intense bombing in Syria. This chapter adapts a perishable inventory–routing problem with a single blood collection center and multiple hospitals with uncertain demand. To solve the problem, we use a newsvendor model in conjunction with a generalized traveling salesman (TSP) model combination to determine optimum quantities of blood and to determine the best possible route to meet the blood demand. We develop a two-stage linear programming representation of the underlying stochastic program. Such a representation is easy to scale up to real –world scenarios of providing humanitarian aid through healthcare supply chains. It also allows us to consider additional complexities such as truncated and correlated demands which can lead to some interesting insights in managing blood products inventory.

Keywords Newsvendor · Stochastic Programming · Blood (or Blood Products) · Traveling Salesman · Routing · Distribution/Allocation · Demand Correlation · Capacity Planning · Integer Programming · Stochastic Models

---

S. S. Kulkarni (✉) · D. R. Nowicki
University of North Texas, Denton, TX 76203, USA
e-mail: Shailesh.Kulkarni@unt.edu

C. G. Kochan
Ohio Northern University, Ada, OH 45810, USA

C. W. Zobel et al. (eds.), *Advances in Managing Humanitarian Operations*,
International Series in Operations Research & Management Science 235,
DOI 10.1007/978-3-319-24418-1_9

9.1 Introduction

Blood bank service operations are central to healthcare systems to meet the overwhelming worldwide demand for blood products. Blood product transfusions are a service provided by blood banks and are frequently required for trauma victims, surgery, and organ transplants; for patients receiving treatments for diseases such as cancer, leukemia, and anemia; and in childbirth (Katsaliaki and Brailsford 2006). Blood products include whole blood, red blood cells, blood platelets, blood plasma, and frozen blood for the specific needs of patients. In this study, we focus on only the whole blood product. The shelf life for a unit of whole blood stored at room temperature is 1 day and if refrigerated the unit of whole blood has a shelf life of 21 to 35 days. (redcrossblood.org 2013).

Distributing blood products in the right quantity, at the right place, and at the right time is challenging, especially in the case of a massive emergency such as a natural or manmade disaster. Since blood products are highly perishable and their humanitarian operational demand is uncertain, matching supply and demand in an effective manner is not straightforward (Beliën and Forcé 2012). The blood product supply-demand mismatch can result in a shortage that can ultimately lead to increases in mortality, imposing a high cost for society. On the other hand, excess supply can lead to expiry due to the blood product's inherently short shelf life. Expired blood products will be discarded leading to increases in the healthcare supply chain cost. Therefore, blood products need to be managed (moved and stored) in a healthcare supply chain to ensure an adequate and appropriate supply is available at hospitals, keeping spoilage to a minimum (Hemmelmayr et al. 2009).

This study is motivated by the Doctors without Borders/Médecins Sans Frontières (MSF) (MSF 2013) report on the medical care and blood needs for civilians under intense bombing in Syria. In March 2011, the first anti-government protests broke out in Syria under the influence of Arab Spring in the Middle East. According to 2013 UN estimates, violent conflicts in Syria have killed more than 70,000 people since the first anti-government protests in 2011 (MSF 2013). The conflict shows no signs of abatement and health workers and medical facilities are still receiving threats and medical structures are being destroyed. According to the MSF's (MSF 2013) report, due to the constant bombing attacks, providing trauma surgery and treating the war-wounded have become very difficult tasks. There is a need for large quantities of blood in order to treat patients with injuries. The only blood bank in the Aleppo area was demolished in an air attack. Donor finding is not the issue in the area since many people volunteer to donate their blood. However, the main issues are: (a) transfusing blood without testing that causes death and (b) keeping blood bags cool since most hospitals are not equipped with refrigerators or even if they have refrigeration, most of them have insufficient power.

Despite the fact that MSF has not received permission from the Syrian government in Damascus to work in the country, MSF manages to provide a fridge, money for fuel, testing supplies and training to set up a new blood collection center (BC). Even though the BC is still at risk of becoming a military target, this facility now supplies three hospitals opened by MSF in the Aleppo region. MSF covers the costs associated with distributing, storing, and testing each unit of blood. Therefore, due to a tight budget MSF, needs to minimize the costs of distributing and storing blood where demand is uncertain.

This study considers a combination of two problems that MSF faces. The first problem is determining and allocating the optimum blood units acquired by each hospital when the capacity is limited. The second problem is delivering the optimum blood units determined in the first problem among several hospitals. The objective of this study is therefore to develop a combined newsvendor-TSP model (see below) to find the optimum quantities of whole blood units that minimize the shortage and excess blood costs for each hospital in each disaster region while also deciding on the best possible route that minimizes the total transportation cost. The contributions of this study are as follows: (a) scalable newsvendor and generalized TSP models are developed and applied to a real and emergent blood distribution problem in Syria; (b) correlated demands are considered that typically do not yield themselves to simple closed-form solutions; and (c) our methodology for managing inventory of blood products can be easily implemented in the field for use by real-world agents.

The organization of this chapter is as follows: First, in Sect. 9.2 we review the literature related to the problem. Then, in Sect. 9.3 we discuss our problem assumptions and develop the models. Next, in Sect. 9.4 we present a numerical example. Finally, in Sect. 9.5 we discuss conclusions and directions for future research.

9.2 Literature Review

Supply Chain Management of Blood Products

There are many relevant blood product studies that exist in the supply chain management literature. These blood product supply chain studies mainly focus on using multi-objective optimization, linear and integer programming techniques to solve the problems of distribution, allocation, location, and inventory routing of blood products with the objective of minimizing the costs, shortages and abundance of blood products (Beliën and Forcé 2012). For instance, Kendall and Lee (1980) develop a goal programming model for redistribution of blood with the objectives of minimizing excess blood, shortages of blood and the costs. Kendall and Lee (1980) show that their proposed model reduces the percentage of unused blood and blood shortages. Şahin et al. (2007) develop integer-programming models to solve the location-allocation problem in the Turkish Red Crescent's regionalization of blood services. Şahin et al. (2007) formulate a model to find the optimal regions in a new area, measure the effectiveness of past location decisions and improve existing location patterns. In this present research, we assume that the location for the blood collection center is already decided. The terms blood collection center and blood center are used interchangeably in the literature. From this point forward we will use the acronym BC. Later, Hemmelmayr et al. (2009) develop an integer programming model to optimize the cost of delivering blood products to Austrian hospitals. Hemmelmayr et al. (2009) investigate two alternative delivery strategies. The first approach uses an integer programming technique to decide which hospital to visit each day to prevent daily shortages, while minimizing spoilage and delivery costs. The second approach combines vehicle routing decisions and delivery regularity. Hemmelmayr et al. (2009) use a variable neighborhood algorithm to solve this particular periodic routing problem. They conclude that integer programming is effective when constraints are tight; therefore, delivery costs can be reduced. They assume that demand for each hospital is known and can be estimated from historical data. However, the assumption of known demand does not apply in the case of a disaster. Thus, in this study we relax this assumption and assume hospital demand is not known with certainty.

In the case of uncertain demand, Nagurney et al. (2012) use a multi-criteria system optimization approach for regionalized blood supply chain network to determine the optimal levels of blood processed in each supply chain network link with the objective of minimizing the total cost associated with blood collection, shipment, processing and testing, storage, distribution, and total discarding costs. In addition, Nagurney et al. (2012) minimize supply chain risk by satisfying the uncertain demand at minimum total cost. In this study, we examine a similar problem, allocating optimal blood units available at a BC among several hospitals (each hospital has a stochastic demand) and determining the best possible delivery route with the objective of minimizing the cost associated with the delivery, spoilage and obsolescence of blood units. To address our problem, we employ newsvendor and TSP solution approaches.

Newsvendor & Generalized TSP with Stochastic Demand

In the literature, studies exist that combine single period-inventory management and routing models with an instant time horizon when the demands are stochastic and not known at the time of planning (Federgruen and Zipkin 1984; Federgruen et al. 1986; Chien et al. 1989). In this case, an instant time horizon refers to a very short planning horizon that requires only one visit per customer (Andersson et al. 2010). This is typically done as an extension of an inventory management problem with the inclusion of a routing component (Andersson et al. 2010). Authors take a wide variety of solution approaches for such problems, such as routing/allocation decomposition (Federgruen and Zipkin 1984; Federgruen 1986), customer selection/routing heuristics (Golden et al. 1984; Dror et al. 1985; Trudeau and Dror 1992), and mixed integer programming (Chien et al. 1989).

One common solution approach for solving inventory management and routing problems with an instant time horizon is to solve both the inventory allocation problem and the TSP together. Federgruen and Zipkin (1984) were the first to develop a combined single period TSP and inventory allocation problem with one warehouse and multiple retailers. They assume that the inventory levels are known to the decision makers who make daily decisions on the optimal inventory quantities for each customer as well as on deliveries utilizing a fleet of vehicles. They then use a newsvendor model, as a part of a nonlinear mixed integer programming model, to determine the optimal quantities delivered to each location, and they solve this nonlinear mixed integer programming model using an interchange heuristic and Benders decomposition approaches.

Later, Federgruen et al. (1986) extend Federgruen and Zipkin (1984)'s model to perishable products. They assume that each delivery size is fixed at the optimal delivery quantity, a fleet of delivery vehicles exists, and customer demands are independent. They then develop an efficient algorithm for two delivery policies: (1) a simple inventory allocation model using a solution procedure in which the deliveries are in individual routes and resulting optimal allocation quantities are utilized as deterministic delivery requirements within a routing algorithm, and (2) an integrated inventory allocation –VRP model in which the deliveries are combined in multistep routes using a repeated solution procedure as a subroutine within a routing algorithm. Both models have the objectives of allocating the optimal delivery quantities and minimizing the number of vehicles. Federgruen et al. (1986) then compare the performance of the integrated solution approach and the inventory allocation solution approach based on the inventory and travel costs. According to the results, they suggest that the integrated solution approach leads to a minor inventory cost savings and significant travel cost savings due to the reduction in the number of vehicles. In traditional commercial supply chains, operational costs such as total costs of resources, inventory cost, and distribution costs are considered as predominant costs (Beamon and Balcik 2008). Unlike commercial supply chains, in humanitarian relief supply chains operational costs are considered as secondary costs since the main concern is to maximize the benefits to the affected population (Perez et al. 2010). In this chapter in addition to the operational costs, loss in human welfare is considered as a social cost/deprivation cost—economic value of the human suffering/life loss due to the shortages in blood units (Perez et al. 2010).

In this current study, unlike in Federgruen et al. (1986)'s inventory allocation problem, we assume a single vehicle and assume that hospital demands are correlated in each delivery region, with the objective of determining the optimal delivery quantities while minimizing the total distance traveled. We present our problem by modeling a combined multi-product newsvendor problem and generalized TSP problem with a finite single period time horizon since the decisions need to be made on a daily basis in a disaster area. We refer to our solution approach as a combined approach since our solution approach uses the optimal results of the multi-product newsvendor problem within the TSP model.

The Multi-Product Newsvendor Problem (MPNP) with Capacity Constraints

Perishable products such as human blood, medical drugs, and food create challenges on inventory management, especially trading off stock outs and on shelf availability against wastage due to expiry (Sebastian et al. 2012). The decision maker has to determine the optimal quantities of a product to stock or order when there is a limited shelf life, while considering the tradeoff between the cost of over stock and stock outs. In blood supply chains, blood stock outs lead to high social costs due to the increased mortality rate (Beliën and Forcé 2012). The newsvendor models and extensions provide an effective framework to study such perishable product problems (Erlebacher 2000).

Since blood products are perishable, limited refrigerated space is another supply chain issue that must be addressed. Allocating refrigeration capacity for blood units to each hospital is a critical blood product distribution decision. In the literature, an extension of the classical newsvendor model, MPNP with capacity constraints, has been extensively used to address such resource allocation challenges. Hadley and Whitin (1963) are the first to discuss a single constraint MPNP. Lau and Hing-Ling Lau (1996) extend Hadley and Whitin (1963)'s MPNP solution procedure to handle general distributions and develop solution procedures for one resource and multiple resource constraints. Later, Erlebacher (2000) presents optimal and heuristic solutions for the MPNP. First, he proves optimality of the order quantities for two special cases. The first case is when each item has a similar cost structure and the demand for each item follows a similar distribution. The second case is when the demand of each item follows a uniform distribution. Then, based on the results of the cases, Erlebacher (2000) proposes three heuristics for several probability distribution functions. Finally, Erlebacher (2000) examines the effectiveness of these three heuristics and shows that they are extremely effective when the optimality conditions are satisfied.

In this chapter we represent the MPNP as a two-stage stochastic problem with recourse since it can be solved optimally (Godfrey and Powell 2001) using the deterministic equivalent representation of the underlying stochastic model. In our problem, the BC acquires the blood units before the actual demand is realized and the decision maker makes an acquisition decision based on the available blood units. Later hospitals report the actual demand. Therefore, the objective of the problem is to obtain an optimal first stage order quantity and second stage order policy for each realization of the demand signal subject to a capacity constraint.

9.3 Problem Definition

Problem Description

In the blood distribution problem, we consider 3 regions (Manbij, As Safirah, and Al-Bab) in the Aleppo Governorate with two hospitals in each region and one BC in the Syrian city of Aleppo. Figure 9.1 shows the healthcare supply chain with its regions (r), hospitals (i_r), BC and its information and blood unit flows. Solid arrows represents flow of blood units and dotted arrows represents flow of information. The amount of blood units delivered on the route is denoted by $n_{i_r}^{r}$, the amount of

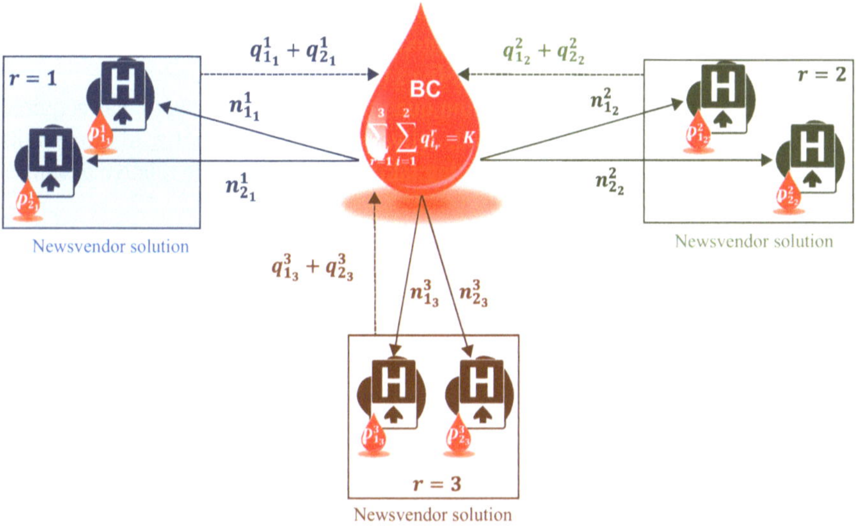

Fig. 9.1 Illustrative model

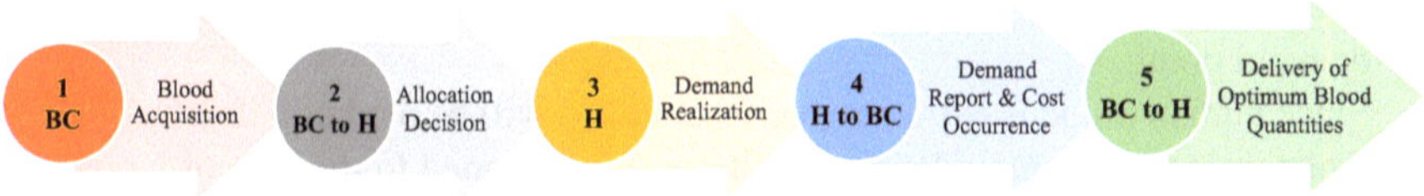

Fig. 9.2 Scenario

acquired blood units is denoted by $q^r_{i_r}$, capacity of BC is denoted by K and demand in each hospital is denoted by $D^r_{i_r}$.

In this problem, we assume that hospitals utilize a Vendor-Managed Inventory (VMI) program. VMI refers to an inventory management policy where the maintenance of inventory levels at a number of locations is controlled by the central decision maker (Kleywegt et al. 2002). VMI is a widely used practice especially by hospitals to solve supply shortages since the 1990s (Kelle et al. 2012). We consider the BC as the central decision maker that makes decisions on the replenishment and the delivery of the blood units. Therefore, the BC has access to the inventory information and historical demand information at all the hospitals. Figure 9.2 depicts the following scenario for our blood distribution problem.

1. At the beginning of the period, the BC utilizes inventory level (safety stock), capacity (refrigeration capacity) and historical demand (injured people) information reported by each hospital to determine the acquisition of the blood units. BC collects the blood units from donors in the face of unknown hospital demands. The number of blood units collected is limited by the BC's capacity.
2. The BC determines the allocation of the blood units.
3. Hospitals see the actual demands (injured people).

4. Hospitals report the actual demands to the BC and are charged with inventory holding and shortage costs represented as social costs—an economic value assigned to suffering and the loss of lives (Perez et al. 2010).
5. The BC, constrained by the optimum quantity of blood units that it has acquired and contingent on demand revealed, delivers the feasible quantity of blood units to each hospital via an optimal routing taking into account the relative distances of hospitals and vehicle capacity.

Problem Assumptions

- Blood units are distributed from a BC to hospitals using a single vehicle with known capacity.
- Vehicle capacity is equal to BC capacity.
- Each blood unit requires the same amount of capacity.
- Cost structure for each blood unit is identical.
- Hospitals are restricted to one and only delivery route.
- Due to the vitality/importance of the availability of blood at the hospitals, a relatively large penalty cost is associated with the shortage of a unit of blood at demand point.
- Since human blood is extremely perishable and will be outdated if not used over a certain period, the outdating penalty cost is assigned to the unit of a possible supply surplus.
- The cost of short-term inventory holding (cold storage-refrigerators), and possibly the discarding cost of the outdated product are added together.
- Throughout the process of blood collection and distribution a fraction of the collected blood may deteriorate, be lost, or be wasted. This study does not consider the cost of wasted whole blood.
- Demand in each hospital is uncertain and continuous with normal distribution. We assume that the demand in each hospital is continuous since the whole blood can be distributed in variety sizes of bags, tubes and bottles. In our problem, each type of container size is calculated as fraction of units.
- Hospital demands in the same regions are correlated.

Problem Formulation

Newsvendor Model

The BC acquires blood units from blood donors in the face of uncertain demand and decides on allocation and distribution of the available blood units for each hospital j in region r. When the demand increases, BC satisfies the demand dependent upon the acquired blood units. We apply the MPNP to find the optimal quantities needed for each hospital in order to prevent cost of shortage and cost of excess whole blood. We assume that demands are correlated in each region but not across regions.

Notation

$r = 0,1,2\ldots|R|$ = index of regions where 0 = BC

$$i_r = 1,2,\ldots|I_r| = i^{th}\text{ hospital of } r^{th}\text{ region}; i_r = \begin{cases} 1,2,\ldots|I_r| \text{ for } r = 1,2,\ldots|R| \\ 1 \text{ for } r = 0 \end{cases}$$

K Capacity limit acquired by BC

$p^r_{i_r}$ Cost per unit of shortage of blood in each hospital i in region r

$w^r_{i_r}$ Shortage quantity of blood units in each hospital i in region $r\left(D^r_{i_r} - q^r_{i_r}\right)$

$v^r_{i_r}$ Cost per unit of holding blood in each hospital i in region r

$z^r_{i_r}$ Excess quantity of blood units in each hospital i in region $r\left(q^r_{i_r} - D^r_{i_r}\right)$

$c^r_{i_r}$ Cost per unit of blood in each hospital i in region r

$q^r_{i_r}$ Quantity of blood acquired by each hospital i in region r

$D^r_{i_r} = N\left(\mu^r_{i_r}, (\sigma^r_{i_r})^2\right), a < D^r_{i_r} < b$ demand with mean μ and standard deviation σ in each hospital i in region r.

In this study, we assume that the demand in each hospital follows truncated normal distribution although it is feasible to use any type of distribution for our scenario. We utilize normal distribution since the normal distribution makes it fairly easy to model large scale stochastic programs. In our problem, we restrict the normal distribution's lower and upper limits because we assume that it is mandatory for hospitals to keep certain amount of blood units as a safety stock which refers to the lower limit of the truncated normal distribution, and each hospital has a limited capacity of refrigerator or refrigerated container that keeps the blood units fresh which refers to the upper limit of the truncated normal distribution.

Two-Stage MPNP with Capacity Constraint

According to the problem, there is a shortage in the refrigerators at the BC that keep blood units fresh. BC must operate within a limited refrigerator size. Therefore, we formulate the newsvendor model with a capacity constraint. Stages in newsvendor model solution procedure follows:

Stage 1: BC utilizes inventory level (safety stock), capacity (refrigeration capacity) and historical demand (injured people) information reported by each hospital to determine the acquisition of blood units constrained by the BC's capacity.

Stage 2: Hospitals realize the actual demand (injured people), report the actual demand to the BC and are charged with the Inventory holding and shortage costs. In addition, we assume that hospital demands in each region are correlated. Aggregated demand in each region is written as $\sum_{i=1}^{|I_r|} D_{i_r}^r$. Each $D_{i_r}^r$ follows a truncated normal distribution $N\left(\mu_{i_r}^r, \left(\sigma_{i_r}^r\right)^2\right), a < D_{i_r}^r < b.$ As is well known, for the two different demands, a bivariate normal distribution can be written as $\left[\mu_{1_1}^1 + \mu_{2_1}^1, \sigma_{1_1}^1 + \sigma_{2_1}^1 + 2\rho\sigma_{1_1}^1\sigma_{2_1}^1\right]$.

The objective function that minimizes the capacity acquired by BC can be written as:

Stage 1:

$$G\left(q^{*r}_{i_r}\right) = arg\ min\,(q_{i_r}^r) \tag{9.1}$$

$$s.t.\quad q_{i_r}^r \geq 0 \tag{9.2}$$

Stage 2:

$$\Pi = Min. v_{i_r}^r z_{i_r}^r + p_{i_r}^r w_{i_r}^r \tag{9.3}$$

$$s.t.\quad \sum_{r=1}^{|R|} \sum_{i=1}^{|I|} q_{i_r}^r \leq K \tag{9.4}$$

$$z_{i_r}^r \geq q_{i_r}^r - D_{i_r}^r \quad \forall r \neq 0 \tag{9.5}$$

$$w_{i_r}^r \geq D_{i_r}^r - q_{i_r}^r \quad \forall r \neq 0 \tag{9.6}$$

$$z_{i_r}^r, w_{i_r}^r, q_{i_r}^r \geq 0, \forall r, i \tag{9.7}$$

where,

$$G(q_{i_r}^r) = E\left[\Pi\right] + \sum_{r=1}^{|R|} \sum_{i=1}^{|I|} c_{i_r}^r q_{i_r}^r \tag{9.8}$$

is the expected (E being the expectation operator) Stage 2 penalty cost, plus the total cost of delivering the acquired blood.

Constraint (9.4) restricts total order quantity to not to exceed BC capacity. Constraint (9.5) ensures that excess quantity for each hospital i in region r equal or greater than the difference between capacity and demand. Constraint (9.6) ensures that shortage quantity for each hospital i in region r equal or greater than the difference between demand and the capacity. Constraint (9.7) indicates that variables $z_{i_r}^r, w_{i_r}^r, q_{i_r}^r$ are non-negative.

To solve formulation (9.8) and thus (9.1), we adopt the solution procedure from Lau and Lau (1996) and Erlebacher (2000). Based on the procedure, we first compute q^*_{ri} for each hospital i in region r and check whether the capacity constraint (9.4) is binding. If it is not binding, then the optimal quantity is $q_{i_r}^{*r}$ given by the typical newsvendor critical fractile.

$$q_{i_r}^{*r} = F^{-1}\left(\frac{p-c}{h+p}\right)\forall_i, \forall_r \tag{9.9}$$

where $\frac{p-c}{h+p}$ represents the critical fractile.

If the capacity constraint is binding, there is no closed form solution. In the case of a tight capacity constraint, which would typically be the case in the real world, one could apply an iterative solution procedure to solve the model (Erlebacher 2000). However, as stated earlier, we solve the large scale linear program which represents the deterministic equivalent version of the stochastic model. We utilize *Xpress-MP* as our solver of choice on a Dell computer using the Windows 8 operating system with an Intel Xeon processor running on a CPU at 2.70 GHz. We typically use 5000 scenarios in our sampling which results in 115,000 constraints and 60,006 variables in the linear programming model. Solution time, 33 s, is quite reasonable and the biggest advantage of solving the stochastic linear program is that it is independent of whether the capacity constraint is binding or not and it can consider truncated correlated demands within and across regions. It is thereby much more generalizable to be used in the field.

Generalized TSP Model

The solution procedure of the generalized TSP model could be construed as stage 3.

Stage 3: The BC, constrained by the optimum quantity of blood units that it has acquired in stage 1 and contingent on demand revealed in stage 2, delivers the feasible quantity of blood units to each hospital via an optimal routing, taking into account the relative distances of hospitals and vehicle capacity. Once the optimal quantity for each hospital has been determined, the optimal routes can be determined based on the following single period single stage model that involves not only selecting the optimal routes, but also how many tours there are to be and the order the hospitals in each tour are visited using a generalized TSP. This model minimizes the total distance traveled and, since each hospital has to be delivered to only once, also minimizes the total cost.

The disaster area is divided into three regions and a single vehicle serves each hospital i. The BC operates vehicles with limited capacity for distributing whole blood units to each hospital i. We formulate the TSP integer program based on Christofides, Mingozzi and Toth (1981).

Notation

$r = 0,1,2\ldots|R| =$ index of regions where 0 = BC

$i_r = 1,2\ldots|I_r| = i^{\text{th}}$ hospital of rth region; $i_r = \begin{cases} 1,2\ldots|I_r| \text{ for } r = 1,2,\ldots|R| \\ 1 \text{ for } r = 0 \end{cases}$

$q_{i_r}^{*r}$ = Optimal quantity of the blood needed in each hospital i in region r

$$m_{i_r,j_r}^{r} = \begin{cases} 1 \text{ if hospital } j \text{ is immediately followed by hospital } i \text{ in a route} \\ 0 \text{ otherwise.} \end{cases}$$

$d_{i_r,j_r}^{r} =$ Distance of hospital i from hospital j

$\sum_{r=1}^{|R|} \sum_{i=1}^{|I|} q_{i_r}^{r} = K$ Maximum capacity of vehicle which is equal to the total blood units acquired by BC

$n_{i_r}^{r} =$ Total amount of blood delivered on the route that includes up to hospital i in region r including hospital i

The objective function to minimize the distance traveled can be expressed as:

$$Z = Min. \sum_{r=0}^{|R|} \sum_{i_r}^{|I_r|} d_{i_r,j_r}^{r} m_{i_r,j_r}^{r} \tag{9.10}$$

$$s.t. \ \sum_{r=0}^{|R|} \sum_{i_r}^{|I_r|} m_{i_r,j_r}^{r} = 1, \ \forall_{j_r} \tag{9.11}$$

$$\sum_{r=0}^{|R|} \sum_{i_r}^{|I_r|} m_{i_r,j_r}^{r} = 1, \ \forall_{i_r} \tag{9.12}$$

$$q_{i_r}^{*r} \le n_{i_r}^{r} \le K, \ \forall_{i_r} \tag{9.13}$$

$$n_{i_r}^{r} \le K + \left(q_{i_r}^{*r} - K\right) m_{i_r,j_r}^{r}, \ \forall_{i_r} \tag{9.14}$$

$$n_{j_r}^{r} \ge n_{i_r}^{r} + q_{j_r}^{*r} - K + K m_{i_r,j_r}^{r} + \left(K - q_{j_r}^{*r} - q_{i_r}^{*r}\right) m_{i_r,j_r}^{r}, \ \forall_{i_r,j_r} \tag{9.15}$$

$$n_{i_r}^{r} \ge 0, \ \forall_{i_r} \tag{9.16}$$

$$m_{i_r,j_r}^{r} \in \{0,1\}, \ \forall_{i_r,j_r}, i_r \ne j_r. \tag{9.17}$$

Constraint (9.11) and Constraint (9.12) ensure that a hospital must be visited exactly once and delivery enters and leaves each region exactly once. Constraint (9.13) ensures that the total amount of blood delivered on the route is as large as the quantity needed by hospital i and within the capacity of the vehicle. Constraint (9.14) states that if hospital i is the first hospital of the tour, then the $n_{i_r}^{r}$ is equal to $q_{i_r}^{*r}$. Constraint (9.15) ensures that the quantity delivered from the BC up to hospital j is no less than the quantity delivered between hospitals i and j on the tour, a quantity that needs to be reduced by the delivery at hospital j. Constraint (9.16) and (9.17) indicate that the variable n_{i_r} is non-negative and m_{i_r,j_r}^{r} is binary, respectively.

Table 9.1 Parameters and initial values for MPNP

Parameters	Initial values
$c_{i_r}^r = c_{1_1}^1, c_{2_1}^1, c_{1_2}^2, c_{2_2}^2, c_{1_3}^3, c_{2_3}^3$	10
$v_{i_r}^r = v_{1_1}^1, v_{2_1}^1, v_{1_2}^2, v_{2_2}^2, v_{1_3}^3, v_{2_3}^3$	55
$p_{i_r}^r = p_{1_1}^1, p_{2_1}^1, p_{1_2}^2, p_{2_2}^2, p_{1_3}^3, p_{2_3}^3$	100
K	400
$D_{1_1}^1 = N\left(\mu_{1_1}^1, \sigma_{1_1}^1\right)$	$N(100,40), 20 < D_{1_1}^1 < 180$
$D_{2_1}^1 = N\left(\mu_{2_1}^1, \sigma_{2_1}^1\right)$	$N(100,40), 20 < D_{2_1}^1 < 180$
$D_{1_2}^2 = N\left(\mu_{1_2}^2, \sigma_{1_2}^2\right)$	$N(80,24), 10 < D_{1_2}^2 < 100$
$D_{2_2}^2 = N\left(\mu_{2_2}^2, \sigma_{2_2}^2\right)$	$N(80,24), 10 < D_{2_2}^2 < 100$
$D_{1_3}^3 = N\left(\mu_{1_3}^3, \sigma_{1_3}^3\right)$	$N(50,10), 30 < D_{1_3}^3 < 70$
$D_{2_3}^3 = N\left(\mu_{2_3}^3, \sigma_{2_3}^3\right)$	$N(50,10), 30 < D_{2_3}^3 < 70$
$\rho_r = \rho_1, \rho_2, \rho_3$	0.5

9.4 Numerical Example & Results

In order to further illustrate the above models, we present a numerical example based on the problem defined above. We solve the problem using *Xpress-MP* as our solver of choice and with 5000 scenarios. Each hospital has the same cost structure. Penalty costs refers to the social costs that occurs from unmet blood demand. In this example, we assign \$100 (a representative weighting intended to cover the cost of suffering and loss of life). Table 9.1 shows the parameter values that are used for the MPNP simulation:

We provide the results of the MPNP model, optimal blood units acquired by each hospitals and optimal total cost comprised of operational and social costs in Table 9.2.

Once the optimal blood units are determined, we solve the generalized TSP to find the best possible route that minimizes the total distance for distributing the optimal blood units acquired, again using *Xpress-MP*. We assume that the distance is equal to the cost of transportation. For each km traveled, transportation cost is \$1. The problem involves 6 hospitals in 3 regions, and a single vehicle. We assume that the vehicle capacity is 400 which is equal to the BC's capacity. Table 9.3 depicts the distance matrix between the hospitals and the BC.

Results of the TSP indicate that the total distance/transportation cost is 589/\$589 and the total quantity delivered is 400 which is equal to the vehicle/BC capacity. The best possible route $\left(\mathrm{BC} \rightarrow \mathbf{1}_1 \rightarrow \mathbf{2}_3 \rightarrow \mathbf{2}_2 \rightarrow \mathbf{1}_2 \rightarrow \mathbf{2}_1 \rightarrow \mathbf{1}_3 \rightarrow \mathrm{BC}\right)$ is shown in Fig. 9.3 below.

Table 9.2 Optimal blood units acquired by each hospital and the optimal total cost

	Blood units acquired
$q_{1_1}^{*1}$	80.1666
$q_{2_1}^{*1}$	81.6829
$q_{1_2}^{*2}$	68.8222
$q_{2_2}^{*2}$	69.2403
$q_{1_2}^{*3}$	50.2752
$q_{2_2}^{*3}$	49.8128
Total cost	*$14,434.9*

Table 9.3 Distance Matrix (in km)

	BC=0	1_1	2_1	1_2	2_2	1_3	2_3
BC=0	*0*	130	145	198	156	116	127
1_1	130	*0*	155	72	169	89	75
2_1	145	155	*0*	42	112	42	136
1_2	198	72	42	*0*	85	120	63
2_2	156	169	112	85	*0*	89	99
1_3	116	89	42	120	89	*0*	61
2_3	127	75	136	63	99	61	*0*

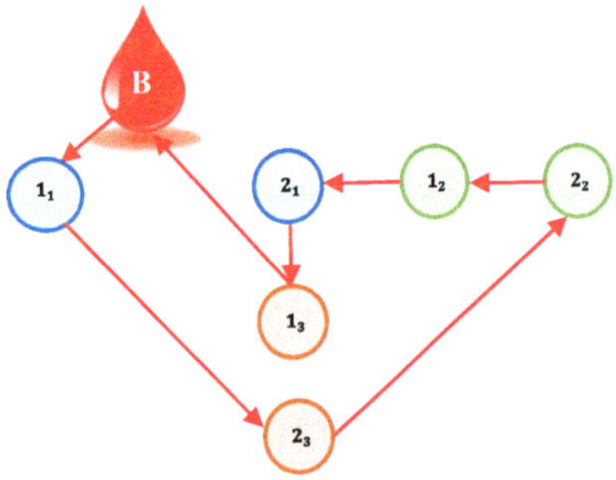

Fig. 9.3 The optimal delivery route

Sensitivity Analysis

In this section, we analyze the sensitivity of optimal blood units to penalty cost and the demand correlation. However, our goal in this study is by no means to do a full experimental design, although that would be a legitimate exercise to understand the nuances of various parameters value combinations.

Table 9.4 Optimal solutions under various penalty cost

$K = 400$	High	Medium	Low
$p_{i_r}^r$	*$500*	*$250*	*$100*
$q_{1_1}^{*1}$	81.998	81.5692	80.1666
$q_{2_1}^{*1}$	83.4250	82.9948	81.6829
$q_{1_2}^{*2}$	69.9944	69.5972	68.8222
$q_{2_2}^{*2}$	70.3037	69.9969	69.2403
$q_{1_3}^{*3}$	47.1903	47.9766	50.2752
$q_{2_3}^{*3}$	47.0886	47.8653	49.8128
Total Costs	*$50,223.5*	*$27,875.3*	*$14,434.9*

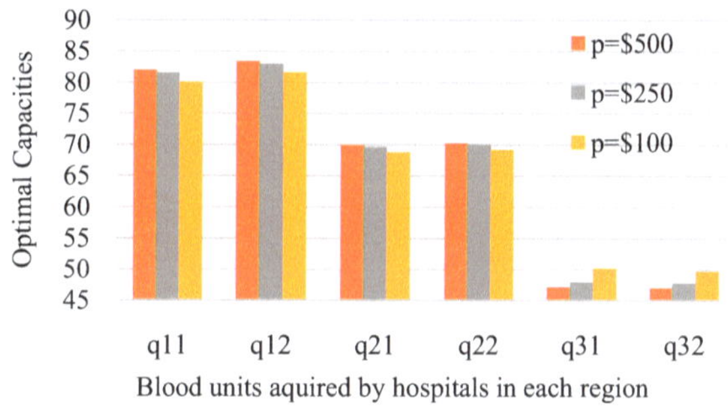

Fig. 9.4 Sensitivity of optimal blood units to $p_{i_r}^r$

First, we consider different levels of penalty costs. We assume that the unit cost $\left(c_{i_r}^r = 10\right)$, the holding cost $\left(v_{i_r}^r = 50\right)$, and the demands are constant. Table 9.4 shows the optimal blood units against three levels of penalty costs $\left(p_{i_r}^r\right)$.

Figure 9.4 demonstrates the optimal blood units against different levels of penalty costs. As seen in Fig. 9.4, the optimal blood units acquired by each hospital are not very sensitive to the changes in penalty costs. However, there is a significant difference between the optimal total costs as expected. The optimal acquired blood units' low sensitivity to the penalty cost can be explained due to the capacity constraint, which is tight, but has K=400. Therefore, next, we consider the effects of penalty costs under a tighter capacity constraint with K=200, holding all else constant. According to Table 9.5 and Fig. 9.5, there are slight changes in the optimal blood units against penalty cost change when the capacity is tighter. A comparison of Table 9.4 and Table 9.5 shows that under tight capacity total costs are higher, which is attributable to the penalty of not meeting demand. What is more interesting however is that in both Fig. 9.4 and 9.5, the blood units acquired in region 3 actually *decrease* in the penalty costs. This represents a subtle substitution effect. The coefficient of variation of region 3 is lower than that of regions 1 and 2.

Table 9.5 Optimal solutions under various penalty cost

$K = \mathbf{200}$	High	Medium	Low
$p_{i_r}^r$	*\$500*	*\$250*	*\$100*
$q_{1_1}^{*1}$	28.879	28.812	27.7661
$q_{2_1}^{*1}$	30.8039	30.4549	30.0912
$q_{1_2}^{*2}$	37.3077	37.0856	36.7422
$q_{2_2}^{*2}$	36.9247	36.6725	36.3329
$q_{1_3}^{*3}$	33.4679	33.9785	34.8181
$q_{2_3}^{*3}$	32.6168	32.9965	34.2495
Total Costs	*\$130,951*	*\$66,512.2*	*\$27,847.2*

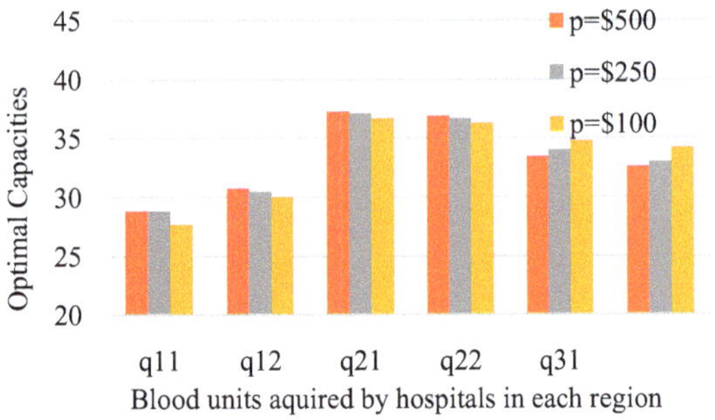

Fig. 9.5 Sensitivity of optimal blood units to $p_{i_r}^r$

Thereby, as the penalty cost increases, the increased blood unit acquisition in regions 1 and 2 is somewhat offset by a decrease in blood unit acquisition in region 3. Note that the overall capacity constraint is still binding, which in essence leads to the substitution.

We also experiment with different levels of correlations with the following levels of: unit cost $\left(c_{i_r}^r = 10\right)$, holding cost $\left(v_{i_r}^r = 55\right)$, and penalty cost $\left(p_{i_r}^r = 100\right)$. Table 9.6 shows how different levels of correlations between demands effect the optimal acquired blood units. Figure 9.6 illustrates the effect of changes in demand correlations on optimal blood units. From Table 9.6 and Fig. 9.6, it seems that optimal blood units are not very sensitive to the changes in the demand correlations. However, there are differences in the optimal total costs. From Table 9.6,

in which capacity is not as tight as in Table 9.7, we observe that *medium* correlation in demand actually yields lesser cost. Whereas, Table 9.7 shows that under tighter capacity, *high* correlation in demand yields less cost. Furthermore the acquisition of blood units within some regions does not necessarily follow a given direction such as monotonically increasing or decreasing in correlation (see Fig. 9.7). Once again, we observe substitution effects that take into account the optimization of the entire system rather than each region alone.

Table 9.6 Optimal solutions under various correlation coefficient

$K = 400$	High	Medium	Low
ρ_r	*0.9*	*0.5*	*0.3*
$q_{1_1}^{*1}$	80.6538	80.1666	80.2706
$q_{2_1}^{*1}$	81.4084	81.6829	81.6108
$q_{1_2}^{*2}$	69.0592	68.8222	68.8803
$q_{2_2}^{*2}$	68.2789	69.2403	69.08
$q_{1_3}^{*3}$	50.4723	50.2752	50.3462
$q_{2_3}^{*3}$	50.1274	49.8128	49.8122
Total costs	*\$14,481.5*	*\$14,434.9*	*\$14,440.6*

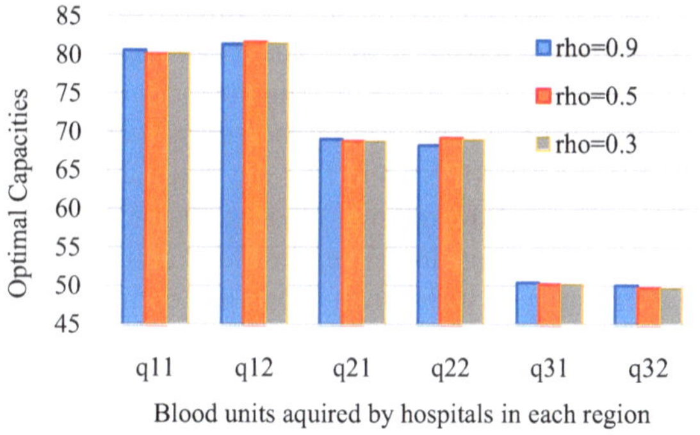

Fig. 9.6 Sensitivity of optimal blood units to ρ_r

Table 9.7 Optimal solutions under various correlation coefficient

$K = 200$	High	Medium	Low
ρ_r	*0.9*	*0.5*	*0.3*
$q_{1_1}^{*1}$	28.4255	27.7661	27.63
$q_{2_1}^{*1}$	29.2645	30.0912	30.0423
$q_{1_2}^{*2}$	36.8244	36.7422	36.1454
$q_{2_2}^{*2}$	36.621	36.3329	37.2513
$q_{1_3}^{*3}$	34.8195	34.8181	34.6957
$q_{2_3}^{*3}$	34.0452	34.2495	34.2352
Total costs	*\$27,830.9*	*\$27,847.2*	*\$27,863.8*

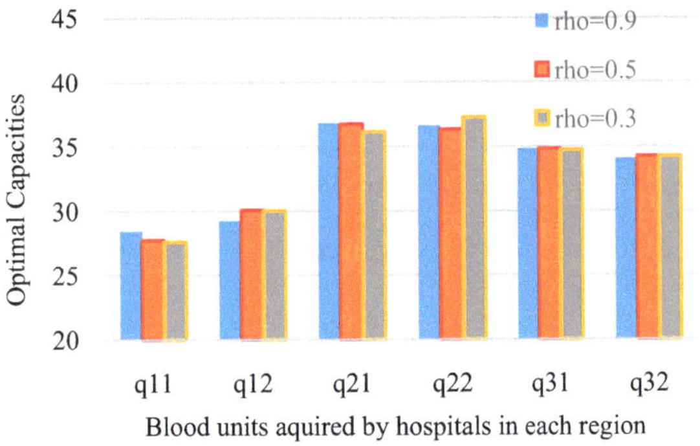

Fig. 9.7 Sensitivity of optimal blood units to ρ_r

9.5 Conclusions and Future Research

In this study, inspired by the real world example of blood distribution under intense bombing in Syria, we develop a two stage MPNP model with capacity constraint and a generalized TSP model. We contribute to the literature by utilizing a combination of extant models in the context of humanitarian aid, specifically the distribution of blood products, while making the models amenable to solution by commercially available solvers and thereby increasing their real-world applicability.

We solve a MPNP and find the optimal blood units acquired by each hospital. Then, we feed the optimal blood units into a generalized TSP model and find the best possible route to deliver the blood units cost effectively. Next, we conduct sensitivity analysis with penalty costs and correlation coefficients being the parameters of interest. The sensitivity analysis sheds light on interesting substitution effects that are a direct outcome of the level of tightness of the capacity constraint. They show that blood acquisitions in a given region may actually decrease with increased penalty costs and that the lowest total costs may be achieved at medium correlation and not necessarily at low or high correlation.

The proposed models in this study could be extended by incorporating other costs such as cost of wasted whole blood. The model of MPNP-TSP can be extended by considering a scenario where there are three regions and three hospitals and the hospital demands in each region are correlated. Another important and interesting extension could be developing an integrated MPNP- generalized TSP model and comparing the results of such an integrated model with the models considered in this chapter.

This study considers a centralized inventory management where the BC controls the inventory and routing. Future studies may consider decentralized inventory management where hospitals themselves control the inventory and the routing. In addition, since this study uses a disaster scenario, future studies could consider optimizing time to delivery as well. Future studies also can include different types of blood products such as blood platelets, red blood cells, plasma, and frozen blood that require different capacities. Finally, another extension could be to consider blood units with various ages (fresh or almost expired) since fresh blood units would be given more storage space than older blood units.

References

Andersson H, Hoff A, Christiansen M, Hasle G, Løkketangen A (2010) Industrial aspects and literature survey: combined inventory management and routing. Comput Oper Res 37(9):1515–1536. doi:http://dx.doi.org/10.1016/j.cor.2009.11.009

Beamon BM, Balcik B (2008) Performance measurement in humanitarian relief chains. Int J Public Sect Manage 21(1):4–25. doi:10.1108/09513550810846087

Beliën J, Forcé H (2012) Supply chain management of blood products: a literature review. Eur J Oper Res 217(1):1–16. doi:http://dx.doi.org/10.1016/j.ejor.2011.05.026

Chien TW, Balakrishnan A, Wong RT (1989) An integrated inventory allocation and vehicle routing problem. (English). Transp Sci 23(2):67

Christofides N, Mingozzi A, Toth P (1981) Exact algorithms for the vehicle routing problem, based on spanning tree and shortest path relaxations. Math Program 20(1):255–282

Dror M, Ball M, Golden BL (1985) A computational comparison of algorithms for the inventory routing problem. Ann Oper Res 4(1):3–23

Erlebacher SJ (2000) Optimal and heuristic solutions for the multi-item newsvendor problem with a single capacity constraint. Prod Oper Manage 9(3):303–318. doi:10.1111/j.1937-5956.2000.tb00139.x

Federgruen A, Zipkin P (1984) A combined vehicle routing and inventory allocation problem. Oper Res 32(5):1019–1037. doi:10.2307/170651

Federgruen A, Prastacos G, Zipkin PH (1986) An allocation and distribution model for perishable products. Oper Res 34(1):75–82. doi:10.2307/170672

Godfrey GA, Powell WB (2001) An adaptive, distribution-free algorithm for the newsvendor problem with censored demands, with applications to inventory and distribution. Manage Sci 47(8):1101–1112. doi:10.2307/822596

Golden B, Assad A, Dahl R (1984) Analysis of a large scale vehicle routing problem with an inventory component. Larg Scale Syst 7(2–3):181–190

Hadley G, Whitin TM (1963) Analysis of inventory systems. Prentice-Hall, Englewood Cliffs

Hemmelmayr V, Doerner KF, Hartl RF, Savelsbergh MWP (2009) Delivery strategies for blood products supplies. OR Spectr 31(4):707–725. doi:10.1007/s00291-008-0134-7

Katsaliaki K, Brailsford SC (2006) Using simulation to improve the blood supply chain. J Oper Res Soc 58(2):219–227

Kelle P, Woosley J, Schneider H (2012) Pharmaceutical supply chain specifics and inventory solutions for a hospital case. Oper Res Health Care 1(2–3):54–63

Kendall KE, Lee SM (1980) Formulating blood rotation policies with multiple objectives. Manage Sci 26(11):1145–1157. doi:10.2307/2631180

Kleywegt AJ, Nori VS, Savelsbergh MWP (2002) The stochastic inventory routing problem with direct deliveries. Transp Sci 36(1):94

Lau H-S, Hing-Ling Lau A (1996) The newsstand problem: a capacitated multiple-product single-period inventory problem. Eur J Oper Res 94(1):29–42. doi:http://dx.doi.org/10.1016/0377-2217(95)00192-1

MSF (2013) In Syria's idlib province, little medical care for civilians living under intense bombing. http://www.doctorswithoutborders.org/news-stories/special-report/syria-two-years-failure-international-aid. Accessed 26 Jul 2013

Nagurney A, Masoumi A, Yu M (2012) Supply chain network operations management of a blood banking system with cost and risk minimization. Comput Manage Sci 9(2):205–231. doi:10.1007/s10287-011-0133-z

Perez N, Holguín-Veras J, Mitchell J, Sharkey T (2010) Integrated vehicle routing problem with explicit consideration of social costs in humanitarian logistics. Rensselaer Polytechnic Institute 19(2):12–19. http://transp.rpi.edu/~HUM-LOG/Doc/Vault/IRP.pdf. Accessed 1 May 2015

redcrossblood.org (2013) Blood components. http://www.redcrossblood.org/learn-about-blood/blood-components. Accessed 11 Jul 2013

Şahin G, Süral H, Meral S (2007) Locational analysis for regionalization of Turkish Red Crescent blood services. Comput Oper Res 34(3):692–704. doi:http://dx.doi.org/10.1016/j.cor.2005.03.020

Sebastian HWS, Wilding R, Yates N, Cotton S (2012) What drives perishable inventory management performance? Lessons learnt from the UK blood supply chain. Supply Chain Manage 17(2):107–123. doi:http://dx.doi.org/10.1108/13598541211212861

Trudeau P, Dror M (1992) Stochastic inventory routing: route design with stockouts and route failures. Transp Sci 26:171–184

Chapter 10
The Vaccine Supply Chain Multathlon: the Reconciliation of Technology, Economy and Access to Medicines

Catherine Decouttere, Nico Vandaele, Stef Lemmens and Mauro Bernuzzi

Abstract Many rigorous models have been developed to support the design of humanitarian supply chains. Supply chains supportive of Access to Medicines (ATM), like vaccine supply chains impose considerable additional challenges on this design process. We embed these models in a broader stakeholder based framework. This will substantially enhance the societal and human impact of the ATM supply chain service delivery. This broader base of stakeholders will deliver a balanced set of Key Performance Indicators, against which the new design options or scenarios will be evaluated. These scenarios will be the outcome of an iterative design and modelling process moderated by a group of key stakeholders. Subsequently, a multi-criteria ranking method will reveal a shortlist of championing scenarios. Finally, a group decision process will decide on the final supply chain design choice. Moreover, the acceptance and buy-in of the (re)designed supply chain will be much more evident for all stakeholders involved. The main purpose of this paper is to present a general framework for ATM supply chain design which constitutes a stakeholder's framework encapsulating a rigorous modelling approach. We build upon our own experience within a vaccine supply chain.

Keywords Stakeholder analysis · Scenario planning · Multi-criteria evaluation · Key performance indicators · Supply chain design · Access to medicines · Vaccine industry · Group decision making · Rotavirus · Data envelopment analysis

---

C. Decouttere (✉) · N. Vandaele · S. Lemmens
Katholieke Universiteit Leuven, Naamsestraat 69, 3000 Leuven, Belgium
e-mail: catherine.decouttere@kuleuven.be

N. Vandaele
e-mail: nico.vandaele@kuleuven.be

S. Lemmens
e-mail: stef.lemmens@kuleuven.be

M. Bernuzzi
GSK Vaccines, Avenue Fleming 20, 1300 Wavre, Belgium
e-mail: Mauro.bernuzzi@gsk.com

C. W. Zobel et al. (eds.), *Advances in Managing Humanitarian Operations*,
International Series in Operations Research & Management Science 235,
DOI 10.1007/978-3-319-24418-1_10

10.1 Introduction

The initial goal of any health care system is not only to address the medical needs of individuals and populations but also involves other factors affecting the general well-being of individuals and societies. The three main goals of a health care system, as stipulated by the World Health Organization (WHO) are: health improvement, responsiveness and fairness in financial contribution (Murray and Frenk 2000; Musgrove et al. 2000). Equally, in a vaccine supply chain design context, these goals constitute the underlying ground for the multi-criteria evaluation of the way a vaccine supply chain is designed: the ATM dimension needs to co-exist with the economic and technological ambitions. This boils down to the observation that a good supply chain design will make the best feasible combination of these multi-criteria evaluation metrics in order to reach as much as possible the aspirations of all stakeholders involved.

Vaccination is generally considered as an important means to enhance the well-being of the earth's population. However, the design and operation of a vaccine supply chain is extremely challenging. It relies heavily on the ability to design the supply chain in such a way that the humanitarian operations are supported in their best conditions. First, like in many other supply chains, there is the permanent striving for balancing supply and demand. Secondly, in the case of vaccine supply chains, some additional features are present which make the supply/demand balancing act more complex. As most of these features will become clear throughout this chapter, at this point it is worthwhile to mention an extreme reliance on quality issues and the presence of a complex stakeholders' structure. Today, like in many other health care systems, medical technology pushes ahead with an ever increasing pace in order to leapfrog global health concerns. However, the cost of health care is also growing rapidly and finally a balancing act from a societal and humanitarian point of view is also mandatory. This results in a complex set of trade-offs, which is clearly very outspoken in a vaccine supply chain. In order to cope with this social responsibility dimension we developed a framework to analyze and support the decision making of the vaccine supply chain. Nevertheless, our findings are applicable to any other supply chain, which is supported by other work (Yadav et al. 2013).

A vaccine supply chain is not only characterized by the typical issues every supply chain is concerned with, but on top of that it also contains some specificities which heavily aggravate the three dimensional balancing act mentioned above (Vandaele and Decouttere 2013). Firstly, technology-wise we have to take the extremely rigorous quality and regulatory requirements, long manufacturing lead times and cold chain operations into account. Also belonging to this category is a complex market structure, characterized by tenders, random demand and repeat orders. All three sources of demand prominently compete for the shared, limited capacity. Additionally, the occurrence of armed conflicts and natural disasters is shaking the market needs almost permanently. Secondly, from the economic point of view, both the inherent high cost of production and complex pricing is not without compromise on a global scale. Profit is not easily brought in line with the delivery and worldwide availability of vaccines. This leads to important considerations like decisions related to the (dis)continuation of the production of vaccines with a decreasing market potential (due to eradication, government policies, competition, donor's policies, etc.), the prioritization on the use of scarce and shared resources and the impact of the vaccine product portfolio decisions on governmental and institutional budgets, only to mention a few important ones. Thirdly, the ATM dimension, embedded in the mission and vision of both private, public and non-governmental stakeholders together with their alliances comes into the picture. This is expressed in terms of saving lives, of religious issues, of public safety, of military and anti-terrorist interest, among others. Similar findings have been reported (Samii et al. 2012; Yadav et al. 2013), where insights have been formulated from other vaccine supply chains as well as from non-vaccine supply chains, for example from food and beverage.

Also it became clear (Thakur et al. 2012; Vandaele et al. 2003) that the effective adoption of any newly designed health care system depends on the support it gets from the key stakeholders involved, e.g. the health care providers on the field may be dissatisfied with the way a vaccine has to be administered. All of these contribute on top of the issues raised from the demand side. They turn the vaccine supply chain design exercise into a very complex problem (Beyan and Baykal 2012).

Typically the modelling of a health-care supply system has been directed in a bottom-up way: adding more and more incremental improvements to the operational models under study (Rechel et al. 2010; Taboada et al. 2011) and measuring their performance by means of a limited selection of tangible performance dimensions such as technical capacity, waiting times, cost of care (Brailsford et al. 2004) etc. In this chapter we look at the vaccine supply chain design problem in a top-down perspective. Starting from a human-centred design approach that aims at a sustainable future supply chain concept and not from a modelling point of view that aims at improving an existing supply chain in an incremental way. With this perspective, starting from the objectives of a global or national health system, the goals for the subsystems such as the manufacturer's supply chains can be derived and the stakeholders' needs can be revealed. This results in a set of system requirements and performance indicators which encompasses the more diverse aspects of health improvement effectiveness, cost efficiency, vaccine availability and fairness. It is clear that some inherently conflicting goals need to be brought into balance. This is the very reason why we propose an integrated approach for the design and modelling of the vaccine supply chain.

Vaccine supply chains need to excel on both technical, economic, and a vast amount of social aspects such as ATM. Due to the multitude of stakeholders involved, it is a real challenge to identify improvements for an existing vaccine supply chain or to design radically new health care systems leading to an overall better societal, economic and technical performance. A good starting point is contained in a patient-centred design approach instead of a disease-centred one. Such an approach may contribute to deliver radical steps forward (Barry and Edgman-Levitan 2012). There are important stakeholder alliances and strategic initiatives in the global vaccine system that surround the vaccine supply chain. As significant budgets are involved, radical changes to the vaccine supply chain cannot be realised by a sole stakeholder on its own. A strategic group-decision will precede the adoption of a new system, where multiple stakeholders' perspectives are taken into account.

In this chapter we will mainly focus on the derivation of the Key Performance Indicators (KPI's) from the stakeholders' needs as this is the basis for balancing the different performance measures in order to be sustainable not only on the technological and economic dimensions but also on the ATM dimensions. We illustrate our findings with evidence from a vaccine manufacturer. The mission of our chapter is to show that reconciliation on technological, economic and humanitarian aspects is feasible and preserves sustainability in the broad sense. Our concern will be how to obtain the relevant performance measures to support this reconciliation effort. The key element is a stakeholder perspective and a scenario based approach. Within this respect, the next section describes a brief literature review focussing on the reconciliation of economical, technological and humanitarian value driven performance measures in model based supply chain network design. Subsequently, Sect. 10.3 describes our suggested integrated stakeholder approach, combining user centred design principles with mathematical models. Section 10.4 concludes the chapter.

10.2 Current Literature on Vaccine Supply Chain Design

This section gives a short literature review on supply chain design, with focus on vaccine supply chain design. Based on the observations in the literature we concluded that there is a considerable added value in terms of adding a stakeholders' view to (vaccine) supply chain design. However, as will be explained in the next section, in our suggested approach, rigorous modelling remains a core element.

To a large extent, the literature on traditional model-based supply chain design falls short on the reconciliation of the technological, economic and ATM performance dimensions of a vaccine supply chain design model (Lemmens et al. 2014). In the latter paper, a recent literature review on supply chain design has been conducted which shows a clear dominance of the single criterion optimization/modelling approach. Despite the existence of a wide range of different extensively elaborated economic performance measures in the supply chain design literature, the presence of technological performance measures is still subordinate while the societal dimension is even more scarce.

Almost every supply chain design or ATM relief distribution manuscript uses at least one of the following economic KPI's: costs, profit, net present value (Guillén et al. 2005; Guillén-Gosálbez and Grossmann 2009; You and Grossmann 2008) or financial risk (Azaron et al. 2008; Franca et al. 2010; Guillén et al. 2005; Hamedani et al. 2013). The yearly facility operating cost, capacity expansion investments, production, transportation, inventory holding costs and shortages costs are the most common drivers that are taken into account for incorporating these KPI's.

For our purposes, the performance of a supply chain design scenario must be measured by multiple criteria. The vaccine supply chain is considered as a slow and unresponsive supply chain (Shah 2004; Shah 2005), motivating us to study the integration of responsiveness, in our view a technological performance metric, into supply chain design. Indeed, a manufacturer with a responsive supply chain can meet the market demand in shorter lead times and react more quickly to the customer needs. Thereby it is a challenge to design a responsive vaccine supply chain, especially when taking unpredictable events into account such as a fast epidemic out-break and other natural or human-induced disasters. The current literature on model-based supply chain design offers various modeling definitions of supply chain responsiveness (Lemmens et al. 2014): the order delivery lead time from plants to customers through distribution centers (Cardona-Valdés et al. 2011; Olivares-Benitez et al. 2012; Naimi Sadigh et al. 2013), the order lead time from distribution centers to customers (Rajabalipour Cheshmehgaz et al. 2013) the expected lead time of a supply chain (You and Grossmann 2008) and the maximum guaranteed service time of the last stage of the supply chain (You and Grossmann 2011). The minimization of these performance metrics is an attempt to ensure a supply chain's responsiveness. Except for the latter two manuscripts mentioned above, the transportation time between the different supply chain's stages is the main driver for the supply chain's responsiveness measures. Nevertheless, the long manufacturing lead times and corresponding variability, due to the inherently low process robustness of the bio-tech processes, are highly relevant for vaccine supply chain design. However, integrating this uncertainty in a supply chain design model in a sophisticated way is, to the best of our knowledge, currently lacking and a challenge for future research (Lemmens et al. 2014).

The distribution of a limited amount of available aid in a humanitarian relief chain is studied extensively and the similarities and differences with the needs of a private supply chain are known (Bhattacharya et al. 2014). Some of the KPI's of these models, besides economic and technological considerations, clearly show the ATM dimension: ensuring a fair distribution of the aid (by fulfilling an equal proportion of each demand node) and prioritizing demand nodes under special conditions (by fulfilling a higher than equal proportion of the demand) (Ortuño et al. 2011; Vitoriano et al. 2009; Vitoriano et al. 2011). Such access performance metrics are non-existent in the literature on supply chain design. However, an increasing attention towards integrating the environmental sustainability dimension is observed instead: minimizing the carbon emissions of facilities and/or transportation modes by integrating pollution costs (Elhedhli and Merrick 2012), the environmental investment level (Wang et al. 2011) and life cycle assessment (Guillén-Gosálbez and Grossmann 2009) are recent modelling efforts that are part of the stream of "green supply chain design". This indicates the increasing attention towards integrating sustainability in supply chain design. We will be focusing on the integration of the social sustainability dimension.

When more than one objective is present, the most popular mathematical programming method to deal with supply chain design is multi-objective optimization (Lemmens et al. 2014): each criterion is modeled as an objective function and these objectives are optimized simultaneously. However, the number of Pareto-efficient points obtained by this method can be enormous and a way to choose a final implementable solution, satisfying the preferences of the decision makers, is often lacking. In the next section, we will propose a broader approach to overcome these difficulties in a generic way.

10.3 Vaccine Supply Chain Design from a Stakeholders' Point of View

The main lessons learned from the mathematical modelling approach in (vaccine) supply chain design are twofold. Firstly, the practical adoption of the newly designed system was not prevalent as many papers remained on the desktop level, a small number of papers used real-life data sets and only a very few were really implemented in practice. We believe that one of the important reasons for this can be found in a (single) stakeholder focus with limited view on other stakeholders' positions. Secondly, the system performance was exclusively measured on technical, operational and financial parameters, hereby strongly neglecting the socially related factors of the vaccine supply chain. It is clear that any health care system, from its intrinsic purpose, has a fundamental humanitarian goal of creating sustainable wellbeing for people (Musgrove et al. 2000). For vaccine supply chains in particular, this is a mandatory issue. Both observations led us to the conclusion that several important stakeholders must be involved. However, this is clearly situated beyond the scope of the mathematical modelling, so a broader perspective is necessary. In order to enable system designers and modelers to take into account these important aspects, we propose a five-step integrated design and modelling process, starting from the stakeholders as shown in Fig. 10.1. The next subsections of Sect. 10.3 are related to each one of these steps. It is also important to state that especially step 1 and 2 are based on structured expert interviews and workshops combined with findings from desktop literature research. We build upon our experience related to the Rotavirus supply chain.

Step 1: Map the Stakeholders Together with Supply Chain Goals and Limitations

In this section we will show how the vaccine supply chain design exercise benefits from a profound stakeholder analysis. Rather than writing a review on stakeholder analysis as such, we will apply the stakeholder analysis immediately to the design aspect of a vaccine supply chain. For the interested reader, we added some relevant references.

The starting point of the system definition is the articulation of the system goal: "what is the core service the system should deliver and for whom?" On the scale of an organization, or an alliance of organizations, this goal is expressed by its vision and mission statement. In the case of our example, the Rotavirus vaccine manufacturer's supply chain acts as a subsystem of the global immunization system as shown in Fig. 10.2. The latter is the worldwide system involved in preventive healthcare by means of immunization and is an integral part of the global health system. For the purpose of design and modeling, the subsystem considered in this chapter is the supply chain of a vaccine manufacturer for the production of a vaccine against Rotavirus. The Rotavirus has been selected on the basis of high quality availability of both capacity and supply related data and expert knowledge from GSK Vaccines. In

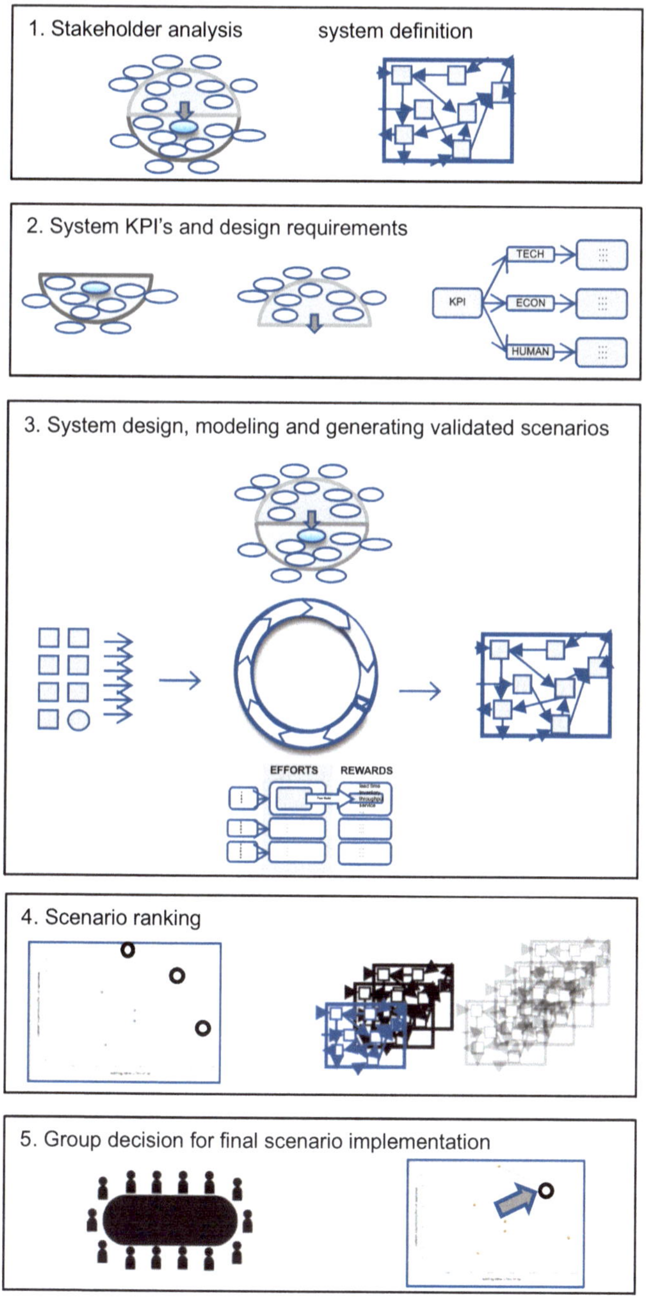

Fig. 10.1 Integrated vaccine supply chain design and modelling process

addition, Rotavirus allows an extended stakeholder analysis and from a supply point of view, presents itself in a rather isolated supply structure which makes modelling independent from the supply structures of other vaccines.

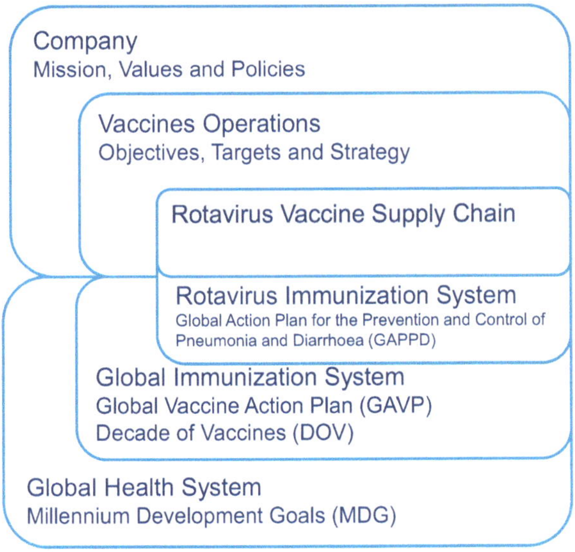

Fig. 10.2 Global immunization system and Rotavirus vaccine supply chain subsystem definition

The globally supported goals of the Global Health System and the Global Immunization System are declared in the Millennium Development Goals (MDG), the Global Vaccine Action Plan (GAVP) and for the specific case of Rotavirus immunizations, in the Global Action Plan for the Prevention and Control of Pneumonia and Diarrhea (GAPPD). On the other hand, vaccine manufacturing takes place in a private company and will be controlled by the company's mission, values, objectives and strategies.

Matching the goals derived from both the global health system and the company's strategy, the goal for the Rotavirus supply chain subsystem can be specified. An example for Rotavirus, in the case of low income countries, is shown in Fig. 10.3, which is based on publicly available communications from GSK Vaccines and subsequently served as company input. Our previously mentioned categories economy, technology and humanity serve as guidelines to preserve the balance in the derivation of KPI's related to the system goals. Obviously, some goals and KPI's may serve more than one dimension. For instance in capacity restricted vaccine production, throughput serves both a financial (turns into sales), a technological (throughput as a result of the production system) as well as a humanitarian goal (more immunized individuals).

The next step is to define the system boundaries together with its elements and to identify the stakeholders involved. The stakeholder concept has been defined in numerous ways since it was originally developed in the context of corporate strategy and defined by Freeman (2010) as follows: *"any group or individual who can affect or is affected by the achievement of the organization's objectives."* We apply this concept to a vaccine manufacturer's supply chain. Different stakeholders have a specific relationship to the organization and this may lead to conflicting interests. Stakeholder theory (Mitchell et al. 1997) proposes a typology and suggests ways

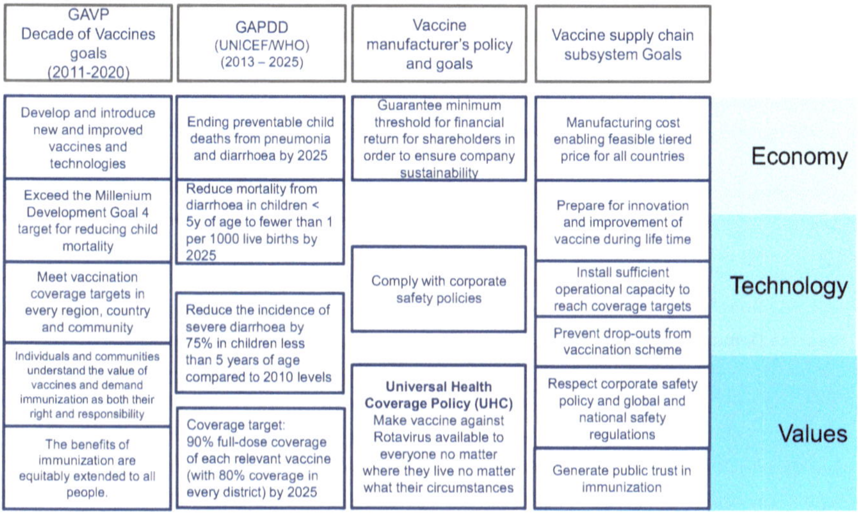

Fig. 10.3 Goal definition for Rotavirus vaccine supply chain subsystem

of working with the different types based on legitimacy, power and urgency. For applications in the field of R&D project management (Elias et al. 2002) and system design, specific stakeholder attributes such as interaction mode (functional, financial, decision power) (Donaldson et al. 2006) and stakeholder dynamics are of great importance (Solaimani et al. 2013).

For the purpose of vaccine supply chain design, we make the distinction between two groups of stakeholders: internal and external stakeholders. Internal stakeholders are those people, roles, activities or organizations directly or indirectly impacting and impacted by the specific vaccine production and delivery to the customer and final administration to the person to be immunized. Internal stakeholders will be actively involved in the design and the decision process of the new supply chain (see steps 3 and 5). External stakeholders are belonging to the close environment of the global immunization system. They have an impact on the performance of the supply chain, but are not impacted by the supply chain themselves. These stakeholders often play a major role in the determination of the available design options for the new supply chain but are typically beyond the control zone of the supply chain decision makers. They can however be influenced by other stakeholders or by the supply chain once it is operational and should therefore be taken into account during the design and decision process. Examples of external stakeholders are advisory and regulatory bodies such as WHO and FDA as well as funding organizations s such as the Global Alliance for Vaccine Immunization (GAVI) or the national governments. Also opinion spreading entities such as universities and religious groups are relevant external stakeholders. A stakeholder mapping for a vaccine supply chain for countries considering ATM as a priority is shown in Fig. 10.4. This stakeholder map is the result of reviewing different sources, among which structured interviews, workshops and literature are key.

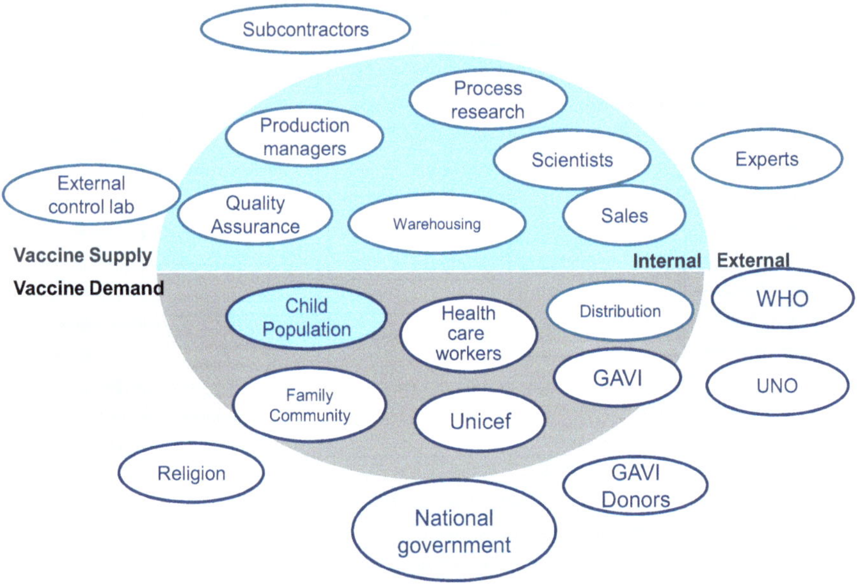

Fig. 10.4 Example of stakeholder mapping for a vaccine supply chain (example for low income countries)

Positive results of stakeholder mapping workshops in health care system design were shown earlier and stakeholder diagrams proved to be well understood by health care professionals (Buckle et al. 2006, 2010; Jun et al. 2009). For instance, by inclusion of the effective users (patients) as stakeholders, new and more ergonomic pharmaceutical packaging was developed in such a way that elderly and/or less abled patients can open the medication package easier, and more importantly, in a safer way. This stakeholder point of view led not only to the inclusion of the user in all its facets, but also introduced the general stakeholder view well known in design and R&D settings. Within the vaccine supply chain system as well as in the global vaccine system, some crucial alliances between stakeholders are present. They appear as stakeholders linked with a strategic interaction, representing their partnerships and common goals. Also on the service demand side, different stakeholders can be linked and networked. Examples are the child and its family, the community and the health care workers, the religious group whereto the child (or its parents) belongs, etc. For each of the stakeholders, the nature of interaction for the vaccine supply chain is determined: experience interaction, financial interaction or decision making interaction, as shown in Fig. 10.5. Please note that this is the stakeholder map for low income countries as Rotavirus is a key vaccine in these countries with a strong capability to save a large amount of lives. The interaction map allows to create an overview of the usually complicated interactions between the stakeholders. Its value lies in the structure, and nature of the interrelationships. However, the

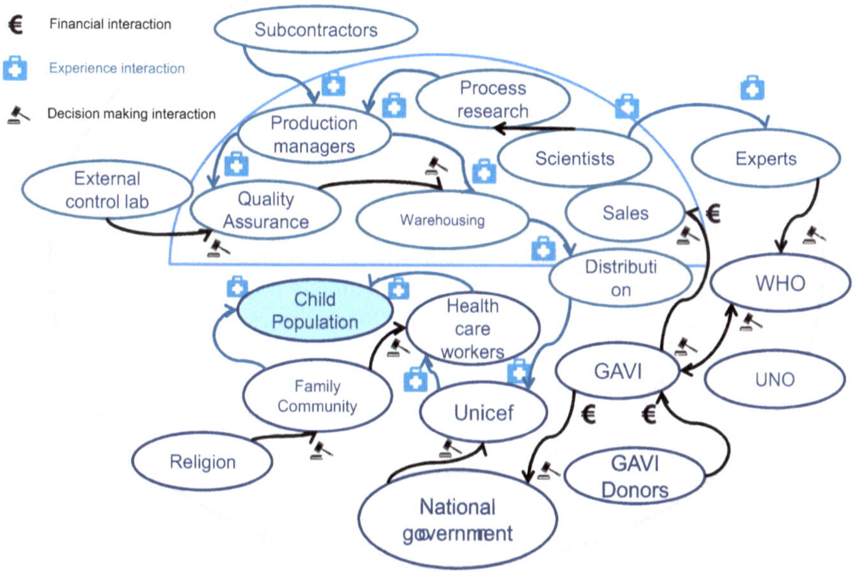

Fig. 10.5 The stakeholder interaction diagram for the vaccine supply chain (example for low income countries)

depth and complexity of the interactions cannot be derived from this type of representation. As an example, the interaction arrow between GAVI and the national governments shows a decision making type and a financial one. This expresses the position of GAVI as a decision maker on a country's eligibility for GAVI support, and on the consequently financial support itself.

The map for high income countries is definitely different, for instance related to the donors, governments, health care workers, etc.

Some of the internal stakeholders will be directly involved in the actual services delivered by the vaccine supply chain. They will participate either in the delivery or the reception of a service. This type of interaction is for the purpose of this chapter named 'experience interaction', and it covers all three levels of interaction: the functional level, the usability level and the emotional level (Norman 2007). The needs of these stakeholders and the experienced performance of the new supply chain designs, deliver the core information for human centred system design. The stakeholders with a functional interaction involve the children and their families, health care workers, clinicians and nurses. Other stakeholders will have a financial interaction with the system and do not necessarily take part in the health care service itself. They set restrictions to the system design in terms of investments and cost of operation e.g. the insurance companies or social security bodies in general. Donors are extremely important in the vaccine supply chain area, and they act pretty interrelated: fund raising bodies like GAVI determine to a large extent not only which vaccination campaign will be supported, but they also have a huge impact on the volume and timing of the required vaccines. UNICEF on the other hand can be considered as acting as the operational execution arm of the GAVI plans, which orders and works out the vaccination campaigns. A third kind of stakeholders are those with a strong decision power on the system design in terms of approving or rejecting a certain solution by the use of laws, rules or standards. They act as gatekeepers or constraints for the vaccine supply chain design and their aim can be the controlled acceptance of new technology with respect to ethics or safety or vaccination priority setting. The respective national health security organizations and WHO are examples of these. Stakeholders can interact on several levels at the same time, e.g. a patient interacts both on the functional as well as on the financial level, but usually not on the decision level. This is definitely true in the case of vaccines. In addition, stakeholders can also act under different roles. For instance, a vaccine manufacturer can also act as an in-kind donor of vaccines.

The manufacturer's vaccine supply chain design will be a subsystem of a global vaccine network as shown in Fig. 10.6. In order to find radically new solutions, the definition of this subsystem and its boundaries ideally does not start from an existing vaccine supply chain which should be improved. On the contrary, this subsystem should find its origin in its actual role and objective in the global system. The physical supply chain system capable of delivering the objectives will be the result of that, this is in contrast to the traditional supply chain design models described in Sect. 10.2. In our vaccine example, that supply chain is typically defined by the existing boundaries of the current as-is supply chain as shown in Fig. 10.6. On the other hand the future supply chain will be designed based on the future vaccination needs of the individuals and populations and the goals defined by the stakehold-

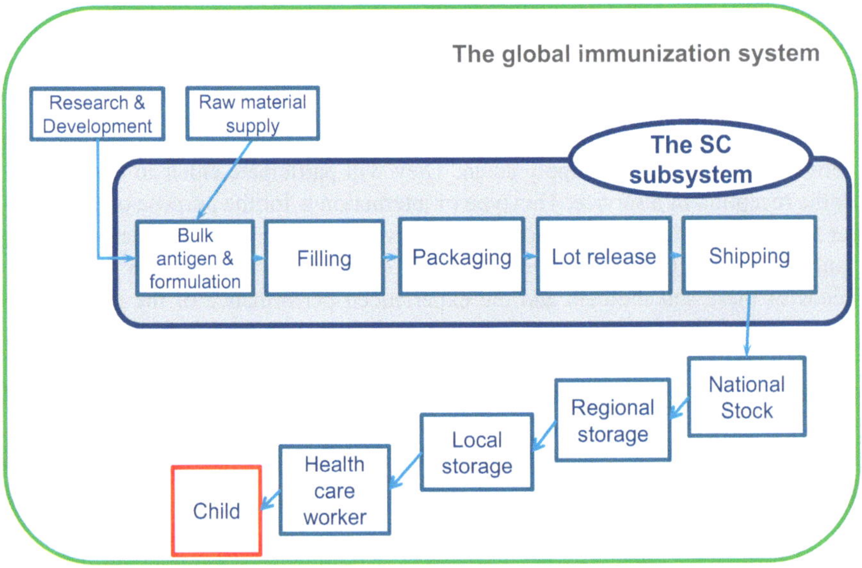

Fig. 10.6 The manufacturer's vaccine supply chain as a subsystem of the global vaccine network

ers involved in the immunization system. From a patient-centered perspective, the vaccine supply chain is a fragment in a broader health care lifecycle in which a patient walks through different stages and where a multitude of stakeholders are found each time the patient gets in touch with the vaccine supply chain. It is generally accepted that vaccination will avoid future health care processes or induces manageable repeat follow-up vaccinations. The supply chain design requirements will include aspects such as health safety, vaccination scheme adherence, cost per person immunized etc. All these are derived from the previous experiences by the NGO's, the immunized people, the manufacturer and other stakeholders. Therefore, the constituting elements and the boundaries of the vaccine supply chain have to be defined for the specific type of vaccine needed and the specific immunization goals for a chosen population or area. As a consequence, different supply chain subsystems may be part of the same global immunization system and/or vaccine operations (company) system. The boundaries will determine the flows, i.e. the kind, the number and timing of vaccines in the supply chain, the process steps taking place between ordering and delivering the vaccination service together with the resources available to support the processes. Both supply chain components, i.e. flows and resources, will constitute the basic building blocks of the modelling effort, exposed in step 3.

Step 2: Derive the Key Performance Indicators and Design Requirements of the Supply Chain

Based on the outcome of the previous step, the stakeholder map and the vaccine supply chain delineation, we need to derive the relevant performance measures together with the apparent thresholds and requirements. In this process step, the design research (Sanders 2008) and analysis is carried out. It involves a mix of qualitative research methods such as expert led workshops, brainstorms and structured interviews, in order to gain empathy with the non-governmental alliances, governments, health care workers and other stakeholders involved. The resulting insights deliver information complimentary to the formal and generally accepted goals set for the global immunization system and to the historical data from the current vaccine supply chain.

In the case that the global immunization goals may aim at eradication of the disease, like for poliomyelitis, the future flows may eventually dry up the vaccine supply chain revenues. Both the global as well as the company specific post-eradication plan is a key input to the design exercise. The same accounts for the introduction of new vaccines. Stakeholder theory, including the principle of stakeholder salience, enables an organization to focus on the most important stakeholder (see Mitchell et al. 1997). We propose an alternative to this approach and will not set an upfront priority to individual stakeholders. Instead, the needs of all relevant stakeholders in the global immunization system are considered to generate design requirements for the vaccine supply chain subsystem. These requirements define the boundaries of the multidimensional solution space for the vaccine supply chain system design. This iterative design mechanism is explained in step 3, where the outcome takes the form of a set of possible, yet validated system design options. A ranking based on a limited set of multi-dimensional KPI's will deliver the best system options for implementation, as shown in step 4. We demonstrate the above with a desktop example related to the immunization against Rotavirus

As it is reflected the WHO-UNICEF Global Immunization Vision and Strategy (Brown et al. 2011), the goal of reducing mortality due to vaccine-preventable diseases by two-thirds by 2015 triggers multi-dimensional objectives and involves many stakeholders. As a consequence, the Global Vaccine Action Plan (GVAP) endorsed in 2012 by 192 member states of the World Health Assembly, had brought together 290 organizations. The resulting plan provides a framework to prevent deaths by 2020 focussing on more equitable access to existing vaccines for people in all communities. The GVAP sets goals, proposes six strategic objectives and defines actions supporting their achievement (World Health Organization 2013). As revealed in the stakeholder analysis from step 3.1, the national governments play an important role on the decision making level. They select implementation strategies and plan their actions via comprehensive multi-year national plans (cMYP) (Kamara et al. 2013). A high impact on the success should also be attributed to local factors such as the behaviour of the parents of the children, the accessibility of the community for the health care workers or the cold chain infrastructure provided in the field.

On the other hand, on the supply side of the global immunization system, vaccine manufacturing companies express their role and intentions in the system through their mission statement, institutional values and strategic initiatives. The vaccine manufacturers may align through federations and they take part in overarching global initiatives from for example WHO or GAVI. From a system perspective, global goals for the immunization system need to be translated into design requirements for the vaccine supply chain subsystem. The example of Rotavirus is shown in Fig. 10.3. For the time being, we consider the Rotavirus vaccine subsystem to behave independently from vaccine supply chains for other vaccines. The latter will be relaxed in future research (Wang 2013).

Notice that the life time for the supply chain to be designed, is expected to be at least 30 years, where the quantitative foresight specified in global targets is only 5–10 years. A thorough understanding of stakeholder needs and scenario analysis allows to mitigate the uncertainty involved. The design requirements define the minimal conditions to be fulfilled in order to qualify as a new vaccine supply chain subsystem in the global immunization system. In order to find the best among the valid supply chain subsystem options, we apply a set of high level aggregate KPI's derived from both the GVAP global strategic objectives and from the company specific strategic objectives. Individual stakeholders' needs from people within the vaccine supply chain are collected by observations, interviews or workshops. Although these needs are often qualitative, their intensity can be measured on a relative scale.

Step 3: Construct Relevant Supply Chain Designs and Scenarios

Without loss of generality, from this point on, our five step procedure is illustrated using a desktop example. In product design, a human-centred approach starts from the user needs, usually the end-user, and takes a selection of other stakeholders into account. The user is preferably actively involved from the earliest stages in the design process, the idea generation phase throughout the concept definition and product development, for concept testing and prototype validation (Martin and Barnett 2012). A nice example of this in the vaccine supply chain context is the decision of the packaging and administration format: single- or multi-dose vials, oral applicators or tubes, etc. The impact of this exceeds the boundaries of the supply chain, and extends to product density, cold chain storage requirements, efficiency and safety of administration, infrastructure for administration, etc. All this clearly has a huge impact on many stakeholders and on the design and expected performance of the vaccine supply chain. When designing a complex product/service system such as a vaccine supply chain, a lot of different stakeholders' needs must to be taken into account simultaneously and conflicting requirements need to be resolved in the new service design (Clarkson et al. 2004).

The vaccine supply chain design process should actively involve the stakeholders both from the demand side and the supply side, and is based on design thinking and participatory design (Brown 2008; Rouck et al. 2008). Stakeholders with experience interaction will be invited to take part in the system design. Their current experience in an existing vaccine supply chain will inspire system and process designers to develop better solutions for unmet and for future needs. In this way, new supply chain concepts created can be radically different from the existing situation, and new technologies or new ways of working can be introduced. The resulting scenario will be tested by these stakeholders and improved accordingly. Concept development methodologies from engineering design and product design, for example Quality Function Deployment (QFD) and Vision in Product Design (ViP), can be applied in this stage (Akao 1990; Hekkert and van Dijk 2011). The concept is iteratively modelled, optimized and tested with stakeholders, including mathematical modelling where possible. A validated concept, i.e. a prototype for a supply chain in a certain context, is here referred to as a *scenario*. Initially, the supply chain concepts are tested by a combination of simulation and mathematical analysis where possible. For the design requirements where no model can be applied, qualitative concept testing techniques are used. They can be carried out as expert reviews or a role play based on a written prototype with allows for simulating the processes, infrastructure and human resources needed to check whether it is feasible with respect to the goals. With the direct or indirect involvement of real stakeholders, feedback on the supply chain concepts is obtained and adjustments are made until the stakeholders validate it or reject it. The validation of a concept by the stakeholders literally means that the resulting scenario is expected to be capable of delivering the anticipated goals, under the conditions (e.g. inputs and assumptions) considered in the scenario. When we go down the road to the modelling effort behind the (re)

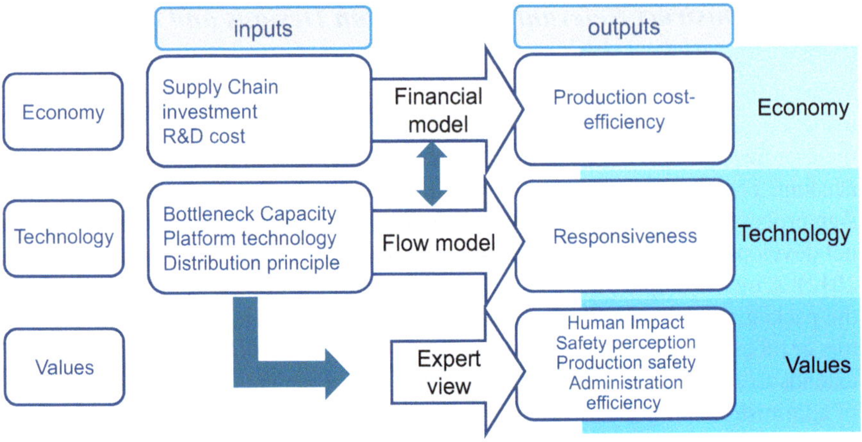

Fig. 10.7 An integrated design approach and assessment

design of the supply chain, we end up with rigorous flow modelling iteratively embedded in a multi-dimensional design approach as depicted in Fig. 10.7.

This figure is an adapted version of a methodology developed for R&D portfolio management (Vandaele and Decouttere 2013).

The design requirements for the vaccine supply chain design problem were obtained from the stakeholders' needs. The creative process of generating new and improved systems led to a set of concepts with improved or new supply chains, an example of which is given in Fig. 10.8. On the left hand side three design dimensions are listed: vaccine dosage, administration complexity and distribution centralization. This leads to 24 design combinations (full factorial), all listed in Fig. 10.8. Clearly some of them will be excluded up front while others will be excluded during the iterative design process by not passing the thresholds of the system requirements.

On the three main dimensions of KPI's, each supply chain scenario is represented by a set of inputs, efforts in the sense of limited resources, and outputs, efforts in the sense of desired outcomes. The mathematical model calculates the relation between a subset of inputs and outputs from the technology-pillar. The other inputs and output variables are the result of design activities. Many of the human-related outputs will be measured by qualitative techniques from user research and concept testing. As shown in Fig. 10.9, some of the design concepts (concept 1 through 3) will survive the iterative design process by passing the thresholds of the system requirements, here denoted by the 'stakeholder evaluation'. Conversely, concept 4 and 5 will be re-examined and modified to improve the concept (here denoted by concept 4' and 5'). In this way step 3 provides a long list of feasible concepts.

At this point we can backtrack on the schematic overview of the broader health care system design and modelling process in Fig. 10.1. The start is the current vac-

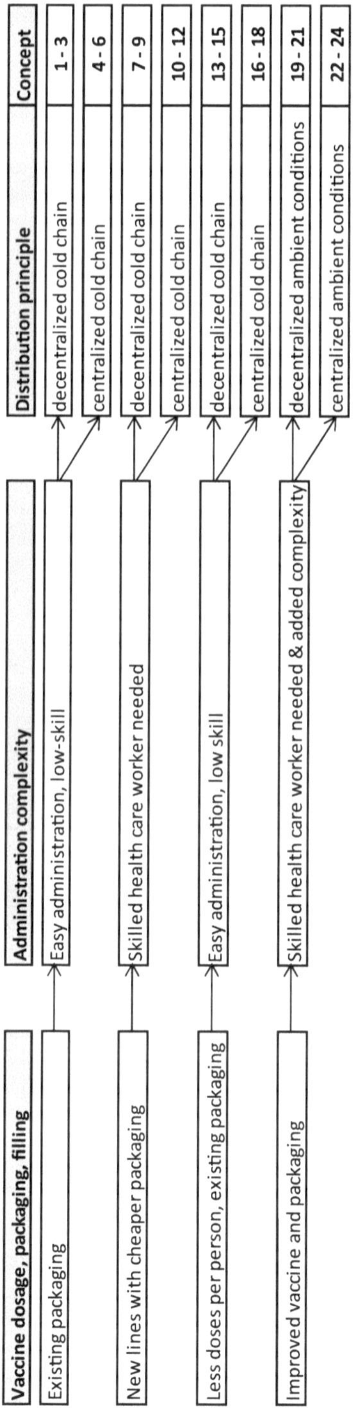

Fig. 10.8 Stakeholders' insights and derived supply chain concepts, desktop example

concept	input			output		Stakeholder evaluation
	bottleneck capacity	technical risk	SC investment	human impact	responsiveness	
	[# doses/yr]	[1 - 10]	[k€]	[# People immunized]	[10^2days-1]	
source	System engineering	Expert evaluation	Financial model	Expert view + Flow model	Flow model	
1	60.000.000	2	50.500	17.800.000	2,174	✓
2	120.000.000	2	170.000	36.400.000	2,326	✓
3	180.000.000	1	240.000	55.200.000	2,381	✓
4	60.000.000	2	600	17.200.000	1,111	✗ → 4'
5	120.000.000	2	109.000	34.400.000	1,250	✗ → 5'
6	180.000.000	1	165.000	51.600.000	1,667	....
7	80.000.000	4	310.500	22.400.000	2,381	...
8	120.000.000	3	370.500	34.000.000	2,381	
9	160.000.000	3	420.500	45.333.333	2,500	
10	80.000.000	4	255.500	21.600.000	1,176	
11	120.000.000	3	309.500	32.800.000	1,429	
12	160.000.000	3	465.500	44.266.667	1,471	
13	40.000.000	6	65.000	17.200.000	2,381	
14	80.000.000	5	175.000	34.400.000	2,500	
15	120.000.000	5	205.000	51.600.000	2,632	
16	40.000.000	6	35.080	16.600.000	1,149	
17	80.000.000	5	135.000	33.200.000	1,205	
18	160.000.000	5	165.000	66.400.000	1,299	
19	40.000.000	10	200.000	10.800.000	3,333	
20	80.000.000	9	210.000	21.600.000	3,571	
21	160.000.000	7	335.000	43.200.000	4,000	
22	40.000.000	10	180.000	10.666.667	1,887	
23	80.000.000	8	193.000	21.333.333	1,961	
24	120.000.000	7	315.000	32.000.000	2,174	

improved concept	input			output	
	bottleneck capacity	technical risk	SC investment	human impact	responsivenes
	[# doses/yr]	[1 - 10]	[k€]	People immunize	[10^2days-1]
	System engineering	Expert evaluation	Financial model	Expert view + Flow model	Flow model
4'	60.000.000	2	600	....	....
5'	120.000.000	2	109.000	34.400.000	....
....					

Fig. 10.9 Integration of mathematical modelling and user validation

cine supply chain where user and stakeholder insights together with the possibilities of new technology, offer opportunities for improvement along the diverse KPI's. The decision making stakeholders establish the limits on the solution space. The development of new supply chain concepts is followed by testing with the stakeholders with experience and financial interaction, in short iterative cycles, each time improving the concept. The technological modelling is used as partial knowledge to model input and output characteristics of the supply chain concepts. The iterative design/model process step continues until a satisfying validated design is reached, called a scenario. Each scenario is characterized by its set of input and output variables. A number of these scenarios is constructed, possibly very diverse in the solution they offer to the design problem. We argue that any other scenario building methodology can be useful here (Brailsford 2008). We also want to point out that this approach can be extended to other vaccine supply chain design exercises and even to supply chain design in general. However, it must be realized that all five steps have to be reviewed as other supply chains may induce other stakeholders, other KPI's and other design/scenario building pathways. This is a major part of our future research.

Step 4: Rank the Scenarios in a Multi-Dimensional Way

As a scenario will unlikely be championing on all dimensions from the diverse set of KPI's, we expect a couple of scenarios to be top of class and thus candidates for implementation. At this point, a multi-criteria ranking method brings great value to give insight into the multiple dimensions of the decision problem. We opt for Data Envelopment Analysis, as described in (Vandaele and Decouttere 2013) where it is used to incorporate sustainability in a R&D portfolio context.

Since no prior importance was attributed by means of stakeholder salience, the diverse scenarios reflect different approaches leading to solutions accepted by the stakeholders. The best solutions are the ones on the efficient frontier in the relative assessment. A partial two-dimensional view is shown in Fig. 10.10 where every dot represents a scenario and 3 and 19 are top ranked scenarios according to these two dimensions. If all dimensions are included, the overall efficient scenarios are visualized by the small squares. Scenarios not on the frontier and not overall efficient are visualized by the dots. For instance also scenario 1 (among others) is optimal when compared on all dimensions simultaneously. In this way a shortlist of 'best' scenarios is handed over to the next step. In order to conduct this ranking step it is necessary to quantify the measurements on all KPI's, even the more intangible ones. Please recall that since Sect. 10.3.3 we explore the procedure with a desktop example. For more details on the mathematical ranking models used, we refer to (Vandaele and Decouttere 2013).

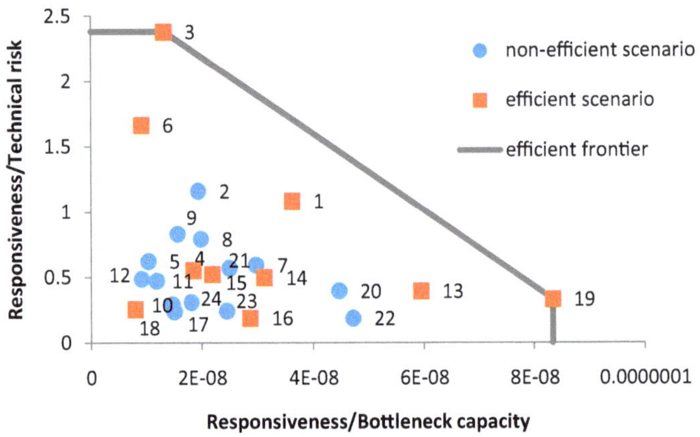

Fig. 10.10 Multi-criteria scenario evaluation

Step 5: The Choice of the Final Scenario for Implementation

With the help of effective and instructive visualisation techniques, such as infographics, insight is gained into the specific profiles of the top ranked scenarios. Each scenario can be considered as equally beneficial with respect to the set of vaccine supply chain goals defined and for the stakeholders identified in the earliest stage of the process (step 1). In the knowledge that each of the scenarios are 'safe' options which are accepted by the stakeholders, the decision group's attention goes to the selection of the best suited scenario in line with the organisation's values and strategic objectives, while fulfilling the short term objectives. The transparency of the scenario strengths with respect to these goals and the early involvement of the decision group in the process, creates openness to consider the different solutions and avoids emotional caveats connected to power play, risk taking, uncertainties and innovating. These create a right environment for better overall decision making in selecting the final scenario to be implemented (Kahneman and Tversky 1979; Milkman et al. 2009).

10.4 Conclusions

In this paper we revisited the (re)design of a vaccine supply chain. Our main contribution is two-fold. First, we propose a broader framework for vaccine supply chain design. In this way, the chances of effective implementation and realisation of the intended results are much more guaranteed. Second, it shows that is possible to reconcile an analytic modelling approach with a more synthetic, descriptive approach. In this way, we aim to fulfil the ATM and humanitarian aspirations together with the financial and technological ambitions. Building on the observation of a very often weak willingness to implement the exclusively model-based supply chain designs, a major reason for this weakness is the ignorant exclusion of major stakeholders in the design process. Therefore a broader approach is put forward based on stakeholder analysis and user-centred design. Building on this analysis, more mind expanding and out-of-the-box design propositions can be generated which will then be dealt with by a multi-criteria decision method in order to select a set of 'best' design options, called scenarios. Additionally, the early involvement of key stakeholders in the design process leads to scenarios with a much better fit and induce a higher willingness to implement the vaccine supply chain design. In this way we believe that health care system design in general will have a much higher probability of reaching the full-fledged implementation benefits for all stakeholders involved.

Future research encompasses a more detailed formalization of the proposed approach and the application of the methodology to other supply chains and health care system design problems. This approach can undoubtedly be extended to other vaccine supply chain design exercises and even to supply chain design in general.

Acknowledgments This research was supported by the GSK Vaccines Research chair in Operations Management.

References

Akao Y (1990) Quality function deployment: integrating customer requirements into product design. Productivity Press, Portland, Oregon

Azaron A, Brown K, Tarim S, Modarres M (2008) A multi-objective stochastic programming approach for supply chain design considering risk. Int J Product Econ 116(1):129–138

Barry MJ, Edgman-Levitan S (2012) Shared decision making: the pinnacle of patient-centered care. N Engl J Med 366(9):780–781

Beyan OD, Baykal N (2012) A knowledge based search tool for performance measures in health care systems. J Med Syst 36(1):201–221

Bhattacharya S, Hasija S, Van Wassenhove L (2014) Designing efficient infrastructural investment and asset transfer mechanisms. Production and operations. Management 23(9):1511–1521

Brailsford SC (2008) System dynamics: what's in it for healthcare simulation modelers. In Simulation Conference, 2008. WSC 2008. Winter, pp 1478–1483. IEEE

Brailsford SC, Lattimer V, Tarnaras P, Turnbull J (2004) Emergency and on-demand health care: modelling a large complex system. J Oper Res Soc 55(1):34–42

Brown T (2008) Design thinking. Harv Bus Rev 86(6):84

Brown D, Burton A, Gacic-Dobo M, Karimov R, Vandelaer J, Okwo-Bele J (2011) A mid-term assessment of progress towards the immunization coverage goal of the global immunization vision and strategy (givs). BMC Public Health 11(1):806

Buckle P, Clarkson P, Coleman R, Ward J, Anderson J (2006) Patient safety, systems design and ergonomics. Appl Ergon 37(4):491–500

Buckle P, Clarkson P, Coleman R, Bound J, Ward J, Brown J (2010) Systems mapping workshops and their role in understanding medication errors in healthcare. Appl Ergon 41(5):645–656

Cardona-Valdés Y, Álvarez A, Ozdemir D (2011) A bi-objective supply chain design problem with uncertainty. Transp Res C Emer Technol 19(5):821–832

Clarkson PJ, Buckle P, Coleman R, Stubbs D, Ward J, Jarrett J, Lane R, Bound J (2004) Design for patient safety: a review of the effectiveness of design in the uk health service. J Eng Des 15(2):123–140

Donaldson K, Ishii K, Sheppard S (2006) Customer value chain analysis. Res Eng Design 16(4):174–183
Elhedhli S, Merrick R (2012) Green supply chain network design to reduce carbon emissions. Transp Res part D Transp Environ 17(5):370–379
Elias AA, Cavana RY, Jackson LS (2002) Stakeholder analysis for R & D project management. R D Management 32(4):301–310
Franca R, Jones E, Richards C, Carlson J (2010) Multi-objective stochastic supply chain modeling to evaluate tradeoffs between profit and quality. Int J Prod Econ 127(2):292–299
Freeman RE (2010). Strategic management: a stakeholder approach. Cambridge University Press
Guillén G, Mele F, Bagajewicz, Espuña A, Puigjaner L (2005) Multiobjective supply chain design under uncertainty. Chem Eng Sci 60(6):1535–1553. Cambridge
Guillén-Gosálbez G, Grossmann I (2009) Optimal design and planning of sustainable supply chains under uncertainty. Am Inst Chem Eng Jl 55(1):99–121
Hamedani S, Jabalameli M, Bozorgi-Amiri A (2013) A multi-objective model for locating distribution centers in a supply chain network considering risk and inventory decisions. Manag Sci Lett 3(4):1077–1088
Hekkert P, van Dijk M (2011) Vision in design: a guidebook for innovators. BIS. Amsterdam, Netherlands
Jun GT, Ward J, Morris Z, Clarkson J (2009) Health care process modelling: which method when? Int J Qual Health Care 21(3):214–224
Kahneman D, Tversky A (1979) Prospect theory: an analysis of decision under risk. Econometrica 47(2):263–292
Kamara L, Lydon P, Bilous J, Vandelaer J, Eggers R, Gacic-Dobo M, Meaney W, Okwo-Bele J-M (2013) Global immunization vision and strategy (givs): a mid-term analysis of progress in 50 countries. Health Policy Plan 28(1):11–19
Lemmens S, Vandaele N, Decouttere C, Bernuzzi M (2014) Integrated supply chain network design models for vaccines: a literature review, FBE Research Report KBI 1414, KU Leuven, Belgium
Martin J, Barnett J (2012) Integrating the results of user research into medical device development: insights from a case study. BMC Med Inform Decis Mak 12(1):74
Milkman KL, Chugh D, Bazerman MH (2009) How can decision making be improved? Perspect Psychol Sci 4(4):379–383
Mitchell RK, Agle BR, Wood DJ (1997) Toward a theory of stakeholder identification and salience: defining the principle of who and what really counts. Acad Manag Rev 22(4):853–886
Murray CJ, Frenk J (2000) A framework for assessing the performance of health systems. Bull World Health Organ 78:717–731
Musgrove P, Creese A, Preker A, Baeza C, Anell A, Prentice T (2000) The World health report 2000: health systems: improving performance. World Health Organization
Naimi Sadigh A, Fallah H, Nahavandi N (2013) A multi-objective supply chain model integrated with location of distribution centers and supplier selection decisions. Int J Adv Manuf Technol 69(1–4):225–235
Norman DA (2007) Emotional design: why we love (or hate) everyday things. Basic books. New York
Olivares-Benitez E, González-Velarde J, Ríos-Mercado R (2012) A supply chain design problem with facility location and bi-objective transportation choices. TOP 20(3):729–753
Ortuño M, Tirado G, Vitoriano B (2011) A lexicographical goal programming based decision support system for logistics of humanitarian aid. TOP 19(2):464–479
Rajabalipour Cheshmehgaz H, Desa M, Wibowo A (2013) A flexible three-level logistics network design considering cost and time criteria with multi-objective evolutionary algorithm. J Intell Manuf 24(2):277–293
Rechel B, Wright S, Barlow J, McKee M (2010) Hospital capacity planning: from measuring stocks to modelling flows. Bull World Health Organ 88:632–636
Rouck SD, Jacobs A, Leys M (2008) A methodology for shifting the focus of e-health support design onto user needs: a case in the homecare field. Int J Med Inform 77(9):589–601

Samii A-B, Pibernik R, Yadav P, Vereecke A (2012) Reservation and allocation policies for influenza vaccines. Eur J Oper Res 222:495–507
Sanders L (2008) On modeling: an evolving map of design practice and design research. Interactions 15(6):13–17
Shah N (2004) Pharmaceutical supply chains: key issues and strategies for optimisation. Comput Chem Eng 28(6–7):929–941
Shah N (2005) Process industry supply chains: advances and challenges. Comput Chem Eng 29(6):1225–1236
Solaimani S, Guldemond N, Bouwman H (2013) Dynamic stakeholder interaction analysis: innovative smart living design cases. Electronic Markets, pp 1–12
Taboada M, Cabrera E, Iglesias ML, Epelde F, Luque E (2011) An agent-based decision support system for hospitals emergency departments. Procedia Comput Sci 4:1870–1879
Thakur R, Hsu SH, Fontenot G (2012) Innovation in healthcare: issues and future trends. J Bus Res 65(4):562–569
Vandaele NJ, Decouttere CJ (2013) Sustainable r & d portfolio assessment. Decis Support Syst 54(4):1521–1532
Vandaele N, Nieuwenhuyse IV, Cupers S (2003) Optimal grouping for a nuclear magnetic resonance scanner by means of an open queueing model. Eur J Oper Res 151(1):181–192
Vitoriano B, Ortuño M, Tirado G (2009) HADS, a goal programming-based humanitarian aid distribution system. J Multi-Criteria Decis Anal 16(1–2):55–64
Vitoriano B, Ortuño M, Tirado G, Montero J (2011) A multi-criteria optimization model for humanitarian aid distribution. J Global Optim 51(2):189–208
Wang F, Lai X, Shi N (2011) A multi-objective optimization for green supply chain network design. Decis Support Syst 51(2):262–269
Wang SA, Hyde TB, Mounier-Jack S, Brenzel L, Favin M, Gordon WS, Shearer JC, Mantel CF, Arora N, Durrheim D (2013) New vaccine introductions: assessing the impact and the opportunities for immunization and health systems strengthening. Vaccine 31(2):B122–B128. (Decade of Vaccines)
World Health Organization (2013) Global Vaccine Action Plan 2011–2020. WHO; 2013. http://www.who.int/immunization/global
Yadav P, Stapleton O, Van Wassenhove L (2013) Learning from coca-cola. Stanf Soc Innov Rev 11(1):51–55
You F, Grossmann I (2008) Design of Responsive supply chains under demand uncertainty. Comput Chem Eng 32(12):3090–3111
You F, Grossmann I (2011) Balancing responsiveness and economics in process supply chain design with multi-echelon stochastic inventory. Am Inst Chem Eng J 57(1):178–192

Chapter 11
Achieving Equity, Effectiveness, and Efficiency in Food Bank Operations: Strategies for Feeding America with Implications for Global Hunger Relief

Irem Sengul Orgut, Luther G. Brock III, Lauren Berrings Davis, Julie Simmons Ivy, Steven Jiang, Shona D. Morgan, Reha Uzsoy, Charlie Hale and Earline Middleton

Abstract One in six Americans (14.3 % of households) reported being food insecure at some time during the year 2013 (i.e., they lacked access to enough food for an active, healthy life for all household members). This translates to 17.5 million food insecure households and 49.1 million Americans, 33.3 million adults and 15.8 million children living in food insecure households (Coleman-Jensen et al. 2011, 2014). This slight decrease from 14.5 % in 2012 was not statistically significant and marks the third consecutive year that the USDA's annual hunger survey has found food insecurity at some of the highest levels since the government started the report in 1995.

There is a growing body of research addressing logistics and supply chain management issues in the area of humanitarian relief. Much of this work has focused in the emergency management domain, addressing critical needs of those affected in the aftermath of a natural disaster. However, the chronic epidemic of poverty and hunger in the United States can also be considered a 'natural disaster'. While this disaster is characterized by being prolonged with a gradual onset (like hurricanes or tornados), it also warrants significant attention given the recent statistics on hunger

---

L. B. Davis (✉) · L. G. Brock III · S. Jiang
Department of Industrial and Systems Engineering, North Carolina Agricultural and Technical State University, Greensboro, NC, USA
e-mail: lbdavis@ncat.edu

I. Sengul Orgut · J. S. Ivy · R. Uzsoy
Edward P. Fitts Department of Industrial and Systems Engineering, North Carolina State University, Raleigh, NC, USA

S. D. Morgan
School of Business and Economics, North Carolina Agricultural and Technical State University, Greensboro, NC, USA

C. Hale · E. Middleton
Food Bank of Central & Eastern North Carolina, Raleigh, NC, USA

C. W. Zobel et al. (eds.), *Advances in Managing Humanitarian Operations*,
International Series in Operations Research & Management Science 235,
DOI 10.1007/978-3-319-24418-1_11

and poverty and the emerging congressional climate to 'remove the safety net' of government programs to support this particular population. Prior research in the hunger relief domain has primarily been conducted in the social sciences addressing such issues as equity and allocation policy and ignoring their relationship to effectiveness and efficiency. A thorough understanding of the partners (private, nonprofit, and governmental) that work together to provide food assistance is needed from a systems perspective.

The opportunity to engage in an overall engineering/operations research effort to improve access to food through equitable, efficient, and effective operations of nonprofits and government agencies has yet to be fully explored. Therefore, in order to effectively advance operations research and inform policy development in this field, this chapter has been written in partnership with collaborators from the Food Bank of Central & Eastern North Carolina (FBCENC) to create a collective vision for the future of research in food bank operations. This chapter creates a framework for operational decision making at food banks, aimed at (1) more effectively translating the ongoing operations research work into practice at a wider range of food banks; (2) identifying synergistic research opportunities for the operations research and public policy communities to collaborate in the area of hunger relief; and (3) identifying critical, open problems in food aid distribution.

Keywords Hunger relief · Humanitarian logistics · Food banks · Food insecurity · Humanitarian supply chains · Food aid distribution · Fairness · Equity · Nonprofit operations · Social welfare systems · Donations management · Poverty · Systems engineering · Emergency food network

11.1 Introduction

Food insecurity is defined as limited or uncertain ability to acquire nutritionally adequate and safe food in socially acceptable ways (e.g., without resorting to emergency food supplies, scavenging, stealing, or other coping strategies) (National Research Council 2006). Instances of food insecurity have been systematically measured in the United States since 1995 (Carlson et al. 1999) and are often associated with persistent poverty and low socioeconomic status (Coleman-Jensen et al. 2014). As shown in Fig. 11.1, the number of people in the United States with inconsistent access to safe and nutritious food rose steadily through 2008 and has remained fairly stable, declining slightly from 2011 to 2013. The U.S. Department of Agriculture estimates that food insecurity affects close to 15% of all U.S. households and recent reports connect food insecurity and economic conditions, showing that food insecurity increased to 14.6% in 2008 with the onset of the recession (Nord et al. 2009; Coleman-Jensen et al. 2011, 2014).

Food insecurity is a global problem. According to the Food and Agricultural Organization (FAO) of the United Nations, approximately one in nine people in the world suffered from chronic hunger between 2012 and 2014 (FAO 2014). Further

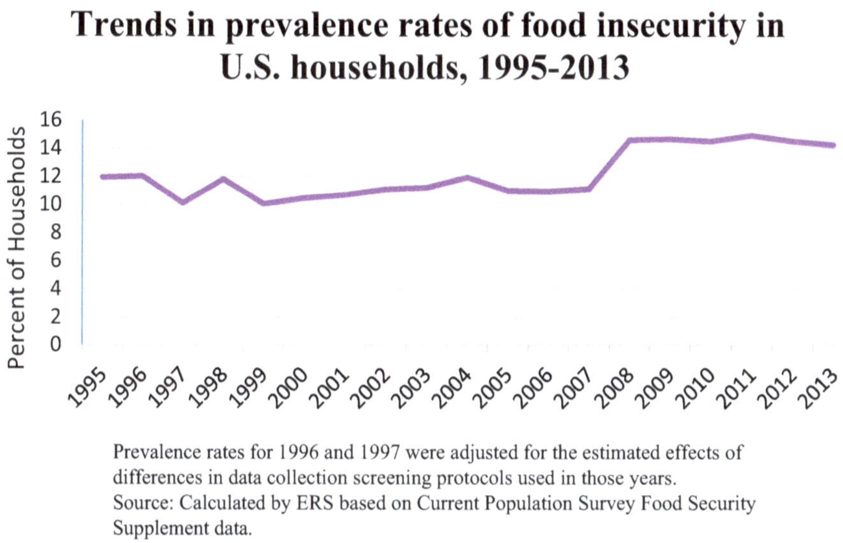

Fig. 11.1 Food insecurity percentages in the U.S., 1995–2013. (Coleman-Jenson et al. 2014)

evidence of the rise in hunger need is documented in high and middle income countries such as the United Kingdom (Lambie-Mumford and Dowler 2014), South Africa (Warshawsky 2011), Australia (Foodbank Australia 2013) and Canada (Tarasuk et al. 2014a) as well as developing countries, e.g., Ethiopia (Goncalves et al. 2013) and Kenya (Rancourt et al. 2013). In Canada, food insecurity in 2012 remained "at or above the levels experienced in prior years" (Tarasuk et al. 2014a). Five percent of the population in Australia experience food insecurity (Butcher et al. 2014), and it is noted that the demand for food relief continues to rise (Foodbank Australia 2013). Warshawsky (2011) states food insecurity in South Africa is a "continuous threat for a third of the population" and "that urban food insecurity has increased to record levels". While measures of hunger need may vary among countries, what is consistent is that experiences of food insecurity are increasing among children, the elderly, and other vulnerable populations that are receiving some form of social assistance from the government (Lambie-Mumford 2014; Tarasuk et al. 2014b).

Food insecurity is mitigated through one or more interventions such as:

1. food vouchers which allow for cash-like acquisition of food items at select retail locations;
2. food literacy initiatives which attempt to address the link between poverty and adverse health outcomes resulting from poor nutrition choices; or
3. direct food distribution by government funded social programs and/or charitable organizations to the food insecure population (Gentilini 2013).

Several countries have used a variety of these interventions to address hunger need, as highlighted in Table 11.1. Many of these interventions involve a collaborative effort between government, public, and private entities comprising a collective safety

Table 11.1 Summary of food interventions by country

Country	Name	Organization (O)/Program (P)	Food provision	Food voucher	Food literacy
Australia	Foodbank Australia	O	•		•
Canada	Food Banks Canada	O	•		•
Egypt	Food Subsidy Reform Project	P	•		
Italy	Fondazione Banco Alimentare Onlus	O	•		
Kenya	Kenya Food Security Steering Group	P	•		
South Africa	Foodbank South Africa	O	•†		
Taiwan	Taiwan People's Food Bank Association	O	•		
	V1[a]	P		†	
United States	Feeding America	O	•		
	Supplemental Nutrition Assistance Program	P		†	
	The Emergency Food Assistance Program	P	†		
United Kingdom	Trussell Trust Foundation	O	•		
	Fareshare	O	•		

[a] Name not specified in Wang and Lyu (2013)
†Government Sponsored, •NGO Sponsored

net for people suffering from insufficient access to food. For example, in the United States individuals can obtain government assistance through food vouchers (Supplemental Nutrition Assistance Program (SNAP)); receive food from charitable agencies (e.g., soup kitchens), some of which are authorized distributors of federal commodities via The Emergency Food Assistance Program (TEFAP); and receive food from supplemental programs targeted for pregnant women and children. In Taiwan, a number of food bank and food voucher programs exist that are managed by non-profit organizations, the government, and partnerships between public and private entities (See Wang and Lyu (2013) for a more complete program listing). However, in Canada, food banks are the only source by which food insecure individuals can receive emergency food, as there are no publicly funded, government-run food assistance programs (Tarasuk et al. 2014a).

At the global level, there are several food provisioning organizations that work to alleviate hunger such as the Global Food Bank Network, the European Federation of Food Banks, and the United Nations World Food Programme. These organizations are networks of food banks and charitable agencies (e.g., soup kitchens, food pantries, and shelters, etc.) which provide in-kind food to those that might otherwise not receive it. They interact with government agencies and local for-profit organizations which produce, manufacture, and distribute food.

A number of studies have discussed the prevalence of food insecurity (McConville and Groot 2013; Lambie-Mumford and Dowler 2014), drivers of food insecurity (Gundersen et al. 2011; Coleman-Jensen et al. 2011, 2014), emergency food relief (Ohls et al. 2002; Webb 2013) and case studies describing specific food bank organizations in select countries (Warshawsky 2011; Tarasuk et al. 2014b; Booth and Whelan 2014; Butcher et al. 2014; Santini and Cavicchi 2014; Wang and Lu 2013). However, a comprehensive examination of the hunger relief supply chain (HRSC) has received little attention from a systems perspective. The intent of this chapter is to provide insight regarding the operational structure of the HRSC with a particular focus on non-profit organizations classified as food banks. We describe the unique challenges inherent within this supply chain and identify opportunities to improve the effectiveness, equity and efficiency of this supply chain in achieving its core mission of increased food access to those in need. We also describe the role that the industrial engineering/operations research/management science community (IEORMS) can play to help these organizations achieve their core mission.

The remainder of this chapter is outlined as follows. We first introduce methods for mitigating food insecurity and describe the structure of the HRSC from a global and domestic perspective. We frame the operational challenges associated with food aid distribution according to three key dimensions: efficiency, effectiveness and equity. These three dimensions of organizational performance are impacted by supply, demand, capacity, and information. The challenges and their corresponding impacts on organizational performance are discussed. Lastly, a review of the related modeling literature is presented highlighting opportunities for future research. Throughout the chapter we interweave the experiences of a particular food bank with a focus on characterizing the operating environment of that organization, which possesses many of the operational complexities inherent in most domestic food banks.

11.2 Overview of Hunger Relief Supply Chain

Global Perspective

The supply chain associated with hunger relief is highly localized and geographically disaggregated. As a result, food aid distribution is a function of the environment in which the distribution is occurring. The socio-economic status of the country within which distribution is occurring plays a particularly important role in the logistics, processes, and implementation of food distribution for hunger relief. The population at risk (i.e. the demand) varies significantly between developing and developed countries. In the developing world (low income countries), the population at risk is estimated based on measures of extreme or chronic under-nutrition or malnutrition. In contrast, in developed countries demand or hunger need is estimated based on food security indicators. For example, the USDA defines two levels of food insecurity:

1. **Low food security** (*old label = Food insecurity without hunger)*: reports of reduced quality, variety, or desirability of diet. Little or no indication of reduced food intake;
2. **Very low food security** (*old label = Food insecurity with hunger)*: Reports of multiple indications of disrupted eating patterns and reduced food intake (Coleman-Jensen and Gregory 2014).

As has been suggested by Gentilini (2013), "the nature and magnitude of food insecurity in developing countries is very different from that of high income countries". As a result, the food distribution supply chain is also quite different. Figure 11.2 summarizes and compares the nature and structure of food aid distribution in developed (high income) and developing countries.

In developing countries, food aid operations are often managed through partnerships between government, the United Nations, and NGOs. These types of partnerships manifest in a variety of ways. For example, in Kenya, the Kenya Food Security Steering Group, a joint group of specialists from the Government of Kenya (national and local), the United Nations, and NGOs, conduct biannual needs as-

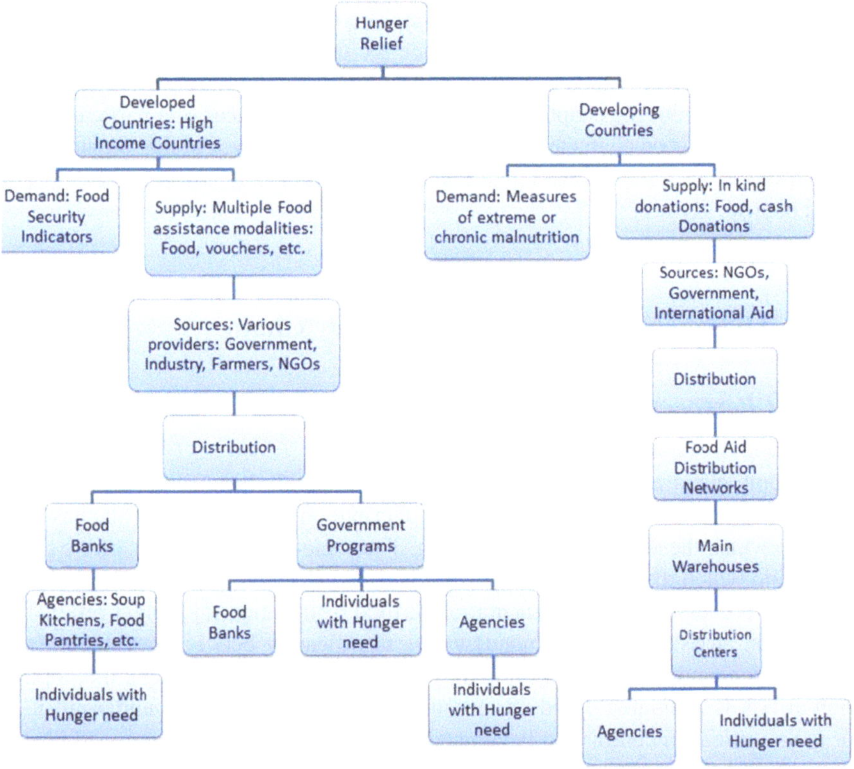

Fig. 11.2 Comparison of hunger relief logistics in high income and developing countries. (based in part on Gentilini 2013)

sessment in order to estimate food need in anticipation of the seasonal droughts and floods. In addition to estimating the number of beneficiaries, the assessments determine the ratio entitlements (i.e., the percentage of a standard food ration that the beneficiaries are entitled to receive based on their nutritional requirements) (Rancourt et al. 2013). In Egypt, the government-run Food Subsidy Reform Project which plays a major role in the Government of Egypt's safety net program receives technical assistance from the United Nations World Food Programme (Quinn 2010). The World Food Programme has also collaborated with the Government of Ethiopia in a similar manner to provide food for its population and assist with the logistics of food aid supply and distribution (Goncalves et al. 2013).

In developed countries food assistance manifests in a variety of ways from government-funded programs that provide food and various types of vouchers (e.g., food stamps) to non-governmental organizations (NGOs) that provide and distribute food. Food banks are NGOs that serve as food collection and distribution centers. "Food banks can be considered an intermediate agent that connects donors and beneficiaries. They raise contributions and support from individuals, private sector corporations, farmers, manufactories, retailors and by governments… food banks help handle, store and deliver these commodities to a web of grassroots organizations" (Gentilini 2013). Most of the national food banks in developed countries are part of networks (e.g., Feeding America, European Federation of Food Banks) for information sharing and capacity building. In the hunger relief supply chain, food banks serve as a central component responsible for supply sourcing and distribution. The use of food banks for the purpose of food distribution to reduce the impact of food insecurity is primarily found in developed high and middle income countries. Foodbank Australia is the largest food bank organization in Australia, operating out of seven regions and distributing more than 29.9 million kg of food. In the United Kingdom, Fareshare recovers surplus food from supermarkets and shops that are distributed to more than 720 charitable organizations in the UK. The Trussell Trust Foundation is a Christian hunger relief organization in the United Kingdom that operates a franchise-type of system. Approximately, 400 affiliated food banks located throughout the United Kingdom provide short-term (3-day provision) emergency food parcels using a voucher referral system.

Domestic (U.S.) Perspective

As depicted in Fig. 11.3, food aid distribution in the U.S. involves the management of food, information, money, and other resources in support of the delivery of food to needy parties.

Food banks collect, store and distribute food to charitable agencies that serve the food insecure population. They serve an important role from a sustainability perspective by recovering surplus food that might otherwise be discarded. Local food producers (i.e., farmers) provide food donations that are leftover from harvested commercial crops. Local food processors and distributors provide items which are

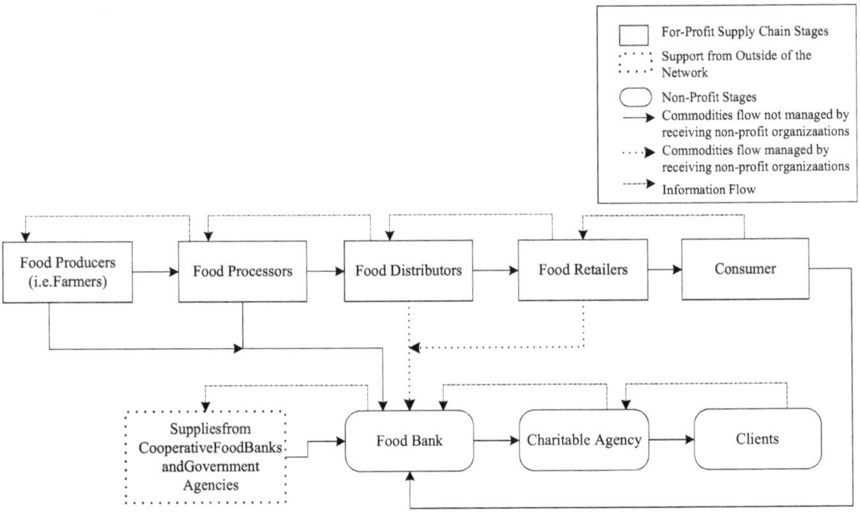

Fig. 11.3 Diagram of material and information flows for the domestic HRSC. (Adapted from Feeding America 2011b)

usable yet unsuitable for sale in local markets. While the majority of high-volume donations are received from food processors and retailers, consumers also serve as a source of supply through local food drives. Charitable agencies may also conduct local food drives to get food. However, their primary source of supply comes from food banks.

Over 200 food banks operating in the United States are affiliates of Feeding America®, a network of food banks, corporations, and community-concerned groups whose objective is to end food insecurity in the United States. Food banks that belong to Feeding America participate in the Choice™ system which was developed in collaboration with the University of Chicago (Houlihan 2006). The Choice system is an online/mobile-enabled auction system that allows food banks to order food based on shares (Campbell et al. 2015). Feeding America posts available products twice a day. Food banks bid for products using shares they hold, which are allocated relative to the need in their service area. The highest bidder wins the product based on the amount of the second highest bid. Shares from the winning bid are then redistributed within the network. The Choice system also facilitates trading and bartering among member food banks. A more detailed discussion regarding the origin of the Choice system can be found in Houlihan (2006).

In addition to these virtual systems, many government-supported programs depend on food banks to transport allocated commodities to the charitable agencies serving local communities (Cabili et al. 2013). These agencies may also rely on food banks to provide additional food allotments generated from local donation sources. Furthermore, food banks are involved with disaster relief, providing millions of pounds of food to individuals in affected areas (Feeding America 2011b).

Local Perspective: Food Bank of Central & Eastern North Carolina

Background

The Food Bank of Central & Eastern North Carolina (FBCENC) is one of the seven Feeding America food banks serving the food insecure population in North Carolina. The FBCENC provides food to over 600 charitable agencies in its thirty-four county service area. Figure 11.4 illustrates the service region of the FBCENC and Fig. 11.5 provides an overview of their food distribution network. While food banks with large service areas often operate from a single warehouse, the structure of this particular food bank is unique in that it uses multiple regional warehouses to support food distribution activities. Each regional warehouse primarily serves a distinct set of charitable agencies located in the counties served, and one serves as the hub (Raleigh) where the bulk of the donations are received. Due to the scale of this network, we illustrate the county distribution for the regional warehouse in Durham, North Carolina. The activities within this supply chain are discussed in terms of food acquisition and food distribution.

Food Acquisition

In the FBCENC's supply chain, food is secured from community, not-for-profit, private, and governmental sources, and Feeding America (FBCENC 2013). As discussed above, Feeding America is a national hunger relief organization that supports food bank members by securing food donations from large manufacturers, retailers, and

Fig. 11.4 The FBCENC's Service Region. (FBCENC 2013)

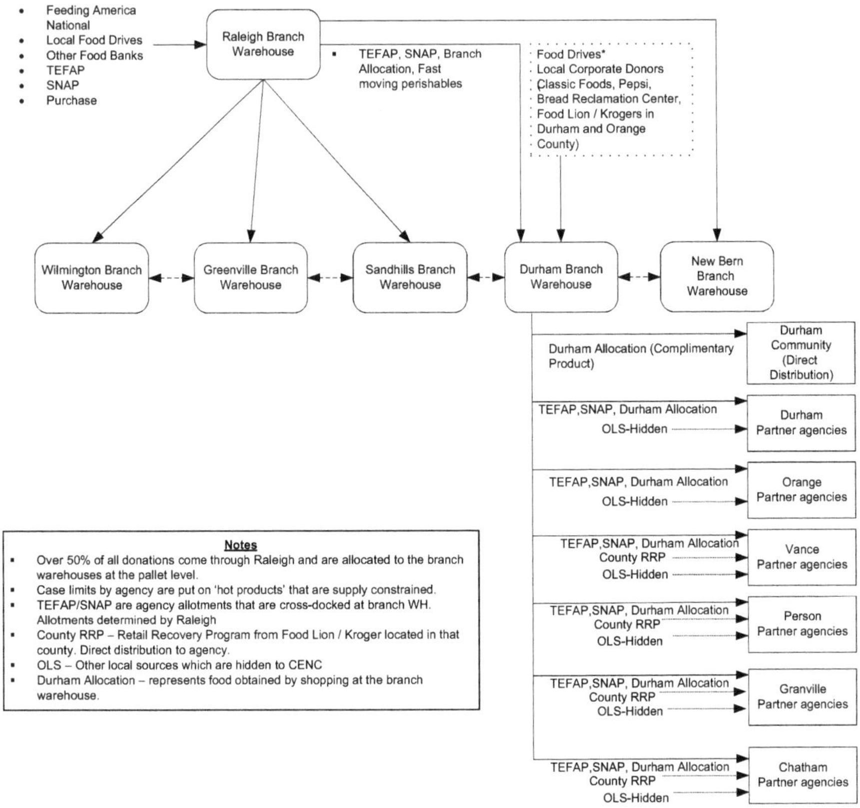

Fig. 11.5 The food distribution supply chain of the FBCENC

foundations. Regional food banks also secure food from local donors. The Choice system facilitates sharing of excess food donations amongst the 200 food banks in the Feeding America network to meet the nutritional needs of people who are food insecure. Food supplies are delivered to food banks from food manufacturers and retailers using commercial trucks.

In addition to the food sourcing partnership available through Feeding America, food banks receive food from TEFAP, a federal program that supplies food to food banks to supplement the diets of the food insecure at no cost. States receive food from the TEFAP program quarterly in proportion to their poverty and unemployment rates. Food from the TEFAP program is fairly standardized and thus provides food banks with a degree of certainty, for any particular calendar quarter, in terms of the type and quantity of their food supply.

Food banks also receive food from civic and faith-based organizations. These donations vary greatly in terms of the product, quantity, and timing. In addition to donations of food and supplies, food banks may also receive monetary donations to procure food. For example, the FBCENC, as a member of a state association of food banks, receives funds from the NC General Assembly. Funds allocated to the

FBCENC by the State of North Carolina have specific stipulations in terms of the items permitted for purchase and the purchasing period. Items purchased using state funds are offered to local soup kitchens and pantries at no cost.

On the other hand, private monetary donations provide food banks with more discretion for purchased items. Typical factors that enter the purchasing decision using private money include item demand, shelf-life, and price. These products are made available to partner agencies for a small percentage markup over the procurement cost. For this reason, items purchased with private funds are distinct from items purchased with state funds.

Food Distribution

Food bank distribution networks can have multiple configurations. For example, food banks can distribute donations to other regional food banks, soup kitchens and pantries, or both. The objective is to distribute food donations to local organizations that service the public in a quick, efficient, and equitable manner. This objective is complicated by several issues. First, food distribution is influenced by the limited shelf life of food products. Therefore, it is critical that the distribution network and associated process assist in preserving the integrity of food. Food safety and quality have to be considered in truck (i.e., refrigerated versus non-refrigerated) and route selection. Second, the food aid distribution, unlike traditional food distribution, is challenged by the unpredictable timing, quantity, and quality of the donated food supply. This in turn makes truck needs, usage, and routes difficult to anticipate. Finally, food banks aim to distribute food equitably amongst their partner agencies. This is particularly challenging since the FBCENC does not have visibility to the true need. For example, they do not know how much, if any, additional food aid is received by a food-insecure person in their service region. Furthermore, there is no system that allows the FBCENC to learn the real-time inventory of the agencies that may be affected by factors like the amount of food they have distributed and the amount of donations that they received from other sources. As a result, they use regional poverty metrics to help establish food distribution practices that provide the beneficiaries with equal access to donated food. It is common to use county-level metrics like poverty population and unemployment statistics as an estimate of the food-insecure population.

The aforementioned challenges point to the need for food bank distribution processes to be flexible. Standard distribution routes and practices may not be feasible. For example, the FBCENC provides food to over 800 charitable agencies in 34 counties utilizing a centralized warehouse and distribution network. Over 50% of all the food flows through the main warehouse in Raleigh and is distributed to other regional warehouses. Centralized operations have advantages particularly in terms of promoting efficiency. The FBCENC secures food from multiple donors and warehouses these donations at the Raleigh hub. Donated products are received by the food bank branches in a number of ways such as food that is transferred from the hub, donations from governmental agencies, and donations from local growers and grocers. Transferring donated product to various food bank branch locations and partner agencies from multiple points of origin presents a significant logistical undertaking particularly for a non-profit organization with limited resources. Centralized distribution and warehousing streamlines record keeping and facilitates equitable distribution objectives. After receiving food donations at the Raleigh hub, the FBCENC determines the amount of food to be allocated to each agency based on a retrospective analysis of equitable distribution. If it is an agency served from the Raleigh hub, food is directly sent there. If it is served from another branch, food is sent to the corresponding food bank branch for final shipment to the charitable agency.

11.3 The Objectives in Nonprofit Food Distribution

What are Food Banks Trying to Do: The Equity-Effectiveness-Efficiency Pyramid

The objectives of equity, effectiveness and efficiency are common in models that consider public services (Savas 1978). However, these objectives may be in conflict with one another resulting in trade-offs for decision making. We propose that this balancing act performed by food banks daily as they attempt to optimize the often conflicting objectives of equity, efficiency and effectiveness is the *Iron Triangle of Hunger Relief* - the equivalent to the "iron triangle of healthcare" (Kissick 1994) which seeks to balance cost, access and quality.

Equity implies serving the needs of the customer fairly. The definition of the term "equity" is subjective and can have different implications for different systems (Stone 1997). For this reason, it is necessary to select the equity measure based on the specific problem being considered (Sen 1973; Marsh and Schilling 1994). Savas (1978) defines equity in the public sector broadly as the fairness, impartiality, or equality of service. For food banks, a common measure of equity is to ensure that each food-insecure person in their service region receives the same amount of food. Fairness has two dimensions in the food distribution network: quantity received per customer and the quality (or type) of food received.

Feeding America evaluates performance of the food banks in their network using different measures. The county-level measure called "**P**ounds of food per **P**erson **i**n **P**overty (PPIP)" is a measure of equity in food distribution. It is the total poundage of food shipped to a county over the total number of food-insecure people residing in that county. The food banks are responsible for reporting the level of PPIP they achieve in their service region to remain affiliates of Feeding America. Feeding America also uses the county-level measure "meal gap" to estimate a county's need, with the current level of donations. Feeding America defines the meal gap as "a conversion of the total annual food budget shortfall in a specified area divided by the weighted cost per meal in that area." Therefore, this measure highlights the regions with highest distress in terms of food insecurity.

In addition to the performance metrics used by Feeding America, the FBCENC uses a metric referred to as the "fairshare". The fairshare measure is used to evaluate equitable food distribution with the goal of ensuring that each county in the FBCENC's service region receives the fraction of total food donations that is equal to the fraction of total demand that resides in that county. This definition is equivalent to each person receiving the same amount of food.

Effectiveness measures the ability to meet the needs of the end customer. As with any supply chain network there are multiple customers depending on their location in the chain. The customer in the HRSC environment is complicated because the immediate customer is the charitable agency but the larger customer is the food insecure population. Effectiveness therefore is measured as a function of meeting hunger need (at the individual level) through food distribution while simultaneously maximizing the yield of the donated supply (i.e. minimize waste). The PPIP not only provides a way to compare if distribution is equitable among counties, but it is provides a way to measure the effectiveness of food distribution activities in a single county.

Efficiency in supply chain management is traditionally defined as "the inverse of the cost of making and delivering a product to the customer" (Chopra and Meindl 2012), where increases in cost lower efficiency. Typical drivers of cost are transportation, facilities, and human capital. In the context of non-profit food distribution, cost is driven by the resources required to collect, manage and distribute donated food. This equates to efficient management of storage capacity, transportation vehicles for collection and distribution activities, and human capital to inspect and process incoming donations.

Equity-Effectiveness-Efficiency Trade-off in Food Distribution

The objectives of equity, effectiveness and efficiency as described above are in conflict with one another. For example, if the only objective of the food bank is to satisfy equity, a trivial solution is to send no food to any agencies: in that situation, everyone will receive the same amount of food (zero), however, no food is distributed hence this distribution is not effective. On the other hand, consider a situation when a food bank has rented many refrigerated trucks but only received dry goods as donations. The food bank may then use those refrigerated trucks to distribute all the donations, resulting in an effective solution. However, since the cost of refrigerated trucks is much higher than the cost of regular trucks, this solution is not efficient.

Although the ultimate goal of food banks is to satisfy hunger need by distributing food in their service region, usually the food donations that they receive are much lower than the demand they serve. For example, the donations received by the FBCENC over the first 6 months of 2009 averaged 3.5 pounds of food per month, per person in poverty. This amount is substantially lower than the nutritional need of an average person for 1 month. Due to this supply limitation, food banks focus on distributing as much food as possible as equitably as possible, as quickly as possible while also minimizing the food waste.

Equity is a primary objective guiding food distribution decisions by food banks, however, it is often contradictory to the other objective of food banks to minimize food waste and their constrained budget. The FBCENC is occasionally faced with the challenging situation where they have some extra food on hand. This food may be in the form of produce which has a lower shelf life and hence, has to be distributed in a timely fashion. If the shelf life of a produce item is ending, they face the conflict of whether they should send that food to waste (which also costs money), or send that food to an agency. Faced with a situation like this, they choose to send food to the agency although it may have a higher cost and may violate equity. Food banks face challenging decisions like these constantly; which makes the use of IEORMS tools relevant for obtaining policies that consider multiple objectives simultaneously.

11.4 Challenges from the Supply Chain Perspective

Supply Constraints

Food banks are impacted by a number of supply-side constraints that make it difficult to balance the multiple objectives of equity, effectiveness, and efficiency. We consider two dimensions of supply: (i) material supply that is obtained via in-kind donations and (ii) monetary supply that is donated and used to make food purchases to supplement material donations. The key characteristics of the supply that impact food distribution are uncertainty, limited availability, perishability, and donor preferences. Each factor is described below.

Uncertain Material Supply Since food banks primarily collect surplus food in the form of donations, there is a significant amount of variability associated with the frequency, type, and volume of donated food. First the donor base consists of food retailers, food manufacturers, farms, and other community groups (e.g. churches) concerned with hunger relief. The motivation for providing the donated food varies by donor group and as a result donation quantities are influenced by external factors. For example, products donated from the manufacturing sector are those which are safe for consumption but possess defects that would impact sales in the primary market (e.g. dented can, damaged labels). Adoption of lean production practices along with improvements in manufacturing quality will only limit the surplus food available for food banks. Donations from food retailers/wholesalers are influenced by customer buying preferences and frequency. Typically food that is close to the sell-by-date is made available to hunger relief organizations. Since the demand experienced by retailers is also uncertain, the amount of surplus food available for donating to the food bank is also uncertain. Likewise, donations from local farmers will be highly uncertain in frequency and food type. Certain crops are only harvested at certain times of the year and are limited by environmental and geographical conditions. While relationships between food banks and retailers can help to reduce the uncertainty associated with donation frequency, there are still a number of one-time or infrequent donors, particularly those associated with community groups, which also affect supply variability. Supply uncertainty impacts all factors of the iron triangle of hunger relief.

Uncertain Monetary Supply Uncertainty in supply is also associated with donated financial resources which are used to supplement donated food. Clearly some donors may only be motivated to give during organized food drives or during the Thanksgiving and Christmas holidays. Therefore, this source of funding is inconsistent and highly variable. In addition, food banks can receive financial allocations from the government to make supplemental food purchases. Given that this allotment can change from year to year as a result of state or federal budget decisions, the amount of food made available through purchasing activity is also uncertain from year to year. This impacts the ability to develop a diverse portfolio of nutritious food products for charitable agencies, and indirectly the individual food recipients. As a result, achieving equity in the type of food distributed is impacted.

Limited Supply Despite the large amount of donated food, there is still insufficient supply to meet the daily demand for meals faced by the food insecure population. Recall that typical measures of food insecurity are linked to census estimates of the poverty population. Figure 11.7 illustrates the relationship between need and distributed supply. Need is estimated based on census estimates for the number of people in poverty in the FBCENC service area. It is assumed that each person has 1 meal supplemented daily by charitable agencies supported by the food bank. The amount of food distributed each fiscal year (in pounds) is determined from

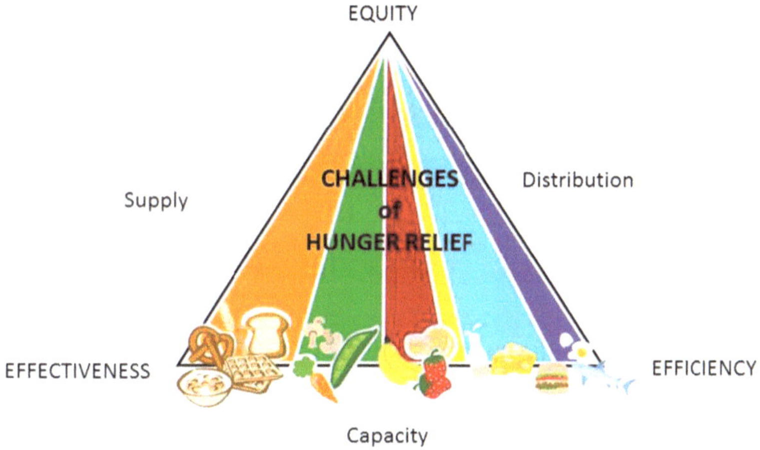

Fig. 11.6 The iron triangle of hunger relief

historical data. An industry standard conversion factor (pounds*16/19) is used to convert pounds to meals. As evidenced by Fig. 11.7, there is insufficient supply to satisfy the demand. It should be noted that charitable agencies represent one source of food supply. Food insecure individuals also have the ability to obtain food from other sources (e.g. federal programs). However, this highlights that supply must be distributed equitably across the food bank network.

Perishability Food banks receive a variety of products both perishable and non-perishable. For example, the FBCENC receives produce, canned goods, prepackaged meals, frozen foods and meats, and dairy products. During the 2010–2011 fiscal year, approximately 26% of the food donations were fresh produce, 12% were refrigerated products like dairy, and 12% were frozen products (e.g. frozen meats and produce). The management of perishable products impacts efficiency since investments must be made in transportation (refrigerated vehicles) and storage (freezers). Given that many products are close to their sell-by-date, the products must be distributed quickly to the agencies that have the ability to take the food, thus impacting equity. For example, it is not uncommon for the FBCENC to receive a large donation of sweet potatoes. Sweet potatoes normally have a shelf life of 3–5 weeks. However, when the FBCENC receives them, they are very near the end of their shelf life. As a result, they will make arrangements to sort, package, and distribute the sweet potatoes as quickly as possible.

Donor Constraints on Supply It is not uncommon for large community donors to place restrictions on who can receive the food. Within the FBCENC structure, regional warehouses receive donations from the community (e.g. private corporations or community groups) where it is located. These donors may request that donations are distributed to the charitable agencies served by the regional food bank. This limits sharing among the network and therefore impacts equity at a regional level. This may result in the some counties in the service area being over-served with respect to the pounds distributed per person in poverty.

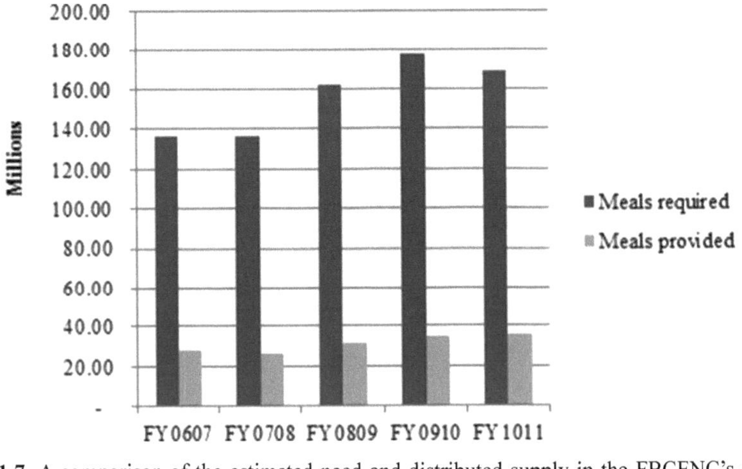

Fig. 11.7 A comparison of the estimated need and distributed supply in the FBCENC's supply chain

Demand Constraints

In the HRSC, demand is associated with the hunger need of the people who are food insecure. Food banks experience constraints in demand in terms of uncertainty and visibility.

Uncertain Demand The population most vulnerable to food insecurity are the elderly and those that suffer from homelessness or chronic poverty (Gentilini 2013). In fact, from a global perspective, many of the users of food bank services are those from single-headed households, the working poor, and "former middle class family members with sporadic income" (Gentilini 2013). The changing economic conditions that result in sudden loss of benefits (social assistance) or income are some of the factors that impact the demand for food aid. It needs to be pointed out that although poverty is an important indicator of food insecurity, there are other factors associated with food insecurity. Economic factors such as unemployment, participation in federal nutrition assistance programs (SNAP, WIC, and National school lunch program), participation in Medicaid, and household assets can have an impact on food insecurity. In addition, geographic locations (urban vs. rural) also have an effect on food insecurity. Estimates of demand based on these indicators is quite complex, especially since many of these indicators are updated at different points in time. Unemployment statistics are updated monthly, while poverty estimates are updated annually.

Limited Demand Visibility Food banks measure effectiveness and equity at the individual population level (e.g. number of people in poverty in a county). However, the individual *demand* is often hidden to the food bank as their primary customers are charitable agencies. While the charitable agencies do provide some information, it is sometimes inconsistent in terms of the information type and frequency. For example some agencies report usage based on households served, while others report based on individuals. In addition, some agencies are required to provide reports on a frequent basis in order to receive access to food. If agencies fail to comply with reporting requirements, they may be temporarily ineligible to receive food. This impacts food equity from a network perspective because loss of an agency (in terms of shopping) is akin to a significant reduction in capacity in a certain geographic area, which is realized as a reduction in the pounds of food distributed.

Capacity Limitations

The capacity limitations constrain the food bank distribution network at different levels of a food bank's supply chain. A food bank's capacity is associated with its ability to receive, store and distribute food donations to the beneficiaries. Campbell et al. (2013) define a food bank's organizational capacity to be a function of their physical infrastructure and equipment, technological infrastructure, staff, and funding. We will follow a similar categorization and suggest that a food bank's network capacity depends on the physical storage capacity at the food bank warehouses and charitable agencies, means of transportation between warehouses and from warehouses to agencies, workforce and budget. We will examine these factors separately.

Physical Storage Capacity The food bank warehouses and charitable agencies that are a part of Feeding America's network show a vast variety in terms of their physical structure. Food bank warehouses vary on their sizes, infrastructures and storage capacities and capabilities. The donated food is initially stored, sorted and prepared for delivery to the agencies in the food bank warehouses. With the increasing demand and increasing amount of donations, food bank warehouses face capacity problems. As a solution to capacity limitations, food bank managers face the challenge of either expanding existing facilities or acquiring a new warehouse. After initial processing in the food bank warehouses, food is transported to charitable agencies. Agencies may be soup kitchens, food pantries, shelters, etc. with varying storage capabilities. 90% of Feeding America's agencies are either faith-based or other nonprofit organizations with limited storage capability (Hunger in America Report 2014). Considering different food types with different storage requirements (such as frozen food, or produce), the physical storage space of an agency becomes a primary constraining factor on its ability to store, and hence distribute, food to the food-insecure population in its service region.

Transportation Food bank warehouses and agencies both face challenges for transporting food from one location to another. Food is commonly collected at the food bank warehouses and then transported from one food bank warehouse to another before being shipped to agencies for distributing to the beneficiaries. Although traditionally it was more common for the agencies to pick up food from the food banks, due to high transportation costs and limited resources, more food banks have started delivering food to the agencies (Balcik et al. 2014). Currently, delivery of food to the agencies is being made either by the food bank vehicles or the means of transportation that the agencies can provide. Similar to their storage capabilities, agencies' means of picking up food from the food banks also affect their ability to serve the people in their region. Lacking the funds to obtain well-equipped transportation vehicles, most agencies only rely on the personal vehicles of the people responsible for picking up food from the food banks; which usually do not provide refrigeration. This becomes even a greater challenge for the agencies which are located in more rural areas and farther away from the food bank warehouses. Considering food quality and spoilage constraints, it becomes even more challenging for these locations to receive food.

To help with the problem of storage and transportation limitations, Feeding America has launched the Mobile Pantry program (Feeding America 2014). A mobile pantry is a vehicle that is managed either by the food banks or agencies allowing for direct food distribution to the food-insecure population. Through the usage of mobile pantries, food banks are able to increase their food distribution capacities and expand the area that they can serve.

Another problem faced by the food banks originates from their desire to collect and distribute nutritious food. Although food banks want to receive donations that consist of more nutritious items such as produce (Campbell et al. 2013), those food types require specific storage requirements (i.e. refrigerated rooms/trucks). For this reason, while trying to promote more nutritious donations, food banks simultaneously have to increase their storage capacity and capability to be able to store those food items.

Workforce According to Feeding America's latest Hunger Study (2014), approximately half of all the partner agencies that are affiliates of Feeding America have no paid staff. This highlights the dependency of the food distribution network on volunteer resources. In an average month, Feeding America's partner agencies get assistance from nearly 2 million volunteers across the entire food distribution network. This irregular work structure is not only a constraining factor of agencies' capacity but is also a primary source of uncertainty since it is difficult to estimate the number of volunteers an agency will receive in a given day. Typically, volunteers help with the sorting and packaging of the incoming food donations. Since both food bank warehouses and agencies have specific reporting requirements, staff that is trained to use computers is also a necessity.

Budget Food banks and agencies in the Feeding America network receive funding from various resources. Some of the sources of their funding are local, state and federal governments and donations from individuals, religious institutions or companies. Food banks use their budget to pay for their overhead expenses, to buy food, to provide transportation of food and pay for their regular staff. Agencies use their funding for purchasing food either from the food banks or other sources or paying for means to store or transport goods. The agencies can receive deliveries from the food banks or they can "shop" food in the "shopping area" of the branches. For the food they receive from the food bank, they pay a "shared maintenance fee". Although the prices of the food they get from the food banks are very low, they still have to work within budget limitations for "shopping" food from the branch. An agency's economic condition also has a significant effect on their staffing levels and transportation capabilities. For these reasons, agencies' budgets constrain their ability to receive and distribute food to serve the population in need.

The charitable agencies, which are the final locations for food distribution, face capacity limitations at different levels. An agency's capacity is directly affected by its ability to receive, store and distribute food. When examining the food distribution network, it is important to consider these capacities and their effect on the equity, effectiveness and efficiency of food distribution.

Balancing Efficiency, Effectiveness and Equity

Table 11.2 summarizes the challenges associated with the food aid supply chain as a function of their impact on effectiveness, efficiency, and equity. Material supply uncertainty impacts efficiency, particularly for activities associated with collecting, processing, and storing food. Limits on available supply as well as perishability affect the ability for food banks to be effective in terms of meeting the hunger need while minimizing waste. Clearly, limited availability of supply will also impact effectiveness as the food bank sets effectiveness targets based on the pounds of food distributed. International food banks also cite supply uncertainty and limited supply as a signifi-

Table 11.2 Impact of food aid supply chain on the iron triangle of hunger relief

	Efficiency				Effectiveness		Equity		
Complexity	Trans-portation	Storage	Labor	Infor-mation system	Waste	Hunger need	Food quantity	Food quality	Access
Limited supply				√		√			√
Uncertain material sup-ply quantities	√	√	√			√	√		
Uncertain sup-ply type	√	√		√	√	√		√	√
Uncertain monetary supply						√		√	
Perishability of donated supply	√	√			√				
Donor preferences							√	√	
Uncertain demand					√	√			√
Distribution network	√		√			√			√
Modes of distribution	√			√		√			√
Volunteer workforce			√	√		√			√
Data		√	√	√		√	√	√	
Reporting			√	√		√			√

cant problem (Booth and Whelan 2014; Food Banks Canada 2013). A national report by Food Banks Canada (2013) states that the amount of food provided to households by food banks was reduced due to supply shortages and several households had to be turned away. Demand mainly affects the level of effectiveness and equity since the uncertainty and limited visibility of demand may cause food to go to waste or may not satisfy the relative hunger need. The factors associated with capacity (such as the modes of distribution, workforce and reporting) also influence the iron triangle of hunger relief through constraining food distribution at different levels.

11.5 Strategies: Review of Current Modeling Literature: What is the State of the Science?

The food distribution modeling literature is dominated by studies focusing on for-profit food supply chains (Akkerman et al. 2010; Ahumada and Villalobos 2009). Although a lot of research has been done for resource allocation problems in for-profit supply chains; little work has been done for nonprofit operations (Lien et al. 2014). This chapter focuses on studies considering nonprofit food distribution supply chains for both emergency food relief and long-term food aid development. A main difference between for-profit and nonprofit food supply chains is their objectives. While studying for-profit supply chains requires considering objectives such as cost minimization and food quality, some common objectives for nonprofit food supply chains are determining optimal food collection and distribution locations (Davis et al. 2014), and assuring equitable, effective and efficient food distribution (Balcik et al. 2014; Sengul Orgut et al. 2015).

The modeling literature for food distribution can be divided into different categories such as stochastic or deterministic; operational, tactical, or strategic; and according to the objectives considered (equity, effectiveness, or efficiency.) We will first consider studies that have deterministic models.

Logistics of nonprofit food distribution is a challenging issue and has been studied by several researchers. Solak et al. (2012) study a nonprofit food distribution system in which food is distributed from warehouses to sites to be picked up by agencies. They build a mixed-integer program to achieve the optimal site selection, optimal allocation of agencies to these sites and an optimal vehicle routing for minimizing total transportation costs. Davis et al. (2014) study the problem of collection and delivery of food donations by a food bank. The food bank vehicles pick up donations from some local sources which provide periodic and large volumes of donations at pre-specified times. These locations can be used as Food Delivery Points (FDPs) where agencies that are located in the rural areas and have limited transportation capabilities can pick up food. The authors first develop a set covering model to determine the assignment of agencies to an FDP. Then, through a periodic vehicle routing problem, they aim to obtain optimal routes and schedules for food delivery and collection considering various constraints such as vehicle capacity and food spoilage.

Some of the modeling literature on food distribution focuses on the Meals on Wheels program which includes distributing cooked meals to people who are incapable of preparing meals for themselves such as the elderly or the sick. Bartholdi et al. (1983) built a user-friendly traveling salesman heuristic to address the routing problem for the Meals on Wheels program. Wong and Meyer (1993) use Geographic Information Systems (GIS) and Spatial Decision Support Systems (SDSS) to increase the efficiency of the publicly funded Meals on Wheels program. They address this problem both at the operational (routing) and strategic (facility location) levels under various constraints such as vehicle capacity and maximum delivery time for food safety.

Sengul Orgut et al. (2015) consider the tactical-level problem of obtaining optimal food distribution strategies. The authors develop deterministic network flow models to achieve equitable and effective food distribution policies under capacity constraints. They find that "*bottleneck counties*" which have low capacity compared to their demand, constrain the total distribution due to the interaction between the equity and capacity constraints. They show that to improve the effectiveness of this system, additional capacity should be allocated to bottleneck counties.

Krejci and Beamon (2010) address the problem of maintaining food security worldwide. They identify critical characteristics of environmentally sustainable food supply chains. They then use these characteristics to develop environmentally sustainable supply chain structures to provide food security for varying customer demand.

Research relevant to food bank operations which utilizes data mining is limited. Brock and Davis (2015) utilize a Multi-layer perceptron—Neural Network (MLP-NN) based approach to estimate the amount of food generated from collections at supermarket branches in isolated collection events. The idea behind this approach is to identify factors which may impact how much unsellable food supermarkets have on-hand for collection. An assumption used in this research is that the type of food available for collection at supermarket locations is known with certainty. Independent variables selected include parameters that describe the date of collection (i.e. weekday, week of year), the financial wellness of the surrounding community, and the outcomes of past operational decisions to collect food at the food bank. The physical characteristics of the food including the product type and its storage requirement are also considered as independent variables. In a computational study, which utilized data from the FBCENC, a MLP-NN model selected through trial and error was proven to be more effective at approximating the amount of food available for collection than multiple linear regression. An additional finding observed through the results is that the MLP-NN model more accurately matched the actual collection amount for food types which had more observations.

Some studies have considered the stochastic aspects of the food delivery problem. As explained in Sect. 11.4, the supply, demand and capacity data for food distribution are highly uncertain and unstable. Lien et al. (2014) aim to maximize the minimum fill rate over all locations to address the objectives of equity (fair distribution) and effectiveness (minimizing waste) in distributing scarce resources by a nonprofit organization under demand uncertainty. They use a dynamic programming approach to achieve the optimal allocation policy to a sequence of customers for a given continuous demand distribution. They use this optimal policy to generate a heuristic when demand follows a discrete distribution. Balcik et al. (2014) extend the work of Lien et al. (2014) from single route to multiple routes and they study the problem of equitable and efficient allocation of food donations through routing and allocation decisions for the donors and agencies of a food bank. They aim to maximize the minimum fill rate among all agencies of a food bank while implicitly maximizing the amount of distributed donations; and hence minimizing food waste. In their models, the allocation decisions are made sequentially where the driver of the food distribution vehicle decides how much food to allocate to an agency upon arrival at that location.

Schweigman et al. (1990) emphasize the difficulty in dealing with food insecurity problems and they address the issue of food insecurity in developing countries. They seek to achieve the optimal amount of land that should be cultivated to prevent any food shortage. Through a discrete-event simulation model, Mohan et al. (2013) address the objectives of efficiency and effectiveness in nonprofit food distribution supply chains. Their aim is to minimize the operational costs and maximize the quantity of food distributed considering the operations of a food reclamation center that is a member of a global hunger relief organization (the Society of Saint Vincent de Paul). They perform what-if analyses to optimize the operations of the reclamation center responsible for collecting, sorting and packaging food.

11.6 Research Opportunities—Open Challenges—How IEs and OR can help

Open Problems Related to Uncertainty

Data Uncertainty There are large amounts of messy, unstructured data generated from multiple sources within the supply chain network. For example, the FBCENC generates more than 0.25 million transaction records annually that record food receipts (of more than 20 product types), distribution, and inventory transfers between branch warehouses. In addition, they collect data from more than 900 charitable agencies in their service area in mixed forms. Based on the classification of the charitable agency (e.g. food pantry, soup kitchen), information may be reported as households served or individuals served. While a significant amount of data is collected, there is also evidence of missing and/or erroneous data. Furthermore, there is a lack of data due to limited visibility to demand, limited visibility to other sources of supply used by charitable agencies, and limited visibility to other sources of supply used by people in need. Methodologies could be developed in the management of big data and in quantifying the value of information particularly as it relates to focusing data collection efforts with respect to supply and demand. Further robust optimization and stochastic programming methods can be explored for capturing the impact of data uncertainty on supply chain management decisions.

Addressing Demand Uncertainty An important source of uncertainty in nonprofit food distribution is the difficulty in accurately forecasting demand. There is very low visibility regarding the actual "need" for the food distribution, and as a result food banks often overestimate their demand (e.g., assuming responsibility for the entire poverty population in a region) and at the same time may underestimate demand (e.g., ignoring populations who though above the poverty level have hunger need). Methodologies could be developed for better estimating the actual need and models that are robust to demand variability could be explored.

Addressing Supply Uncertainty There are also research opportunities associated with better prediction of donated supply. This research could be directed towards approximating the donation quantity as a function of donation source, type, and environment. For example, researchers in operations management and agriculture could collaborate with food producers to identify factors which can be helpful in estimating crop yields at different phases of harvesting season. These factors could be integrated into historic donation records, allowing researchers to study whether these variables are helpful in estimating food availability. Continued research could be conducted which evaluates in-kind food availability at supermarket branches. Marketing research techniques could also be explored to model donor relationship management strategies for other solicited donation sources. Using information obtained through supply forecasting methods, stochastic optimization models could be formulated which promote the presence of sufficient inventory to satisfy immediate and future demand.

Addressing Capacity Uncertainty Methods could be developed for estimating the capacities of food bank warehouses and agencies. Capacity, in this context, represents a food bank's or an agency's ability to receive, store or distribute the food that they receive, and hence it is affected by many factors such as storage space and budget as discussed in Sect. 11.4. Usually capacity is challenging to estimate because of the complexities associated with these different attributes that come together to create an agency's or a food bank's capacity. Although food bank warehouses can control their own inventory, they have very low visibility regarding the real-time inventories of the charitable agencies that they serve. Therefore, since the capacity is highly variable and depends on multiple factors, it is a primary source of uncertainty in the nonprofit food distribution chain. This uncertainty in capacity may be addressed using stochastic operations research tools.

11.7 Open Problems Related to Supply Chain Network

Coordination and Information Sharing Coordination mechanisms have been utilized by for-profit agencies to ensure that profits are shared among all stages of the supply chain. Domestic hunger presents a unique perspective as many of the high-volume suppliers either provide unsolicited donations in response to their awareness of food insecurity concerns or make solicited donations available for collection as a result of poor forecasting or operational inefficiencies.

Tools are being developed to increase visibility between agencies, food banks and donors. Aidmatrix is the first global non-profit organization aiming to develop such tools to increase the efficiency of the delivery of humanitarian aid (Aslanzadeh et al. 2009). Two of the tools they developed in collaboration with Feeding America are AgencyExpress and DonorExpress. AgencyExpress is a web-based software application developed to help agencies place food orders on the food bank and have higher visibility to the food banks' resources; thus increasing the efficiency of the agencies' food ordering process (Aslanzadeh et al. 2009). DonorExpress is also a web-based solution that aims to improve the supply chain operations of Feeding America through providing communication between donors and food banks.

In addition to these tools, warehouse management systems (WMS) can facilitate the management of internal data and product flow. For this reason, automation of internal food bank operations is ripe for investigation. Although there are significant infrastructure costs to deploy WMS, the benefits in terms of keeping track of inventory, space, and equipment usage may in the long term outweigh the costs. WMS can also provide rules/algorithms for storage location assignment and optimal routings for food bank shoppers. Further research is necessary to equip food banks with the ability to make proactive inventory management decisions through information sharing and coordination.

Equity-Effectiveness-Efficiency Trade-off Deterministic and stochastic operations research methods should be used to address the objectives of equity, effectiveness and efficiency. The trade-offs between these objectives should be explored and guidelines which can directly be implemented by the food banks should be developed.

Scheduling Challenges Scheduling is one of the fundamental challenges of operations. It involves understanding demand and providing supply in the form of workforce or resources (i.e. volunteers or vehicles) to ensure that demand is fulfilled. In terms of operations scheduling, there are opportunities for integrated models that incorporate supply, demand, and capacity uncertainty and their impact on scheduling.

Vehicle Routing Strategies While there is prior work addressing routing related challenges in non-profit food distribution, there are still opportunities to develop operations research and systems engineering techniques to enhance vehicle routing policies. The structure of the distribution environment in food bank supply chains is varied and complex. Food banks can operate out of a single warehouse or multiple warehouses. There is uncertainty in the quantity of food donations which impacts collection frequency and vehicle capacity. Charitable agencies can be located in areas that are urban, rural, or a hybrid of both which also has implications on vehicle capacity and food delivery to charitable agencies. Food banks collect both shelf-stable and highly perishable products which has implications on the size and capability (e.g refrigerated vs. non-refrigerated) of the vehicle fleet. Lastly, many food bank networks serve hundreds of high-volume donors and charitable agencies. In addition, the network is frequently changing as new charitable agencies or donors are added. Routing strategies must be adaptable to the dynamic environment of food distribution and capable of identifying solutions for problems which may not be solvable using exact solution approaches. These problems may be further complicated by the existence of customer-specific service time intervals, heterogeneous fleets and vehicle departures from multiple depots.

Warehouse Operations Efficient and effective warehouse design needs consideration to improve food bank warehouse operations. A common issue with food bank warehouses is insufficient/poorly used storage space to respond to variability in donation levels. Thus, warehouse dimensioning should be studied in the context of food banks to understand the tradeoffs between the warehouse size, locations of warehouse departments, and the dimensions of each department on overall warehouse operations. Warehouse layout issues in turn impact equipment selection and utilization, material handling costs, space utilization, and construction and maintenance costs (Gu et al. 2010).

Supply Chain Network Design The distribution network significantly affects equity and access for food banks and can vary based on the service area. For example, meeting the needs of charitable agencies that are located in rural, hard to reach locations can be very different from those located in urban areas. Methodologies could be developed to examine the cost of using multiple modes of food delivery and multiple delivery locations to increase food access.

References

Ahumada O, Villalobos JR (2009) Application of planning models in the agri-food supply chain: a review. Eur J Oper Res 196(1):1–20

Akkerman R, Farahani P, Grunow M (2010) Quality, safety and sustainability in food distribution: a review of quantitative operations management approaches and challenges. OR Spectr 32(4):863–904

Aslanzadeh M, Rostami EA, Kardar L (2009) Logistics management and SCM in disasters. Supply Chain and Logistics in National, International and Governmental Environment. Physica-Verlag HD, pp 221–252

Balcik B, Iravani S, Smilowitz K (2014) Multi-vehicle sequential resource allocation for a nonprofit distribution system. IIE Trans 46(12):1279–1297 (just-accepted)

Bartholdi III JJ, Platzman LK, Collins RL, Warden III WH (1983) A minimal technology routing system for meals on wheels. Interfaces 13(3):1–8

Booth S, Whelan J (2014) Hungry for change: the food banking industry in Australia. Br Food J 116(9):1392–1404

Brock LG, Davis LB (2015) Estimating available supermarket commodities for food bank collection in the absence of information. Expert Syst Appl 42(7):3450–3461

Butcher LM, Chester MR, Aberle LM, Bobongie VJA, Davies C, Godrich SL, Begley A (2014) Foodbank of Western Australia's healthy food for all. Br Food J 116(9):1490–1505

Cabili C, Eslami E, Briefel R (2013) White paper on the emergency food assistance program (TEFAP) (USDA, Trans.). In USDA (Ed.). Washington, DC

Campbell EC, Ross M, Webb KL (2013) Improving the nutritional quality of emergency food: a study of food bank organizational culture, capacity, and practices. J Hunger Environ Nutr 8(3):261–280

Campbell E, Webb Kl, Ross M, Crawford P, Hudston H, Hecht K (2015) Nutrition-Focused Food Banking. http://www.iom.edu/~/media/Files/Perspectives-Files/2015/FoodBanking.pdf. Accessed 14 May 2015

Carlson SJ, Andrews MS, Bickel GW (1999) Measuring food insecurity and hunger in the United States: development of a national benchmark measure and prevalence estimates. J Nutr 129(2):510–516

Chopra S, Meindl P (2012) Supply chain management. Strategy, planning & operation, 5th edn. Prentice Hall, New Jersey

Coleman-Jensen A, Gregory C (2014, September 3) Definitions of food security. United States department of agriculture economic research service. Web. http://www.ers.usda.gov/topics/food-nutrition-assistance/food-security-in-the-us/definitions-of-food-security.aspx#.VElFbPl4ob0. Accessed 3 Nov 2014

Coleman-Jensen A, Nord M, Andrews M, Carlson S (2011) Household food security in the United States, 2010 economic research report No. ERR-125. USDA Economic Research Service, Washington, DC

Coleman-Jensen A, Gregory C, Singh A (2014, September) Household Food Security in the United States in 2013, ERR-173, U.S. Department of Agriculture, Economic Research Service

Davis LB, Sengul I, Ivy JS, Brock III LG, Miles L (2014) Scheduling food bank collections and deliveries to ensure food safety and improve access. Soc-Econ Plan Sci 48(3):175–188

FBCENC (2013) Food Bank of Central & Eastern North Carolina. http://www.foodbankcenc.org/site/PageServer?pagename=FBCENCHome. Accessed 2013

Feeding America (FA) (2011b) Our food bank network. http://feedingamerica.org/how-we-fight-hunger/our-food-bank-network.aspx. Accessed 26 Oct

Feeding America (FA) (2014) http://www.feedingamerica.org/. Accessed 25 Oct 2014

Food and Agricultural Organization of the United States (FAO) (2014) The state of food insecurity in the world: strengthening the enabling environment for food security and nutrition. www.fao.org/3/a-i4030e.pdf. Accessed 17 Oct 2014

Foodbank Australia (2013) End hunger in Australia report 2013. www.foodbank.org.au/wp-content/uploads/2013/10/foodbank-End-Hunger-Report-2013.pdf. Accessed 17 Oct 2014

Food Banks Canada (2013) HungerCount 2013, Food Banks Canada, Toronto. http://www.foodbankscanada.ca/FoodBanks/MediaLibrary/HungerCount/HungerCount2013.pdf. Accessed 29 Oct 2014

Gentilini U (2013) "Banking on food. Examining the scale, features and implications of food banks in high-income countries", Draft, Institute of Development Studies, Brighton

Goncalves P, Leiras A, Chawaguta B, Yoshizaki H (2013) Stochastic optimization for humanitarian aid supply and distribution of World Food Programme (WFP) in Ethiopia. 2013 International Production and Operations Management Society (POMS) Conference Proceedings. http://www.pomsmeetings.org/ConfProceedings/043/FullPapers/FullPaper_files/043-0635.pdf. Accessed 11 Jan 2014

Gundersen C, Kreider B, Pepper J (2011) The economics of food insecurity in the United States. Appl Econ Perspect Policy 33(3):281–303

Gu J, Goetschalckx M, McGinnis LF (2010) Research on warehouse design and performance evaluation: a comprehensive review. Eur J Oper Res 203(3):539–549

Houlihan P (2006) How food banks came to love the free market. http://www.chicagobooth.edu/magazine/spr06/YJ_34-39_Front.pdf. Accessed 14 May 14 2015

Hunger in America (2014) Feeding America. http://www.feedingamerica.org/hunger-in-america/our-research/the-hunger-study/. Accessed 25 Oct 2014

Kissick WL (1994) Medicine's dilemmas: infinite needs versus finite resources. Yale University Press New Haven, CT

Krejci CC, Beamon BM (2010) Environmentally-conscious supply chain design in support of food security. Oper Supply Ch Manage 3(1):14–29

Lambie-Mumford H (2014) Food bank provision & welfare reform in the UK., SPERI British Political Economy Brief No. 4. http://speri.dept/shef.ack.uk/publications/policy-briefs. Accessed 25 Oct 2014

Lambie-Mumford H, Dowler E (2014) Rising use of 'food aid'in the United Kingdom. Br Food J 116(9):1418–1425

Lien RW, Iravani SMR, Smilowitz KR (2014) Sequential resource allocation for nonprofit operations. Oper Res 62(2):301–317

Marsh MT, Schilling DA (1994) Equity measurement in facility location analysis: a review and framework. Eur J Oper Res 74(1):1–17

McConville A, Groot S (2013) Food insecurity and urban poverty in New Zealand: a scholar-activist engagement. Aust Community Psychol 25(2):38–52

Mohan S, Gopalakrishnan M, Mizzi PJ (2013) Improving the efficiency of a non-profit supply chain for the food insecure. Int J Prod Econ 143:248–255

National Research Council (2006) Food Insecurity and hunger in the United States: an assessment of the measure. Panel to review the U.S. Department of Agriculture's Measurement of Food Insecurity and Hunger. In: Wunderlich GS, Norwood JL (eds) Committee on national statistics, division of behavioral and social sciences and education. The National Academies Press, Washington, DC

Nord M, Andrews M, Carlson S (2009) Household Food Security in the United States, 2008 Economic Research Report, Vol 83: United States Department of Agriculture, Economic Research Service

Ohls J, Saleem-Ismail F, Cohen R, Cox B (2002) The emergency food assistance system-findings from the provider survey, volume ii: final report (No. 33797). United States Department of Agriculture, Economic Research Service

Quinn E (2010) Logistics for food assistance: delivering innovations in complex environments. In: Omamo S, Getilini U, Sandstrom S (eds) Revolution: from food aid to food assistance: innovations in overcoming hunger. World Food Programme, Italy, pp 307–328

Rancourt M, Cordeau J, Laporte G, Watkins B (2013, February) Tactical network planning for food aid distribution in Kenya. CIRRELT-2013-09

Santini C, Cavicchi A (2014) The adaptive change of the Italian Food Bank foundation: a case study. Br Food J 116(9):1446–1459

Savas ES (1978) On equity in providing public services. Manage Sci 24(8):800–808

Schweigman C, Bakker E, Snijders T (1990) Operations-research as a tool for analysis of food security problems. Eur J Oper Res 49(2):211–221

Sen A (1973) On economic inequality. Clarendon Press, Oxford

Sengul Orgut I, Ivy JS, Uzsoy R, Wilson JR (2015) Modeling for equitable and effective distribution of donated food under capacity constraints. IIE Trans (in press)

Solak S, Scherrer C, Ghoniem A (2012) The stop-and-drop problem in nonprofit food distribution networks. Ann Oper Res 1–20

Stone DA (1997) Policy paradox: the art of political decision making. WW Norton, New York

Tarasuk V, Mitchell A, Dachner N (2014a) Household food insecurity in Canada, 2012. Research to identify policy options to reduce food insecurity (PROOF), Toronto. http://nutritionalsciences.lamp.utoronto.ca/

Tarasuk V, Dachner N, Loopstra R (2014b) Food banks, welfare, and food insecurity in Canada. Br Food J 116(9):1405–1417

Wang KYT, Lyu LC (2013) The emergence of food bank/voucher programs in Taiwan: a new measure for combating poverty and food insecurity?. Asia Pac J Soc Work Dev 23(1):48–58

Warshawsky DN (2011) Food bank Johannesburg, state, and civil society organisations in post-apartheid Johannesburg. J South Afr Stud 37(4):809–829

Webb K (2013) Food banks of the future: organizations dedicated to improving food security and protecting the health of the people they serve. J Hunger Nutr 8(3):257–260

Wong DWS, Meyer JW (1993) A spatial decision support system approach to evaluate the efficiency of a meals-on-wheels program. Prof Geogr 45(3):332–341

Chapter 12
Understanding Resilience: A Spatio-temporal Vulnerability Assessment of a Population Affected by a Sudden Lack of Food

Thomas Münzberg, Marcus Wiens and Frank Schultmann

Abstract Natural disasters such as massive floods and severe storms can lead to disorganizations of the food supply chain. A cutoff of the food supply chain is a worst case scenario in which the population is not able to buy food in stores for a couple of days. In this situation, the private food stock in households is the key self-helping capacity to cope with the effects of a sudden lack of food. In this paper a spatio-temporal vulnerability model is introduced to facilitate disaster preparedness and to enhance the understanding of resilience of the population in Germany. The model is indicator-based and uses the empirical data about the stock piling behavior of the population, statistical data about the population structure of different city districts, and the results of an expert survey to estimate the time-dependent vulnerability of a population derived from the continuous consumption of their private food stock. The paper includes a review about studies recently published about the empirical quantifications of food stocks in households in Germany and an evaluation of the expert survey. The model was applied to two cases of German cities to demonstrate and discuss its benefits.

Keywords Resilience · Coping capacity · Indicator-based vulnerability assessment · Spatio-temporal assessment · Disaster relief · Disaster management planning · Lack of food · Food supply disruption

---

T. Münzberg (✉)
Institute for Nuclear and Energy Technologies, Karlsruhe Institute of Technology, Karlsruhe, Germany
e-mail: thomas.muenzberg@kit.edu

M. Wiens · F. Schultmann
Institute for Industrial Production, Karlsruhe Institute of Technology, Karlsruhe, Germany
e-mail: marcus.wiens@kit.edu

F. Schultmann
e-mail: frank.schultmann@kit.edu

C. W. Zobel et al. (eds.), *Advances in Managing Humanitarian Operations*,
International Series in Operations Research & Management Science 235,
DOI 10.1007/978-3-319-24418-1_12

12.1 Introduction

Although the German food supply chain is robust against many natural and man-made disasters, it has been shown in previous examples, such as the Elbe floods in 2002 and 2013 as well as the Münsterland power outage in 2005, that problems in food supply are possible in Germany. In these cases, abrupt cutoffs in the food supply chain led to temporarily limited food shortages. Usually, it was not possible to buy food in the affected areas for a couple of days.

To cope with such short-term lacks of food, stock piling in private households is the key self-helping capacity of the population. However, our review of recent studies revealed that the German population is not sensitized well for the consequences of food shortages. Only a small number of households have stored food to cope with emergency situations that last longer than 1 week. Even short disturbances of the food supply chain could lead to critical situations in which the population has to rely on advanced assistance by disaster management authorities. In order to prepare for these kinds of humanitarian operations, disaster response activities have to be adjusted to the temporal needs of the population.

To facilitate the disaster preparedness in Germany, we introduce a spatio-temporal vulnerability assessment model. The model highlights the spatio-temporal sufferings and the resilience of a population potentially affected by a sudden food shortage. The model aims at providing German disaster management authorities on a local level with a better understanding of the resilience of a population in case of such an event.

Assessing vulnerabilities is a well-established approach to risk management and also applied in the area of operations research. The spatio-temporal assessment of vulnerabilities for understanding disasters is a relatively new field of research. There is an increasing number of models addressing different hazards like floods or earthquakes, but there is no approach considering food shortages.

Our spatio-temporal vulnerability assessment model proposed addresses the suffering of a population during an ongoing lack of food. We review and compare empirical and statistical results of studies published most recently with respect to food stock quantifications in households in Germany. Based on an appropriate data base selected, we develop a completely new model to understand the escalations of a lack of food as a time function. The model is indicator-based and allows for analyzing the spatio-temporal development of the problem taking into account the duration of a food supply cutoff, the empirically confirmed behavior of stock piling of households of different sizes, and the number of households of various household sizes in potentially affected city districts. In addition, we conducted an expert survey to better understand estimations of the vulnerability of people with different food stock capacities at different points in time during an ongoing food shortage. The results of the model enable decision makers to identify crucial city districts and provide insight into the escalation of a lack of food. The effects of alternative disaster measures to enhance the resilience of a population can be simulated. This additionally supports decision making in preparing for and responding to such incidents.

This paper is structured as follows. First, we give a brief introduction to the core challenges in disaster management while coping with a sudden lack of food availability. For this purpose, we distinguish different kinds of lack of food supply and derive a working definition in the context of the German disaster management environment. The vulnerability model proposed should have an empirical basis. To ensure this, we reviewed current empirical studies of the stock piling behavior in Germany and selected the most appropriate one for further model development. Subsequently, we explain the principles of the spatio-temporal vulnerability assessment proposed and an approach to deriving the resilience of a city. Finally, we apply the assessment proposed to two cases using real data of household sizes of the German cities of *Cologne* and *Mannheim*. We present and discuss the results for each example. The paper is completed by a critical remark and a conclusion.

12.2 A Sudden Lack of Food: A Challenge for Disaster Management

Several causes for a sudden lack of food are known. The Pan American Health Organization (2001) distinguishes between (1) the destruction of food stocks and (2) the disorganized distribution system which may not give rise to localized scarcity of food.

In the context of destroyed food stocks, food shortages are often associated with long-term famines caused by e.g. loss of crops, droughts, calamities or poverty in developing countries (see more e.g. in Anderson and Woodrow 1991; Patt and Schröter 2008; Wright and Vesala-Husemann 2006). These forms of food shortages are also understood to be a form of lack of food security (for more information see e.g. Dilley and Boudreau 2001). They are recurrently caused by major long-term or chronic problems in fundamentally unstable food systems.

These types of lack of food need to be distinguished from those associated with distribution systems that become disorganized. Such a disorganization causes a sudden and localized lack of food. The initial events causing such a state of disorganization may be natural disasters like hurricanes, floods, power outages or widespread infrastructure destruction of, for example, warehouses, streets or bridges. Massive food aid usually is not required, as these disasters only rarely lead to a loss of crops (Jacob et al. 2008). These forms of food problems generally are of short-term character but result in situations where it is no longer possible to buy food in supermarkets and discounters. In this paper, lack of food, food shortages and food scarcities are used synonymously for situations in which no food is available on the market and affected populations rely on their private food stock capacity or on advanced assistance from disaster management authorities.

An unprecedented example of a short-term lack of food is the disaster caused by the hurricane Katrina. When reviewing the impact, Jacob et al. (2008) noted that people who were not evacuated endured several days without food. They relied on outside assistance such as trucks that brought food into the affected region.

Furthermore, shops were even closed in relatively unaffected areas (Jacob et al. 2008). Although the occurrence of hurricanes in Europe is unlikely, natural disasters such as massive floods, heavy snow fall and severe storms can also cause local destructions and disruptions of the food supply chain system.

In recent years, the risk of food insecurity in Germany is increasingly discussed (Akinyemi 2010). Like in other countries, there are food banks which collect usable but unsellable food from local businesses for those who have difficulty in buying food. The system how the food is distributed can be distinguished in (a) warehouse systems in which food banks store foods and distribute it to non-profit agencies (Brock and Lauren 2015) and in (b) direct systems in which food banks collect and directly distribute food to deprived persons. Mostly all food banks in Germany operate as direct systems which usually do not store any food for disaster situations.

Some German federal states have disaster management concepts in place to supply a greater number of affected and vulnerable people during a disaster situation ("Betreuungsplatz-Bereitschaft BTP-B 500"). The concepts are based on modular combinations of response resources from different organizations. Each module combination should enable disaster management authorities to cater 500 affected people with food and beverages for at least 24 hours. Longer supply situations with a great multitude of vulnerable people are hard to manage through these response resources.

In addition, there are legal instruments available to manage long-term adverse impacts on the German agricultural and nutrition system ("Ernährungsnotfallvorsorge" referring to the Emergency Food Shortage Preventive Act, "Ernährungsvorsorgegesetz" from 1990). These instruments are embedded in emergency laws that can only be applied in situations in which the German food market is no longer able to absorb the adverse effects and thus a risk for famine occurs. In this case, the German authorities are allowed to intervene in the operational processes of food companies and to take over of food production and distribution. At the moment, this legal instrument is being politically discussed regarding its practicability.

However, the key player in coping with food shortages is the local disaster management authorities that have to consider at which point in time an official interaction is necessary to provide the population with advanced assistance or to consider evacuations. Especially in the beginning of a scarcity of food, response resources are limited and, hence, priorities need to be defined to concentrate on the most severely affected regions. One example of an alternative measure may be evacuation which requires long preparation. While considering this, private food stockpiling of the population is a key impact driver which helps to delay escalations of food shortage situations. This driver also influences the vulnerability and the resilience of the population.

In Germany, there are no explicit and binding provisions for food stockpiling in households. Due to the general principle of subsidiarity, food stockpiling is recommended by the German Federal Agency for Civil Protection and Disaster Assistance (BBK). Until 2013, the BBK recommended a food stock for 1 to 2 weeks in its well-distributed leaflet for disaster preparation for the German citizens. In 2013, the leaflet was fundamentally revised. The recommended capacity for food stockpiling was described as follows: "It must be the aim to have a food stock for surviving 14 days without going shopping." Actually, the German population stores much less food than recommended by the BBK. To better understand the stock piling behavior of the citizens in Germany and to select an appropriate data base for our model proposed, we reviewed several empirical surveys. The findings obtained will be presented in the following section.

12.3 Review of Studies Focusing on the Quantification of Food Stocks in Households in Germany

Several studies and surveys have been performed since 2000 to investigate the actual storage of food in households in Germany. As a result of this review, we identified four studies and surveys to be most relevant. Table 12.1 lists these studies and characterizes their methodological approaches regarding the used survey method, the number of participants and the status of representativeness, which indicates how the study results and especially the statistical evaluations accurately reflect the entire German population.

Table 12.1 Overview of the most relevant surveys focusing on the quantification of food stocks in households in Germany

Author	Year	Kind of method survey	Number of participants	Status of representativeness
Rasche, Schmidt, Schneider, and Waldtmann	2000	Telephone poll	$n=360$	Representative study
Menski and Gardemann	2008	Standardized questionnaire	$n=567$	Cross-sectional non-representative study
Rhein	2013	Computer-aided telephone interviews	$n=2000$	Representative empirical study
Ohder, Röpcke, Sticher, Geißler, and Schweer	2014	Standardized questionnaire	$n=800$	Randomly selected group of Berlin citizens

In 2000, Rasche, Schmidt, Schneider, and Waldtmann made a telephone poll regarding private food stocks in households in Germany. The results are shown in Table 12.2. The authors described their study as representative although the number of 360 random samples is very small.

Table 12.2 Degree of preparedness of telephone poll participants. Selected result for the quantification of food stock piling by Rasche et al. (2000) ($n=360$)

Degree of preparedness	Percentage of the total participants
Have no food stock	15.7
Have food stock	71.6
Those who have a food stock and do additional planning for stockpiling	69.4

In the aftermath of the long-term power outages in Münsterland in 2005, a region in North Rhine-Westphalia, Germany, Menski and Gardemann (2008) investigated the impact on the food supply sector. This study was characterized by the authors to be cross-sectional and non-representative. It was unique, as the participants recently had experienced a food supply disruption. This may result in a more realistic evaluation of the participants' stock piling behavior compared to the participants of the other studies who actually did not experience a food shortage. A standardized questionnaire was used to evaluate food stock piling before and after the power outage. The authors studied how the experience of the power outage changed the awareness and the practice of food stock piling for disaster situations. Menski and Gardemann (2008) pointed out that although the preparation level was bad before the power outages, no significant changes were noticed in the preparation behavior of the affected population afterwards. Based on the published results we calculated the percentages for each specific degree of preparedness (see Table 12.3).

Table 12.3 Degree of preparedness of telephone questionnaire participants. Selected result for the quantification of food stock piling by Menski and Gardemann (2008) ($n=567$)

Degree of preparedness	No stockpiling	1 day	2 days	3–5 days	6–8 days	More than 8 days
Percentage of total participants	2.43	7.01	32.05	40.57	11.56	6.29

After the former study of Menski and Gardemann (2008), the BBK conducted a representative empirical survey in 2012. The aim of this survey was to build a scientific and reliable basis for quantifying coping capacities of the population in the situation of power outages. The results were evaluated and published in the study of Rhein (2013). Additionally, the study provided information about social parameters that potentially influence the stock piling behavior. Table 12.4 lists the results obtained for the degree of preparedness due to the stock piling of food. The stock piling of water was recorded separately and is excluded from the table.

Table 12.4 Degree of preparedness of telephone interview participants. Selected result for the quantification of food stock piling by Rhein (2013) (n=2000)

Degree of preparedness	No stock piling	1 day	2 days	3 days	4 days	5 days	6 days
Percentage of total population	19.2	0.8	2.7	9.2	7	6.6	1.1
Degree of preparedness	7 days	2 weeks	More than two weeks	Self-sufficient			
Percentage of total population	24.8	18.4	9.8	0.5			

Embedded in a German national research project on the improvement of disaster preparedness for power outages of the German states of Brandenburg and Berlin, Ohder et al. (2014) conducted a survey using standardized questionnaire with 800 randomly selected citizens of Berlin. This study is not representative, but provides useful information about tendencies of the disaster preparation behavior regarding the stock piling of water, food, and medicine of citizens of Berlin. Table 12.5 displays the degree of preparedness of the participants regarding the stockpiling of food which can be consumed immediately and does not need any additional processing. This specification had not been made in the studies before.

Table 12.5 Degree of Preparedness of telephone interview participants. Selected result for the quantification of food stock piling by Ohder et al. 2014 (n=768)

Degree of preparedness	1–2 days	3–4 days	5–6 days	More than 6 days
Percentage of all surveyed participants	98.7	76.3	31.1	12.4

The initial representations of the study results show that different scales were used to investigate the degree of preparedness of the population. Additionally, the designs of the questions varied. For instance, the storage of drinking water was excluded or the focus was placed on food that does not need any preparation for consumption. Remarkably, only the study of Rhein (2013) presents a representative evaluation based on empirical data, with the number of samples being the highest of all studies.

Generally, it became obvious, that stock piling for 14 days as recommended by the BBK is not done by the majority of the German population.

12.4 A Vulnerability Assessement Regarding the Impacts of a Lack of Food

Introduction to Spatio-temporal Vulnerability Assessments

Vulnerability models are used to analyze risks (Birkmann and Wisner 2006) and to identify weaknesses in disaster management (Bogardi 2004). Several definitions of the term vulnerability can be found depending on context, discipline, hazard, and assessment purpose. Reviews on the terminology and the embedding of vulnerability into risk approaches are provided by e.g. Thywissen (2006), Manyena (2006), and Hufschmidt (2011).

A commonly used understanding is that vulnerability describes the inherent character of a system to suffer from adverse effects under the impact of hazards (e.g. see Thywissen 2006; Cardona 1999; Birkmann and Wisner 2006). Usually, indicator frameworks are used to assess the vulnerability of a system (see e.g. Kasperson et al. 1995; Brooks 2003; Adger 2006). Indicators describe aspects of vulnerability generation and their selection is tailored to the purpose of the assessment and the field of application.

The vulnerability of a system can be assessed statically for a defined point in time or dynamically by considering time-dependent changes in the system itself or its environment. There are a growing number of temporal assessment models which analyze a changing vulnerability status of a system over time. The models address different temporal references which can be distinguished between frames that focus on (1) the short-term changes of vulnerability of a system during the exposure to an ongoing hazard and (2) the vulnerability development of a system during long-term changes of the environment or the system. The first kind of model provides insights into temporal effects and escalation during disasters. This enhances the understanding of the benefits and limits of, for instance, buffer capacities and (counter) measures. The second kind of model is mostly associated with adaptation management and the consideration of long-term environmental, social, and political processes (Bankoff et al. 2004; Kienberger et al. 2013). Typical temporal vulnerability assessments are related to:

- climate change (e.g. Metzger and Schröter 2006; Aubrecht et al. 2013; Gaillard 2010),
- sea level rise (e.g. Sahin and Mohamed 2010; Sobiech 2013),
- earthquake risks (e.g. Debnath 2013),
- flood risks (e.g. Rodríguez-Gaviria and Botero-Fernández 2013; Giupponi et al. 2013; Chen and Chen 2012), or
- industrial risks (e.g. Belliveau et al. 2006; Münzberg et al. 2014).

Spatio-temporal Vulnerability in the Context of a Sudden Lack of Food

Our proposed spatio-temporal vulnerability model should provide support for disaster management authorities in Germany. As Germany is a federal country, the German disaster management system is strictly organized in a decentralized way. Regarding the legal responsibility for the operational response, the local disaster management authorities at the level of administrative districts ('Landkreise') and independent cities ('kreisfreie Städte') are the key actors in coping with disaster events. To support the preparedness for a sudden lack of food, the proposed model provides spatial insights about the potential suffering of the population that lives in the communities of administrative districts and independent cities. In the following, we focus on communities in city districts although the concept is also applicable to administrative districts.

In this context, we define the spatial vulnerability of a city district to be the inherent character of its inhabitants to suffer from adverse effects of a sudden lack of food. Time-dependent, dimensionless vulnerability values of a city district are used to compare the sufferings of the inhabitants in different city districts for several points in time during an ongoing lack of food.

As mentioned before, the behavior of the population in stocking food is a key driver that influences the ability to cope with a sudden lack of food. We use this as a key indicator to measure vulnerability. To operationalize this indicator, we use the empirical findings of the studies mentioned above to illustrate the development of vulnerability as a function of the duration of a food shortage. In addition, we use statistical data about the number of households with different household sizes to consider the population structure of the city districts. In this way, we ensure that the key assumptions of our model have a strong empirical and statistical basis.

The evolution of vulnerability is determined by assuming constant consumption of stockpiled food during a food shortage. The more food of the stockpile is consumed the lower the ability to cope with an ongoing lack of food. Hence, the vulnerability of the population increases with time. Those affected people who prepared a stockpile for some days have a level of vulnerability of zero at the beginning of a cutoff of the food chain which we define as t_0. The highest degree of vulnerability is reached at the point in time when the food stock is (fully) exhausted, which is defined as t_c. In this way, we understand vulnerability as a temporal and inherent character of a population to suffer from adverse affects of a sudden unforeseen food shortage that is depending on the time of an ongoing food shortage and the depletion of their emergency food stock. How the vulnerability increase could be measured is discussed in the next section.

Increasing Vulnerability as a Result of Ongoing Consumption of the Private Food Stock

The model proposed is based on the concept that the capacity of a disturbance-affected system to absorb adverse effects is limited and depletes with time during the disturbance (Brauner et al. 2015; Münzberg et al. 2015). In the context of a food shortage, the level of vulnerability of a household increases by the consumption of its private food stock. The less food is available in the private stock, the less time is left to react and to refill the stock. The increasing vulnerability can be defined by different kinds of time functions. The simplest form could be a linear function, but other non-linear functions are also possible.

We conducted an expert survey to select an appropriate function. The purpose of the survey was to look for references of a non-linear time function that describes this time dependency. To ensure the trust of decision makers in the proposed model in which this function is to be integrated, an easily understandable function is required.

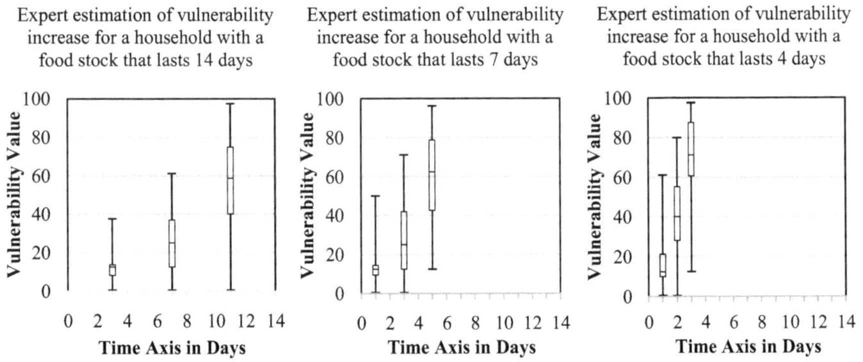

Fig. 12.1 Expert estimation of time-dependent vulnerability values for households with different food stock capacities ($n=31$). The expert estimations are displayed as box plots with median, quartiles, and maximum and minimum values

The survey was conducted with 31 German experts in the field of disaster management and planning. These experts are responsible in authorities for disaster preparedness and response at their respective governmental organizations. We asked them to estimate the time-dependent vulnerability of three fictitious households with various degrees of preparedness (food stocks for 14 days, 7 days and 4 days). For each household, the experts were asked to estimate the vulnerability for three points in time. We assumed that at the beginning, t_0, of a period of a lack of food the vulnerability is zero. Additionally, we supposed that the value of vulnerability is 100 at t_c in which the food stock is fully exhausted.

To interpret the data collected we determined the statistical parameters of median, quartiles and maximum and minimum values as displayed by the box plots in Fig. 12.1.

Because of the different stock capacities, the results from the expert estimations are difficult to compare on a time scale. For a better comparison we standardized the degree of preparation on a depletion scale that expresses the percentages of the depleted food of each food stock on a notionally common scale between 0 and 100 (see Fig. 12.2). The results show that the dependence of the time-dependent vulnerability, the food stock capacity and the duration of a food shortage are not understood to be linear by the experts. The evaluations on the vulnerability level that depends on the percentage of depleted stock are very similar independent from the capacity of the stock. However, especially for small stocks the expert estimations vary considerably (see Fig. 12.1).

As mentioned above, the non-linear increase of vulnerability needs to be formalized to take it into account for the proposed model. Based on the results from the expert survey, statistical data is available to provide three individual regression or interpolation analysis of stocks that last 4, 7, and 14 days (see Table 12.6).

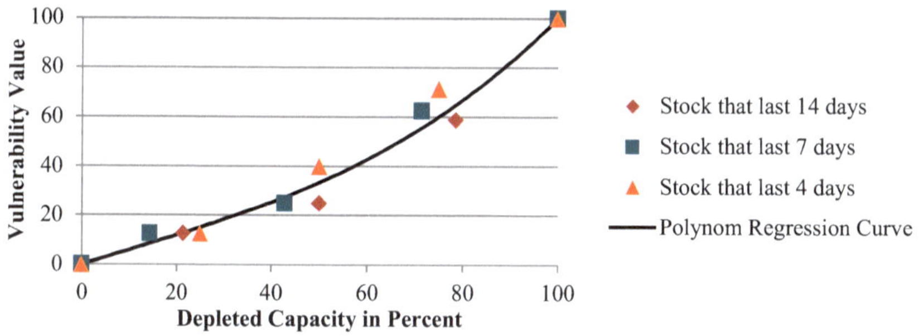

Fig. 12.2 Result from the expert estimations displayed on a common scale of depleted capacity in percent. The considered stocks for 4, 7, and 14 days were standardized to this scale. The points express the median of all expert estimations (N=31) of the vulnerability level depending on the depleted capacity percentage for the considered stocks

Table 12.6 Regression functions for food stock that last 4, 7, and 14 days. The functions are based on the statistical medians of all expert evaluations for each capacity (N=31) and express the course of vulnerability depending on the depletion of food stock capacity

Food stock capacity (t_c)	Regression function $C_v(t_f)$ ($t_f \leq t_c$)
4 Days (96 h)	$C_v^{t_c=96h}(t_f) = -\frac{729}{500}t_f^3 + \frac{7087}{625}t_f^2 + \frac{372}{125}t_f - \frac{3571}{5*10^4}$
7 Days (168 h)	$C_v^{t_c=168h}(t_f) = \frac{4269}{5*10^5}t_f^3 + \frac{4187}{3125}t_f^2 + \frac{5139}{1000}t_f + \frac{3783}{2000}$
14 Days (336 h)	$C_v^{t_c=336h}(t_f) = \frac{6447}{156,250}t_f^3 - \frac{3817}{10^4}t_f^2 + \frac{4399}{1000}t_f + \frac{3991}{10^4}$

However, the previous discussed surveys about the stockpiling behavior of the German population provide percentages of the population with different stock volumes. This also includes stock volumes that were not evaluated by the experts, for instance, stock volumes that last shorter as 4 days and bigger volumes that last longer than 7 days. A common polynomial function is necessary to generalize and to apply the expert evaluation results to any stock capacity. This could be derived by further statistic evaluations based on the data of the expert estimations and the available regression function for the stocks that last 4, 7, and 14 days (Table 12.1). Aiming at a high level of comprehensibility for decision makers, such advanced sophisticated regression are much too complex and do not improve the meaning. Therefore, we believe that it would be much more useful to reduce the complexity. We propose the predefining of a small set of easy-to-understand reference points that are very similar to those provided by the expert survey. Based on these reference points, a common polynomial function or vulnerability function can be derived that can be easily applied to any capacity of a food stock.

For better insight and transparency, we assume a vulnerability function $C_v(t)$ based on the level of the depleted food stock capacity that fits exactly to the reference points $P_0(0;0)$, $P_1(25;15)$, $P_3(75;60)$, and $P_4(100;100)$. Based on these reference points, we determine a polynomial regression function $C_v(t)$ that describes the vulnerability value for a person or household at a point in time of an ongoing food shortage t_f and for its specific food stock capacity that lasts until a point in time t_c. The increasing vulnerability between t_0, indicating the start of a food shortage, and t_c, is described by

$$C_v(t_f) = \frac{1}{18750}(100\tfrac{t_f}{t_c})^3 - \frac{1}{750}(100\tfrac{t_f}{t_c})^2 + 100\frac{3}{5}\frac{t_f}{t_c} \quad (12.1)$$

This polynomial regression curve is shown in Fig. 12.2. It slightly overestimates the vulnerability at the beginning of a food shortage compared to the data points provided by the experts. In this way, continued consumption that would begin at t_0 can be taken into account much better. With a higher depletion rate, the slope of the vulnerability function is similar to the estimations of the experts. In our opinion, the selected regression is absolutely sufficient for our purpose. However, other functions also could be considered and selected based on the decision maker's preferences.

Detailed Comparison of the Study Results Regarding the Quantification of Private Food Stocks

The vulnerability function $C_v(t)$ provides a notionally common scale that allows for a comparison of the results of the empirical studies mentioned before. To do so, the function needs to be applied to each fraction of the total population (or participant) P_i with i number of fractions. Each fraction has its specific food stock t_{c_i}. Summing up these products results in the increased vulnerability of the total population:

$$C_{study}(t) = \Sigma_{k=1}^{p} u P_i \;;\quad u = \begin{cases} C_v(t)_i, & t_f < t_{c_i} \\ 1, & t_f \geq t_{c_i} \end{cases} ;\quad i = 1,2,3,\ldots p \quad (12.2)$$

To apply this to the studies introduced in the section before, some slight transformations of the original numbers are necessary. The results of the studies by Menski and Gardemann (2008) and by Ohder et al. (2014) were recalculated as absolute percentages of the participants or population for each category of a food stock. The quantification by Menski and Gardemann (2008) is limited to shortages that last a maximum of 8 days and the quantification by Ohder et al. (2014) to shortages of 7 days. The results of Rhein (2013) can be directly used and do not need a transformation. The results of Rasche et al. (2000) are not detailed enough and, hence, were skipped.

Applying the vulnerability function to the three study results allows a comparison on a common time scale which presents the duration of an ongoing food shortage. Because the studies of Menski and Gardemann (2008) and Ohder et al. (2014) do not quantify food stocks that last longer than 8 and 7 days, respectively, the period under review is restricted to 7 days (168 h) (Fig. 12.3). It could be assumed that after this initial period of a food shortage first responses are in place and further help requests are submitted. The comparison shows a similarity of the results of Menski and Gardemann (2008) and Ohder et al. (2014) although both studies are non-representative. There are differences in the study participants as those of

Menski and Gardemann (2008) lived in a rural area and had recently experienced a food scarcity, whereas those of Ohder et al. (2014) lived in an urban area and had no experiences with a food shortage at all.

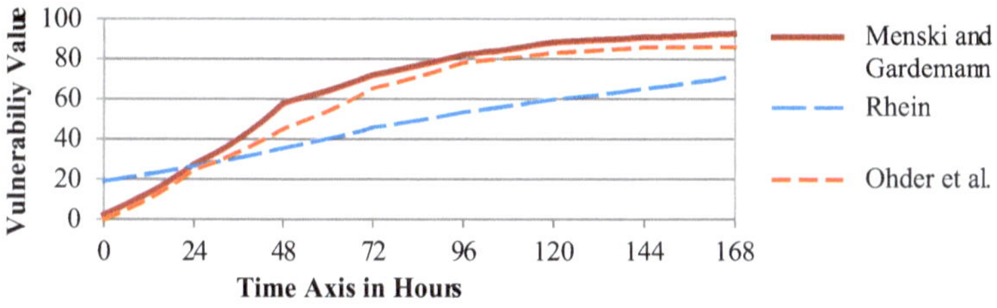

Fig. 12.3 Detailed comparison of the study results relating to the quantification of private food stocks

Compared to these two non-representative studies, the representative results by Rhein (2013) show higher vulnerability values in the initial phase of a food shortage and lower vulnerability values later on. These deviations may be caused by the smaller number of participants in the studies of Menski and Gardemann (2008) and of Ohder et al. (2014), however, objective reasons for these effects are not clearly visible. Due to its representative evaluation and the very high number of participants, the study of Rhein (2013) seems more significant than the others, which may underestimate the initial impacts and overestimate the long-term consequences. Consequently, the quantifications provided by Rhein (2013) will be used in the next sections.

Spatio-temporal Assessment of a Population Affected by a Sudden Lack of Food

We assume that a city, a, is compromised of n total districts ($b=1,2, \ldots, n$). To take the different population structure of each city district into account, we use the number of households of different household sizes in a city district.

Let $h_{total}^{r}(a_b)$ represent the total number of households of size r in district b ($b=1,2, \ldots, n$). The study of Rhein (2013) provides quantitative information about specific household sizes r that have no food stockpile ($h_0^r(a_b)$) and a stockpile for 1 week ($h_{168}^r(a_b)$). Although Rhein pointed out that there is no statistical proof of a correlation between household sizes and preparedness, we used his empirical and representative data as a first guess for a temporal vulnerability estimation that takes into account the city district's population structures. The following household sizes are considered: Single ($r=1$), two-person ($r=2$), three-person ($r=3$), four-person ($r=4$), and five-or-more-person household ($r=5$).

The time-dependent vulnerability of one household of a specific household size $H_v^r(t)_{a_b}$ in a district a_b is presented by:

$$H_v^r(t)_{a_b} = \frac{h_0^r(a_b) + C_v(t) * h_{168}^r(a_b)}{h_{total}^r(a_b)} \tag{12.3}$$

Before the time-dependent vulnerability value of a city district $V_{(t)}^{a_b}$ can be derived, the population of a city district has to be brought into relation to the total population that lives in the whole city. In this way, it is possible to compare the final vulnerability values of the city districts and to sum them to an overall value that expresses the temporal vulnerability of the whole city. To take into account the population structure of a district, we define a ratio $z_{a_b}^r$ of the number of people who are living in households of a specific household size in the city district a_b to the total number of people who are living in the whole city. The ratio $z_{a_b}^r$ is scaled between [0,1]. We derive:

$$z_{a_b}^r = \frac{r * h_{total}^r(a_b)}{\sum_{r=1}^{5} h_{total}^r(a_b) * r} \quad \text{with} \quad \sum_b^n \sum_{r=1}^{5} Z_{a_b}^r = 1 \tag{12.4}$$

The time-dependent vulnerability value of a city district then is determined by the sum of all products of these relative household size ratios, $z_{a_b}^r$, with their corresponding vulnerability functions, $H_v^r(t)_{a_b}$:

$$V_{(t)}^{a_b} = \sum_{r=1}^{5} H_v^r(t)_{a_b} * z_{a_b}^r \tag{12.5}$$

The resulting function describes a time-dependent vulnerability profile of a city district. Spatio-temporal assessments then are possible by comparing vulnerability profiles of different districts. Additionally, the overall time-dependent vulnerability of the whole city $V_{(t)}^{city}$ can be calculated by summing the values for $V_{(t)}^{a_b}$ over all city districts:

$$V_{(t)}^{city} = \sum_{b=1}^{n} V_{(t)}^{a_b} \tag{12.6}$$

12.5 Deriving Resilience From Vulnerability Profiles

Depending on the discipline and the context, different definitions of resilience exist. In the context of a sudden lack of food, resilience is understood to be the ability of a population to cope with a disruption of food supply. This also includes an independence from advanced disaster management assistance mainly due to prepared food stocks. Other resilience definitions may also include recovery and lessons learned. To focus on the initial impacts and to support decisions of first responders, such additional definitions are excluded.

To quantify system resilience over time, time functions are widely used in engineering (see e.g. Bruneau et al. 2003; Zobel and Khansa 2014; Pant et al. 2014). In the context of the resilience to cope with a sudden lack of food, the resilience score $R(\alpha)$ is defined to be an integral of the city-specific vulnerability values for the period under observation that starts at t_0, the beginning of the food shortage, and ends at t_p, the end of the observation:

$$R(\alpha) = \int_{t_0}^{t_p} (1 - V_{(t)}^{city})dt \tag{12.7}$$

where α represents either the baseline scenario or other scenarios in which the preparation of the population is changed through the implementation of measures. In this way, the effects of a higher level of preparedness of households with specific sizes can measured, and the effects of different changes of such input parameters can be compared. This allows sensitivity analysis regarding the effects of alternative ways to improve the resilience of a city and makes it possible to estimate the benefit of single measures or combinations of multiple measures. Considering the same observation period for all scenarios, a normalization of the resilience score is defined to bring the relative results from the resilience score to a common scale between [0,1]:

$$R'(\alpha) = \frac{R(\alpha)}{\int_{t_0}^{t_p} 1dt} \tag{12.8}$$

In the following section, exemplary alternatives based on the two cases are analyzed.

12.6 Examples

For demonstrative purposes, the model proposed is applied to real use cases of two German cities, namely, *Mannheim* and *Cologne*. *Mannheim* comprises six central and eleven peripheral city districts. The city of *Cologne* comprises one central and eight peripheral city districts. Approx. 300.000 inhabitants are living in *Mannheim* and approx. 1.000.000 inhabitants are living in *Cologne*.

The population structure of the city districts is determined by the number of households of a specific household size. For this purpose, census data relating the structure of city districts are used. For the case of the city of *Mannheim*, the data provided by the municipal office for statistical information are used (Kommunale Statistikstelle Stadt Mannheim 2014). For the case of the city *Cologne*, the data provided by the department of city development and statistics of the city of *Cologne* are applied (Amt für Stadtentwicklung und Statistik Stadt Köln 2014).

We define the period under observation to be the first 7 days of a food shortage (t_p=168 h). This is the initial shock phase of a food shortage. Its progress is of greatest interest for the disaster management authorities because they have to know at which time of a food shortage the population is no longer able to help itself. The defined observation period is appropriate to analyse the baseline development that expresses the progress without any measures taken by the disaster management authorities. It can be assumed, that during this period the majority of the affected people run out of food and then request help.

City District Specific Vulnerability Values

The model results are displayed in different ways below. For each city we plot the increasing vulnerability for the individual city district (Figs. 12.4 and 12.5). Both

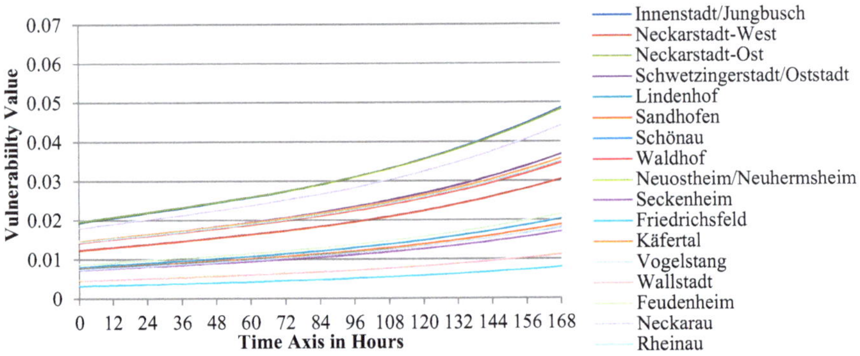

Fig. 12.4 The increasing vulnerability values of the city districts of Mannheim, Germany

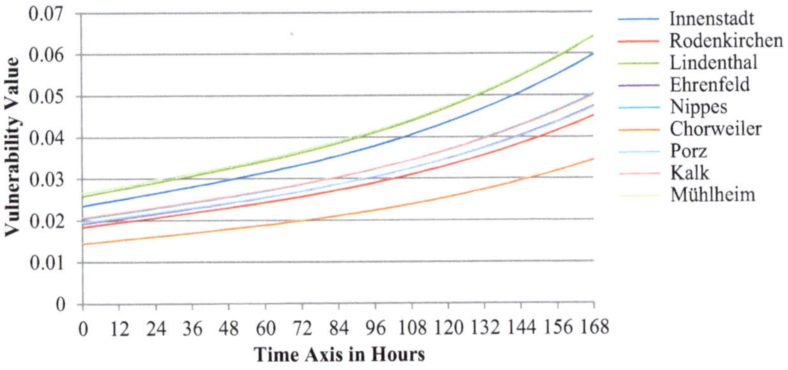

Fig. 12.5 The increasing vulnerability values of the city districts of Cologne, Germany

diagrams allow for an interpretation as to which city district is the key driver and, hence, could be a main focus for preparedness and response measures of disaster management authorities.

Based on these results, the city districts of *Mannheim* can be clustered in three vulnerability categories. Independent of the point in time during an ongoing food shortage, the city districts *Innenstadt/Jungbusch*, *Neckarstadt-Ost*, and *Neckarau* have the highest vulnerability (category 1). The city districts *Schwetzingerstadt/Oststadt*, *Rheinau*, *Waldhof*, and *Neckarstadt-West* are characterized by a moderate vulnerability (category 2). All other city districts have a relatively low level of vulnerability (category 3).

Vulnerability values can also be classified for the city districts of *Cologne*. The results show that the city districts *Innenstadt*, *Lindenthal*, and *Mühlheim* are highly vulnerable independently of the point in time during an ongoing lack of food (category 1). Except for *Chorweiler*, all city districts are characterized by moderate vulnerability values (category 2). Compared to the other city districts, *Chorweiler* has only a low vulnerability level (category 3).

Based on this categorization, the initial response can be prioritized to those districts that reach higher vulnerability values. With the illustration of the vulnerability courses disaster management authorities have a better insight about how the situation in a single district, and the overall situation in a city, develop during a food shortage. In both of our examples it becomes obvious that the overall situation will dramatically change after 2 days. Because our modelling is essentially based on an exponential vulnerability evaluation of food stock depletion, the escalation is not expressed as a sudden shock. The escalation is gradual and has a continuous development without any visible steps.

Although the ranking of the categories does not change over time, it becomes obvious how the individual level of vulnerability of a district increases during a food shortage. The cause of this homogenous rise can be found in the very similar population structure of the districts in both cities. The deviations between the clusters slightly increases during the food shortage. A possible reason for the similar population structure in the districts of cities in Germany may be seen in the German Spatial Planning Act (Raumordnungsgesetz from 2008). The law strongly claims to ensure equivalent living conditions in the districts of cities in Germany. It would also be interesting to analyze and compare areas that do not have similar population structures. This would probably result in much more heterogeneous rises of districts' vulnerabilities with possible graph intersections.

For the city of Cologne, the vulnerability gradient of the districts appears slightly steeper than for Mannheim and they reach higher values. This effect may be induced by the selected geographic boundaries and the total number of districts. Due to the selected standardization of $z^{r}_{a_b}$, the temporal vulnerability of a population in one city district is normalized to the total number of people in the city. The district's vulnerability reaches higher values with a smaller number of considered districts. Hence, a high number of city districts results in relatively small vulnerability values that are hard to compare to each other. This is the case for Mannheim in that it has 17 city districts, whereas there are only nine city districts in Cologne that are much easier to compare.

The results are additionally mapped in Geographic Information Systems (GIS) illustrations for each city (Figs. 12.6 and 12.7). For a better demonstration we normalized the time-dependent vulnerability values of the city district on a scale between 0 and 100 using the value 0 for a vulnerability value of 0 and 1 for the highest vulnerability value that is reached by one of city district in the period under consideration.

The illustrations show the vulnerability values for each city district at the beginning of a food scarcity, after 2, 4, and 7 days. It is possible to clearly highlight those city districts that potentially suffer most, however, the increasing of the vulnerability is very homogeneous.

Resilience and Sensitivity Assessment

To perform a sensitivity analysis and to provide a better understanding of resilience, several alternatives that change the model parameters are defined (see Table 12.7). These alternatives may be caused by measures like advertisements or training courses that raise the awareness and motivate the population to improve their preparedness level for food shortages.

The preparedness level may be enhanced, for instance, by (a) an increasing number of people who stock food for 1 week or more. It is also possible to (b) concentrate on those people who do not stockpile and to try to reduce the number of those people. Another approach may be to (c) concentrate on specific household sizes and to try to reduce the number of households which are not prepared or to increase the number of households that stock food for more than 1 week.

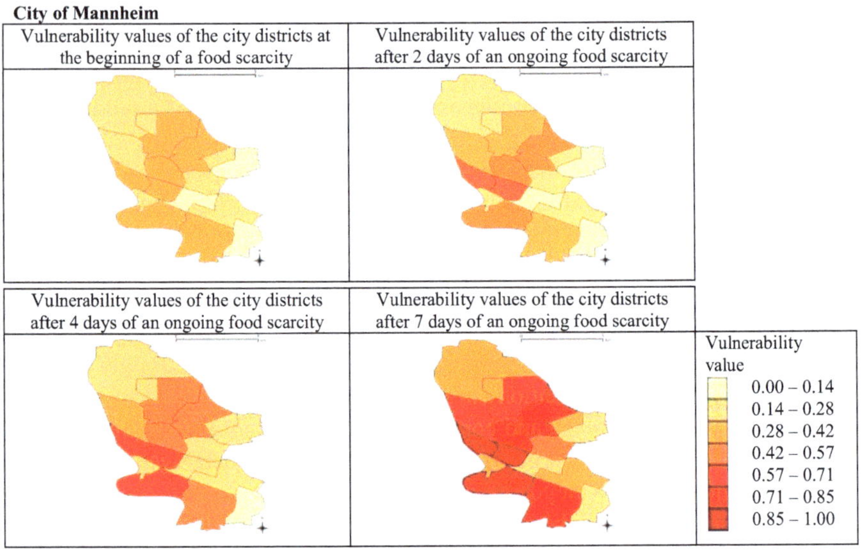

Fig. 12.6 A GIS illustration of the vulnerability values of the city district of Mannheim, Germany, at different points in time during an ongoing food scarcity

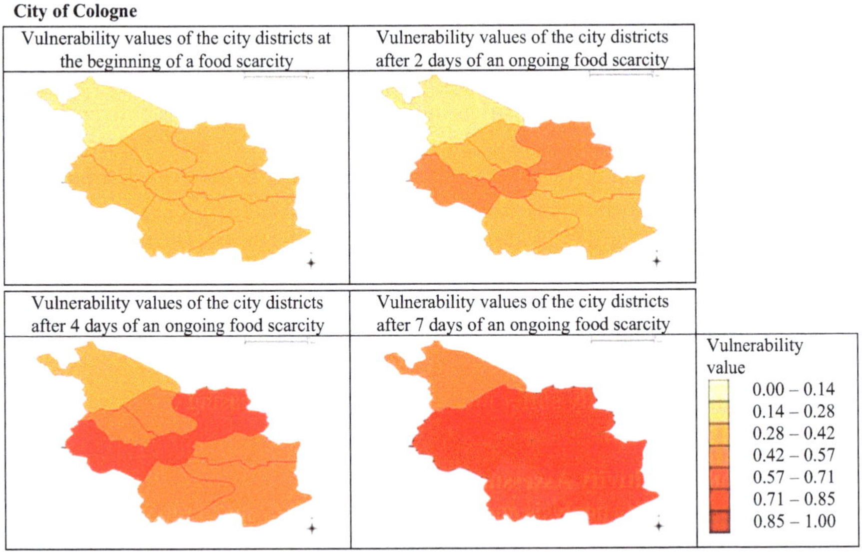

Fig. 12.7 A GIS illustration of the vulnerability values of the city district of Cologne, Germany, at different points in time during an ongoing food scarcity

We calculated the resilience score $R(\alpha)$ for the baseline scenario with the current estimated vulnerability increasing and for each of the twelve scenarios under review (Table 12.7 and 12.8). For each scenario the normalized resilience score $R'(\alpha)$ was derived to compare the effects on a scale between [0,1] (Fig. 12.8).

Table 12.7 Overview about the measures under review

Category	Description	Scenario
Baseline	Current preparation status without the implementation of any measure	$\alpha=1$
a) Improving food stock piling in general	2.5 % points *more households have a food stockpile* that last 1 week	$\alpha=2$
	5 % points *more households have a food stockpile* that last 1 week	$\alpha=3$
	7.5 % points *more households have a food stockpile* that lasts 1 week	$\alpha=4$
b) Improving the preparedness levels of those who are unprepared	2.5 % points *less households have no food stockpile*	$\alpha=5$
	5 % points *less households have no food stockpile*	$\alpha=6$
	7.5 % points *less households have no food stockpile*	$\alpha=7$
c) Improving preparedness levels for specific household sizes	2.5 % points more *single households* have a food stockpile that last 1 week and 2.5 % points less of them have no food stockpile	$\alpha=8$
	2.5 % points more *two-person households* have a food stockpile that last 1 week and 2.5 % points less of them have no food stockpile	$\alpha=9$
	2.5 % points more *three-person households* have a food stockpile that last 1 week and 2.5 % points less of them have no food stockpile	$\alpha=10$
	2.5 % points more *four-person households* have a food stockpile that last 1 week and 2.5 % points less of them have no food stockpile	$\alpha=11$
	2.5 % points more *five-person households* have a food stockpile that last 1 week and 2.5 % points less of them have no food stockpile	$\alpha=12$

Table 12.8 Overview of the normalized resilience score $R'(\alpha)$ of each measure under review for Mannheim and Cologne

	Baseline Scenario				
	R′(1)				
Mannheim	0.702				
Cologne	0.704				
	a) Improving food stock piling in general				
	R′(2)	R′(3)	R′(4)		
Mannheim	0.711	0.721	0.731		
Cologne	0.714	0.723	0.733		
	b) Improving the preparedness levels of those who are unprepared				
	R′(5)	R′(6)	R′(7)		
Mannheim	0.717	0.732	0.747		
Cologne	0.719	0.734	0.749		
	c) Improving preparedness levels for specific household sizes				
	R′(8)	R′(9)	R′(10)	R′(11)	R′(12)
Mannheim	0.711	0.711	0.709	0.708	0.706
Cologne	0.711	0.711	0.709	0.708	0.706

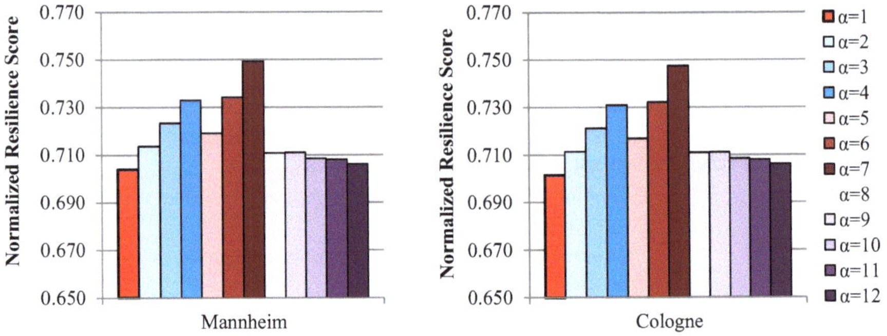

Fig. 12.8 Visualization of the normalized resilience score $R'(\alpha)$ of the baseline scenario ($\alpha = 1$) and each measure under review for Mannheim and Cologne

The result of the sensitivity analysis reflects improvements resulting from single parameter changes. It is obvious how the overall resilience of a city changes by modifying parameters in different degrees. In both examples, the impacts are similar although the individual absolute values are slightly different. The similarity is due to the very similar population structure of both cities. A comparison with e.g. a more rural area with a different population structure would probably show more differences. In our examples the differences caused by the population structure are more obvious on the level of the city districts.

By comparing the parameter modifications it is evident that measures reducing the proportion of households that do not stockpile any food enhance the resilience of a city more effectively than measures that increase the number of households with a stock of food for 1 week. Hence, it may be recommended to concentrate on the reduction of the number of households with no preparation and to stabilize the proportion of households with food stocks for one or more weeks. Additionally, the results show that a change of the preparedness level of single and two-person households has the greatest impacts on the resilience. Awareness and training courses should therefore concentrate on these two categories.

Our new model can provide disaster management authorities with a better understanding of the resilience and the key parameters determining the resilience of a city. The above studies exemplarily demonstrate possible actions to raise the preparedness level of a city. They may also be of help when developing disaster response plans as the spatio-temporal understanding of potential problems in the preparedness of the population is enhanced.

12.7 Critical Remark and Conclusion

A lack of food can bring a population into a situation which requires advanced assistance by disaster management authorities. To support addressing this, our new spatio-temporal vulnerability model enables decision makers to assess time-depending vulnerability values of a population that is affected by an ongoing lack of food. The two cases of *Mannheim* and *Cologne* demonstrate the potential benefit for disaster management. The model results provide a comprehensive insight into the escalation of food shortages, and allow for a comparison of vulnerability values of city districts for different times of a food scarcity and an assessment of alternative preparedness measures by analyzing their influence on the overall resilience level of a city.

The spatio-temporal vulnerability model is based on empirical data relating to food stock capacities in households in Germany. Having reviewed several studies, only the one by Rhein (2013) seems to be representative. That is why this study was used as basis for our model. In addition, statistical data about the number of households with specific household sizes were used.

Like other approaches, our vulnerability model is based on general assumptions and subject to limitations. The calculation of time-dependent vulnerability values does not consider emergency measures that are taken during an ongoing lack of food. Realistically, stocks of food might exist in general stores at the beginning of a food shortage. It may be assumed that this stock will be sold to the population. Hoarding might be considered. These effects must be taken into account in the final interpretation of the model results.

The current model does not take into account any ad-hoc reactions and further abilities of the population. It is conceivable that some people may travel long-distances to non-affected areas to buy food. Even uncoordinated self-evacuation into non-affected areas could be envisaged. These reactions are highly dependent on soft factors such as the availability of information, the way how the individual person assesses its situation and whether alternative supplies are available. These factors are hard to model using an indicator-based approach.

Further research is need with respect to social parameters that characterize the resilience of the affected population. The proposed model considers the population structure of city districts by taking into account the number of households, household sizes and food stock capacity. The resilience of different social groups that are living in a city is not considered. Statistical data on the stock piling behavior and on the resilience to cope with a lack of food of different social groups, such as unemployed, ill, migrated, old people or single-parents, would be of high value. Although parts of these groups were considered in the studies reviewed, no appropriate statistical evaluation suitable for spatio-temporal vulnerability modeling was made.

The current consideration of households of different sizes leaves room for (mis-) interpretation. On the one hand, single and two-person households might be considered very flexible in coping with a lack of foods. For a single or two-adult household it might be easier to travel to a non-affected area than for households with more persons. Additionally, children or older people who may live in such households may need special food. On the other hand, single-parent families with one child are also two-person households having limited resilience to cope with the effects of a lack of food. These examples illustrate the limits of the proposed model in considering social parameters. Hence, more social empirical data is required to define better resilience indicators.

Furthermore, food shortages are often secondary impacts and caused by other initial disaster events. Side and interdependent effects are not considered. In addition, the point in time t_0 which describes the beginning of a food shortage depends on this initial event. Sometimes, a food shortage situation emerges slowly and an explicit start is difficult to determine. This may be a significant source of uncertainty.

The time-dependent vulnerability value indicates the food demand of a population. To derive concrete measures to address this demand, absolute quantifications for the required food capacity are needed. The model proposed allows for a rough and approximate estimation only, because the model focuses on displaying the time-dependent sufferings of a population. It may be concluded that the current development state of the model clearly indicates its added value for the preparedness phase. During the response phase, however, more sophisticated approaches might be needed. Future research work might also be directed towards coupling the demand perspectives with modeling and forecasting of freight flows that simulate food supply chains and, hence, ensure a bridge between demand and supply.

Acknowledgement The research reported in this paper is part of the SEAK project that aims at developing a decision support framework to manage food supply chain disruptions. The SEAK project is funded by the security research program of the German Federal Ministry of Education and Research (BMBF) and we thank the BMBF for the funding support.

We would like to thank all representatives from the German authorities that participated on our workshop on "Disaster Management Planning for Food Shortages and Power Outages" in October 2014 at the Karlsruhe Institute of Technology (KIT), Germany.

References

Adger N (2006) Vulnerability. Global Environ Chang 16:268–281

Akinyemi OM (2010) The Status and Barriers to the implementation and continuation of the school gardening projects in Germany, Nigeria and the United States of America: Obstacles, Strengths and Weaknesses. Doctoral Dissertation, Witzenhausen

Amt für Stadtentwicklung und Statistik, Stadt Köln (2014) Haushaltsgröße bezogen auf die Stadtbezirke der Stadt Köln zum 31.12.2012. http://www.offenedaten-koeln.de/dataset/c9f17c17-a8b5–4f73–8710-76db2cfec601/resource/c9f17c17-a8b5-4f73-8710-76db2cfec601. Accessed 20 Nov 2014

Anderson MB, Woodrow PT (1991, March) Reducing vulnerability to drought and famine: developmental approaches to relief. Disasters 15(1):43–54. doi:10.1111/j.1467-7717.1991.tb00426.x

Aubrecht C, Steinnocher K, Köstl M, Züger J, Loibl W (2013) Long-term spatio-temporal social vulnerability variation considering health-related climate change parameters particularly affecting elderly. Nat Hazards 68:1371–1384

Bankoff G, Frerks G, Hilhorst D (2004) Mapping vulnerability, disasters, development, and people. Earthscan Publications, London

Belliveau S, Smit B, Bradshaw B (2006) Multiple exposures and dynamic vulnerability: evidence from the grape industry in the Okanagan Valley, Canada. Global Environ Chang 16:364–378

Birkmann J, Wisner B (2006) Measuring the un-measurable. The challenge of vulnerability. UNU Institute for Environment and Human Security (UNU-EHS), Source 5, Bonn

Bogardi JJ (2004) Hazards, risks and vulnerabilities in a changing environment: the unexpected onslaught on human security? Global Environ Chang 14:361–365

Brauner F, Münzberg T, Wiens M, Fiedrich F, Lechleuthner A, Schultmann F (2015) Critical infrastructure resilience: a framework for considering micro and macro observation levels. Proceedings of the international conference on information systems for crisis response and management—kristiansand, Norway, May 24–27, Palen, Büscher, Comes & Hughes, eds

Brock III LG, Lauren BD (2015) Estimating available supermarket commodities for food bank collection in the absence of information. Expert Syst Appl 42(7):3450–3461

Brooks N (2003) Vulnerability, risk and adaptation: a conceptual framework, Tyndall centre for climate change, working paper 38, 16.

Bruneau M, Chang S, Eguchi R, Lee G, O'Rourke T, Reinhorn A, Shinozuka M, Tierney K, Wallace W, von Winterfelt D (2003) A framework to quantitatively assess and enhance the seismic resilience of communities. EERI Spectra 19(4):733–752

Cardona OD (1999) Environmental Management and Disaster Prevention: Two Related Topics - A Holistic Risk Assessment and Management Approach. In: Ingleton J (ed) Nat Disaster Manag. Tudor Rose, London

Chen P, Chen X (2012) Spatio-temporal variation of flood vulnerability at the Poyang Lake ecological economic zone, Jiangxi Province, China. Water Sci Technol 65(7):1332–1340

Debnath R (2013) An assessment of spatio-temporal pattern of urban earthquake vulnerability using GIS: a study on Dhaka City. Ann GIS 19(2):63–78.

Dilley M, Boudreau TE (2001) Coming to terms with vulnerability: a critique of the food security definition. Food Policy 26(3):229–247

Gaillard J (2010) Vulnerability, capacity and resilience: perspectives for climate and development policy. J Int Dev 22:218–232

Giupponi C, Giove S, Giannini V (2013) A dynamic assessment tool for exploring and communicating vulnerability to floods and climate change. Environ Model SC 44:136–147

Hufschmidt G (2011) A comparative analysis of several vulnerability concepts. Nat Hazards 58(2):621–643

Jacob B, Mawson AR, Payton M, Guignard JC (2008) Disaster mythology and fact: hurricane Katrina and social attachment. Public Health Rep 123(5):555–566

Kasperson JX, Kasperson RE, Turner BL (1995) Regions at risk: comparisons of threatened environments. United Nations University Press, New York

Kienberger S, Blaschke T, Zaidi RZ (2013) A framework for spatio-temporal scales and concepts from different disciplines: the 'vulnerability cube'. Nat Hazards 68:1343–1369

Kommunale Statistikstelle Stadt Mannheim (2014) Anzahl und Struktur der Mannheimer Privathaushalte in kleinräumiger Gliederung Statistische Daten 2/2014, Mannheim, 2014. https://www.mannheim.de/sites/default/files/page/12213/d201402_privathaushalte_2013.pdf. Accessed 20 Nov 2014

Manyena SB (2006) The concept of resilience revisited. Disasters 30(4):433–450

Menski U, Gardemann J (2008) Auswirkungen des Ausfalls Kritischer Infra-strukturen auf den Ernährungssektor am Beispiel des Stromausfalls im Münster-land im Herbst 2005, Empirische Untersuchung im Auftrag der Bundesanstalt für Landwirtschaft und Ernährung (BLE). http://www.hb.fh-muenster.de/opus/fhms/volltexte/2011/677/. Accessed 20 Nov 2014

Metzger MJ, Schröter D (2006) Towards a spatially explicit and quantitative vulnerability assessment of environmental change in Europe. Reg Environ Change 6(4):201–216

Münzberg T, Wiens M, Schultmann F (2014) Dynamic-spatial vulnerability assessments: a methodical review for decision support in emergency planning for power outages, humanitarian technology: science, systems and global impact 2014, Hum Tech 2014. Procedia Eng 78:78–87

Münzberg T, Wiens M, Schultmann F (2015) The effect of coping capacity depletion on critical infrastructure resilience. Proceedings of the international conference on information systems for crisis response and management—kristiansand, Norway, May 24–27, Palen, Büscher, Comes & Hughes, eds

Ohder C, Röpcke J, Sticher B, Geißler S, Schweer B (2014) Hilfebedarf und Hilfebereitschaft bei anhaltendem Stromausfall, Ergebnisse einer Bürgerbefragung in drei Berliner Bezirken
Pan American Health Organization (2001) Humanitarian supply management and logistics in the health sector. PAHO, Washington, D.C.
Pant R, Barker K, Zobel C (2014) Static and dynamic metrics of economic resilience for interdependent infrastructure and industry sectors. Reliab Eng Syst Safe 125:92–102
Patt AG, Schröter D (2008) Perceptions of climate risk in Mozambique: implications for the success of adaptation strategies. Global Environ Chang 18(3):458–467
Rasche J, Schmidt A, Schneider S, Waldtmann S (2000) Organisation der Ernährungsnotfallvorsorge, Schriftenreihe der Schutzkommision beim Bundesmi-nister des Innern, Neue Folge B and 47. http://www.bsl-mb.com/fileadmin/user_upload/pdf-beitraege-mitarbeiter/Zivilschutzforschung_Band_47.pdf
Rhein S (2013) Kapazitäten der Bevölkerung zur Bewältigung eines lang anhalten-den flächendeckenden Stromausfalles, Empirische Untersuchung für das Bezugs-gebiet Deutschland. http://www.bbk.bund.de/SharedDocs/Downloads/BBK/DE/Publikationen/Praxis_Bevoelkerungsschutz/Band_12_PraxisBS_Stromausfall.pdf?__blob=publicationFile
Rodríguez-Gaviria EM, Botero-Fernández V (2013) Flood vulnerability assessment: a multiscale, multitemporal and multidisciplinary approach. J Earth Sci Eng 2:102–108
Sahin O, Mohamed S (2010) Coastal vulnerability to sea level rise: a spatio-temporal decision making tool. Proceedings of the 2010 IEEE international conference on industrial engineering and engineering management, 29 Oct—31 Oct 2010, Xiamen, China
Sobiech C (2013) Agent-based simulation of vulnerability dynamics, a case study of the German North Sea Coast, 2013: Springer Theses
Thywissen K (2006) Components of risk: a comparative glossary, publication series of UNU-EHS, Source 2
Wright M, Vesala-Husemann M (2006) Nutrition and disaster preparedness: focusing on vulnerability, building capacities. J Issues Nurs 11(3) Manuscript 5. doi:10.3912/OJIN.Vol11No03Man05
Zobel C, Khansa L (2014) Characterizing multi-event disaster resilience. Comput Oper Res (special issue on MCDM in emergency management) 42:83–94

Index

A
Access to Medicines (ATM), 5, 73, 83, 206, 207

B
Blood products, 5, 186
    management of, 187, 188
    types of, 202

C
Capacity planning, 18, 233, 253
Case studies, 4, 88, 100
    analysis of, 89
Cash-based interventions, 62, 80, 82, 85, 101, 103
    approaches, 81
    indicators for, 102
Cash-based response, 61, 80, 82
Cash programs, 81, 83
Cash transfers, 4, 60, 64, 65, 69, 70, 72
    concerns of, 61
    programmes, type of, 62
Change adoption, 5, 136, 138, 139, 146, 150–152, 154
    barriers to, 149
    factors affected on, 143
    technology design, role of, 149
Change management, 135, 136, 144
    aspects of, 150
    external, 141
    organizational and cultural, factors, 155
Cold chain, 175, 206
    storage, 163
    transport, 176
Coping capacity, 126, 262
Corruption, 4, 22
    risk of, 71–73
Costs, 14, 15, 187
    external, 44
    logistic, 41, 42, 44, 84
    social, 192
    transportation, 15, 41, 246, 249

D
Data envelopment analysis (DEA), 24, 223
Delivery mechanisms, 73, 75
Demand correlation, 198, 200
Digital food, 60
Disaster collaboration life cycle, conformity model, 125, 126
Disaster management, 4, 67, 112, 113, 115
    life cycle, 120
Disaster management planning, 278
Disaster plans, 34, 47, 50, 53, 276
Disaster preparedness, 258, 263, 266
Disaster relief, 3, 10, 65
Disaster response, 4, 35
    activities in, 37
    vehicle routing problem, 19
Disaster response logistics (DRL), 3, 34, 35, 54
Distribution, 5, 14, 178, 190, 233, 239
    blood, 187, 191, 192, 202
    control of, 163
    food, 19, 61, 97, 233, 235, 249
    goods, 19
    networks, 48, 49
Donations management, 16

E
Efficiency metrics, 10, 14, 23
Emergency food network, 230, 232, 233, 249
Emergency support functions(ESFs), 35
EMMA analysis, 81

C. W. Zobel et al. (eds.), *Advances in Managing Humanitarian Operations*,
International Series in Operations Research & Management Science 235,
DOI 10.1007/978-3-319-24418-1

Equality, 14, 17, 19
Equity, 5, 10, 14, 17–19

F
Fairness, 18, 19, 206, 207, 240
Financial flow, 66, 71
Food aid distribution, 6, 233, 235, 239
Food banks, 6, 232, 233, 235, 238, 239, 245, 253
  acts, 240
  goal of, 241
Food insecurity, 60, 98, 99, 230, 245
  definition of, 230
  issue of, 250
  prevalence of, 232
Food supply disruption, 262
Funding, 67, 68, 138, 243, 247
  and corruption, 71

G
Group decision making, 207, 214, 224

H
Humanitarian culture, 5, 136–138, 140
Humanitarian logistics (HL), 17, 37, 41, 137, 153, 154
  case study, 4
  challenges, 115
  research, 138
Humanitarian organizations, 64, 85, 136, 154
  characteristics, 137
  culture of, 136
  managing change in, 139
  research on, 138
Humanitarian supply chains, 64–66
  charactristics, 179
  features of, 69
  measuring performance of, 84, 85
  measuring visibility of, 164
Hunger relief, 5, 233
  iron triangle, 240, 243, 248
  supply chain, overview of, 233

I
Indicator-based vulnerability assessment, 258
Information, 5, 26, 34, 70, 73, 163, 170, 173, 217, 251, 277
Input metrics, 14, 15
Integer programming, 187
  model, 188, 189

K
Key performance indicators (KPIs), 208, 217

L
Lack of food, 6, 60, 258, 259, 265
  causes, 259
  spatio-temporal vulnerability in, 264
    population affected, 269
  vulnerability assessement, 264
Logistics, 37, 67, 82, 153
  of non profit food distribution, 249
Longitudinal research, 136

M
Managerial practices, 10
Market systems, 4, 86, 87, 94, 97, 102
  definition of, 81
Material convergence, 38, 41, 47, 49, 52, 53, 55
Material flow, 61, 66, 67
Mobile phone projects, 168
Modeling approaches, 3
Multi-criteria evaluation, 206

N
Newsvendor, 187–190, 193
Nonprofit operations, 3, 9–12, 14–17, 27

O
Output metrics, 14, 16, 17

P
Partner proliferation, 4, 112, 115–119, 128
Performance measures, 18, 83, 101, 103, 208, 217
Performance metrics, 10, 11, 13, 14, 28, 209, 241
Poverty, 17, 22, 98, 239, 241, 259
Purchasing power, 60, 64, 65, 92, 96

R
Resilience, 25, 69, 258, 259
  definition of, 270
Rotavirus, 210, 214, 218
Routing, 5, 17, 39, 187, 189, 202, 249

S
Satellite technology, 172
Scenario planning, 25
Short-term collaboration, 121, 124, 126–128
Social welfare systems, 60
Spatio-temporal assessment, 258, 269, 270
Stakeholder analysis, 210, 211, 218, 225
Stochastic models, 5, 190, 195
Stochastic programming, 5, 195, 251

Supply chain design, 75, 206–210, 213, 215, 216, 219, 223, 224
Systems engineering, 253

T
Technology, 5, 149, 164, 175, 206, 223
Tracing, 161, 163, 170
Tracking, 154, 161, 167
  of inventory, 162
  of lacation, 162
Traveling salesman (TSP), 187–189, 195, 202, 249

V
Vaccine industry, 5
Value for money, 82, 85
Virtual organization, 4, 113, 117, 120–129
Visibility, 5, 69, 159, 160, 162, 164, 179
Vouchers, 60, 61, 64, 81, 93

MIX
Papier aus verantwortungsvollen Quellen
Paper from responsible sources
FSC® C105338

If you have any concerns about our products,
you can contact us on
ProductSafety@springernature.com

In case Publisher is established outside the EU,
the EU authorized representative is:
Springer Nature Customer Service Center GmbH
Europaplatz 3, 69115 Heidelberg, Germany

Printed by Libri Plureos GmbH
in Hamburg, Germany